RADIOACTIVITY IN THE ENVIRONMENT:
Sources, Distribution, and Surveillance

RADIOACTIVITY IN THE ENVIRONMENT:
Sources, Distribution, and Surveillance

Ronald L. Kathren

University of Washington
Joint Center for Graduate Study
Richland, Washington

 harwood academic publishers
chur . london . paris . new york

Harwood Academic Publishers

Poststrasse 22
7000 Chur
Switzerland

P.O. Box 197
London WC2 4DL
England

58, rue Lhomond
75005 Paris
France

P.O. Box 786
Cooper Station
New York, New York 10276
United States of America

Library of Congress Cataloging in Publication Data

Kathren, Ronald L., 1937–
    Radioactivity in the environment.

    Includes indexes.
    1. Radioactive pollution. 2. Radioecology.
3. Environmental monitoring. I. Title.
TD196.R3K38 1984   363.7'38   84-3838
ISBN 3-7186-0203-2

DEDICATION

To the memory of

HERBERT M. PARKER

(1910–1984)

whose numerous and lasting contributions to medical and health physics spanned more than a half century.

Contents

Preface

Although radioactivity and ionizing radiations had been known to man for a half century prior to the detonation of the first atomic bomb in the New Mexico desert in 1945, that explosion kindled new interest in the study of the radiological environment. The importance of the study of the radiological environment was underscored by the testing of nuclear weapons in the atmosphere during the 1950s and '60s; this study took on a new dimension with the coming of the environmental protection movement during the late 1960s. Interest in the radiological environment has been further heightened by the proliferation of environmental protection legislation in recent years and by the controversy surrounding the generation of electricity by nuclear reactors. For many, however, it was the accident at the Three Mile Island nuclear generating station in the spring of 1979, that provided the first real awareness of the need to better understand radioactivity and its potential impacts on our world.

This book seeks to provide the background necessary for understanding and monitoring the radiological environment. It is thus meant to serve as both a text and reference book for scientists, engineers, regulators, and others, both within and outside the nuclear field, who may be involved with the application and control of nuclear power or other sources of radioactivity. It is also intended for use as a textbook in graduate or advanced undergraduate courses devoted to the study of the radiological environment from a physical rather than a biological point of view, and is thus complementary to standard works on radioecology. Its level presupposes a basic understanding of modern physics, chemistry, biology and mathematics through calculus, although the nonmathematical reader may skip over the derivations of equations without loss of continuity or diminished understanding of the basic concept.

The book grew out of a graduate-level course in environmental radioactivity taught as part of the interdiciplinary program in Radiological Sciences of the University of Washington at the Joint Center for Graduate Study in Richland. Over the years, this evening course has been taken primarily by degreed persons actively working full time in health physics and related nuclear fields; thus, it reflects their interests and needs. I am deeply indebted to my colleagues and friends: David L. Waite, Battelle Columbus Laboratories, who established the course bearing the title Environmental Radioactivity at the Joint Center for Graduate Study and served as its first instructor; Monte J. Sula, Senior Research Scientist, Occupational and Environmental Protection Department, Battelle Pacific Northwest Laboratories, for his complete and thorough reading of the manuscript and his candid and incisive comments; Naomi Pascal, University of Washington, for her aid and counsel and arranging for review of the draft manuscripts; Gene L. Woodruff, Professor and Chairman of the Department of Nuclear Engineering at the University of Washington; Geoffrey J. Eichholz, Professor of Nuclear Engineering at the Georgia Institute of Technology, whose numerous comments and suggestions greatly improved the final product; and to Dale Fancher, Joint Center for Graduate Study, for his arrangement of the line drawings and other general support. Despite their efforts and those of the editor and proofreaders, errors will inevitably appear, and for these I take full responsibility.

Ronald L. Kathren

A Note on Units

For many years, certain radiological units and quantities have been in common use within the scientific community, in accordance with the rigorous definitions put forth by the International Commission on Radiological Units (ICRU). Thus, the special unit of activity of a radioactive nuclide has been the curie, which is defined as that quantity of activity producing exactly 3.7×10^{10} spontaneous nuclear transformations per second; the term nuclear transformation has been defined as a change of nuclide or an isomeric transformation.

Similarly, exact definitions have been provided for three other quantities: exposure in a special sense, dose, and dose equivalent. The most recent (1983) definitions for these quantities are given in ICRU Report 33, "Radiation Quantities and Units," and are paraphrased here. Exposure, as used in its special sense, is measured in units of roentgens. One roentgen is that quantity of electromagnetic radiation which produces a charge of 2.58×10^{-4} coulombs in one kg of air. As defined, this special unit of exposure, the roentgen, applies only to photons in air; it therefore cannot be used for any other radiations nor for any other absorbing media.

Radiation dose is a measure of the energy imparted by ionizing radiation to a unit mass of an absorbing medium. The special unit of absorbed dose is the rad, which has been defined as 100 ergs/g or 0.01 J/kg. The rad may be used for any ionizing radiation or any medium. Note that in past years, prior to the special and limited definition of the roentgen as the special unit of exposure, the roentgen was used as the unit of dose and will appear as such in the older literature. The practice regretably was erroneously continued to some extent even after exposure was defined in the special sense. Thus, references to the roentgen in the scientific literature should be carefully examined.

xiii

The dose equivalent, H, measured in units of rem, is the product of the dose, D, the quality factor, Q, and the product of any other modifying factors symbolized by N. The dose equivalent is generally defined by the mathematical relationship

$$H = DQN.$$

For external radiations, N has been assigned a value of unity; Q is dependent on the linear energy transfer of the exposing radiation in water, which has been tabulated in simplified form for application to radiation protection.

In the older literature, usage of the radiation units described above differs from current usage, largely as a result of different and less precise definitions as well as less sophisticated means of measurement. Thus, the roentgen was used to characterize dose not only from photons but also from corpuscular radiations. The older literature should thus be carefully interpreted, although in general, for most penetrating photons, 1 R is approximately equal to 1 rad and to 1 rem. The situation for other radiations is similarly confused. An older unit, the rep (also known as the parker) was originally used to quantify dose or absorbed energy. The rep was defined as 93 erg/g, roughly equal to the present value of 100 erg/g for the rad.

The above system of units and quantities, hereinafter referred to as the old units, is currently being used in parallel with a new system of units based on fundamental physical quantities. By international agreement, the old units will be replaced by the new SI (International System) units in 1985—although there has been some resistance to the changeover, particularly in the United States—and can no longer be properly used in the scientific literature after this year. The becquerel (Bq) will thus replace the curie as the unit of activity; one becquerel is exactly equal to one nuclear transformation per second. The gray (Gy) will supplant the rad as the unit of dose; one Gy = 1 J/kg = 100 rad. The seivert (Sv) will replace the rem as the unit of dose equivalent; 1 Sv = Gy $\times$ Q $\times$ N = 100 rem. No special unit has been adopted to supplant the roentgen, which presumably will no longer be a valid scientific quantity after 1985; hence, units of C/kg must be used to express exposure in the special sense with 1 R = 2.58 $\times$ 10^{-4} C/kg.

Preparation and publication of a book during the transition period raises the vexing question of which system of units to use. The problem is exacerbated by the fact that the literature of environmental radioactivity is voluminous and has almost exclusively and quite comfortably used the old units. In many cases, specific multiples and submultiples that provided easily stated numerical values have been used; for example, concentrations of radioactivity in various media are often expressed in units of pCi/l.

In some respects, the SI system does not have the flexibility of the old system, and the numerical expression of quantities appears somewhat distorted to those used to the old system or to converting from the units used in the older literature. However, as 1985 is almost here, the decision was made to specify quantities in both systems of units. Thus, with the exception of the roentgen, for which an equivalent SI unit does not exist, and a few well-known and established American units, quantities taken from the literature have been given in the original units, along with a conversion to the units of the other system. Hopefully the conversions were made without error. This dual presentation of units, although redundant, possibly somewhat confusing, and inconsistent in the use of multiples and submultiples, seems to strike the best balance between the impending use of the SI system and the old units which have been in active use for so many years.

CHAPTER 1

Historical Introduction

Radioactivity and ionizing radiations have been part of our world since the beginnings of time. Our earth is perpetually bathed in a sea of ionizing radiations, and it and its atmosphere contain many different radioactive species. Everything—earth, air, water, plants, and animals—contain radioactivity, both from natural and anthropogenic sources. The amounts and kinds are constantly changing and are of interest both from the standpoint of pure science as well as from an environmental protection and public health point of view.

Knowledge of the existence of ionizing radiations and radioactivity is very recent, even when considered within the span of recorded history. Mankind has been aware of these phenomena for less than a century, but in that relatively short time span has uncovered a great deal about them. The story of the discovery of x-rays and radioactivity and of the people involved is among the most exiting in the history of science and technology, and appropriately serves as an introduction to the study of the radiological environment.

The Discovery of X-Rays

The discovery of x-rays preceded the discovery of radioactivity by only a few weeks, with the stage being set by the numerous and intensive studies of the cathode ray phenomena in the latter half of the nineteenth century. The study of cathode rays was greatly aided by the invention of the mercury air pump about 1865 by a German technician named Herman Sprengel. This device enabled experimenters to quickly obtain a high degree of vacuum and led to the developmment of highly evacuated glass tubes by Heinrich Geissler, a glassblower at the University of Bonn. Geissler tubes, as they came to be called, were small cylindrical glass tubes containing two electrodes that could be filled with various rare gases or air at very low pressures, and were widely used by experimenters studying the effects of electrical discharges in gases.

The first studies with the Geissler tubes were, quite predictably, made by Geissler himself in collaboration with Julius Plucker, then professor of mathematics at the University of Bonn. In 1865, these two experimenters observed a pale green fluorescence in the wall of an electrically energized Geissler tube opposite one of the electrodes. Unknown to them, this effect was attributable to the cathode rays—electrons freed from the cathode striking the glass wall of the tube. Plucker continued and expanded these early studies, showing that the direction of the rays emerging from the cathode could be altered by application of a magnetic field.

A few years later, in 1869, Johann Wilhelm Hittorf, a former student of Plucker and then a professor at the Academy of Munster, used these same evacuated tubes to establish that the fluorescence was attributable to the cathode rays. Hittorf inserted a small piece of metal in the shape of a Maltese cross into the path of the cathode rays within the tube. The metal blocked the rays, producing a fluorescent glow on the wall of the tube that clearly showed the outline of the Maltese cross. This work was verified and expanded by Eugen Goldstein, who named the cathode rays, and led British physicist G. F. Varley to correctly theorize that the cathode rays consisted of a stream of negatively charged particles of matter ejected from the cathode.

Many researchers were occupied by studies of the cathode rays. Sir William Crookes, a towering figure in British science of the late nineteenth century, performed numerous experiments, including a classic demonstration before the Royal Institute in London. On August 22, 1879, Crookes demonstrated the solid nature of the cathode rays by showing how they could drive a low friction wheel incorporated into an evacuated tube. Crookes theorized that the cathode rays were a fourth state of matter, and came close to making the discovery of x-rays himself, for he observed fogging of unused photographic plates in his laboratory, which he frequently returned to the manufacturers as defective. X-rays, inadvertently produced by Crookes but never recognized as such, were, of course, the cause of the fogging.

The work of Crookes was followed by that of the great German physicist Heinrich Hertz, who constructed a glass tube with a thin window to bring the cathode rays out into the air where they could be more easily studied. In 1891, Hertz was joined in his researches in Bonn by Philipp Lenard, a Hungarian physicist with whom he had a brief but productive association. The 37 year old Hertz was to die tragically in 1894 of blood poisoning, and in the space of a year and a half his pupil and associate Lenard moved on to Breslau and then Aachen. Although he made many observations of cathode rays during this time, Lenard failed to recognize that he was also creating a new and more pentrating radiation—the x-ray—as well.

Crookes and Lenard were not the only ones who unknowingly witnessed the effects of x-rays prior to the discovery. As early as 1700, Francis Hauksbee and the Abbe Nollet had seen extraordinary effects when experimenting with electrical discharges. Hauksbee later wrote ". . . the shape and figure of all parts of the hand could distinctly be seen." Whether this effect was in fact attributable to x-rays is uncertain; more certain is what took place in 1784 when William Morgan, a Welsh mathematician and actuary, inadvertently and unknowingly was experimenting with electrical discharges in evacuated tubes. In one series of experiments witnessed by Benjamin Franklin, the evacuated tube cracked, permitting a gradual inleakage of air to the still electrified tube. The magnificent color changes they observed—yellow-green, blue, purple, and finally red—are suggestive of the production of x-rays. Nearly a century later, Crookes observed similar effects in addition to the fogging of his photographic plates. And, in 1890, an American professor at the University of Pennsylvania, Arthur W. Goodspeed, accidentally made an x-ray photograph of some coins his assistant planned to use for carfare and had temporarily placed on a photographic plate near some Crookes tubes. When developed, the plate clearly showed the circular outlines of the coins, and as this was an unexpected and inexplicable event, was put aside in a drawer with an eye towards future study.

Thus the stage was set for the discovery of x-rays by Wilhelm Conrad Roentgen, a physics professor at the University of Wurzburg. On November 8, 1895, Roentgen was working alone in his laboratory, repeating Lenard's cathode ray experiments. Accordingly, he took an all glass Hittorf-Crookes tube (Figure 1-1), wrapped it carefully in black cardboard, and energized it with a Ruhmkorff coil. After establishing that the cardboard was indeed lighttight, Roentgen noticed a faint glow atop a laboratory table about a meter from the energized tube. Striking a match in his darkened laboratory, he was amazed to find the source of the light was a small barium platinocyanide screen. He quickly realized that this effect could not be caused by the cathode rays, which penetrated but a few millimeters in air, and in a flash of insight realized that what he was seeing was caused by a wholly new and as yet undiscovered kind of penetrating radiation.

Roentgen spent the next weeks in feverish activity, studying the prop-

FIGURE 1–1   Hittorf-Crookes tube of the type used by Roentgen to discover x-rays. This specific tube dates from the 1890's; the dark spot in the glass wall at the large end is a result of irradiation with cathode rays and x-rays.

erties of his newly discovered rays. The last few days in December were spent assembling his notes and preparing a paper on his findings which he delivered to the secretary of the Wurzburg Physical Medical Society with the somewhat unusual request that it be published in the *Proceedings* even though it not yet been presented orally at one of the Society meetings. The secretary, himself a scientist, recognized the significance of the discovery, and the report was printed in the next issue bearing the date December 28, 1895. On New Year's Day, 1896, Roentgen sent off reprints of his report along with some x-ray pictures to several of his colleagues, one of whom, Franz Exner in Vienna, broke the news to the press. The story first appeared in the Viennese newspapers on January 5, notifying the world of this extraordinary new discovery.

All in all, Roentgen published but three papers, dated December 28, 1895, March 9, 1896, and March 10, 1897, totalling 34 pages of text on the x-rays. Each was a model of scientific writing, both in clarity and content. So well had he done his work that little of a fundamental nature regarding the x-rays was uncovered for more than 15 years after the discovery. He was also gracious in recognizing the work of those who had preceded him, clearly acknowledging Lenard's work in his first paper, and specifically noting that Lenard and Hertz had observed that there were different kinds of cathode rays. But it was Roentgen who first had correctly interpreted the signals, and in so doing made the discovery, although an embittered Lenard attempted to claim the discovery for himself. The recognition and acclaim quite properly went to Roentgen, who in 1901 received the first Nobel Prize in Physics for his discovery.

The Discovery of Radioactivity

The discovery of radioactivity followed that of x-rays by only a few weeks. Henri Becquerel, a third generation French physicist, had undertaken a systematic study of the effects of sunlight on various phosphorescent substances, initiating the work with a sample of potassium uranyl sulfate that had previously been used for studying phosphorescence by his father. The experiment performed by the junior Bequerel consisted of placing the mineral on a photographic plate well wrapped in black paper to exclude any light, and then exposing the combination to the sun. Upon development of the plate, the outline of the mineral could readily be seen.

Initially, Becquerel concluded that the effect was attributable to the activation of the potassium uranyl sulfate by exposure to sunlight, and duly reported this to the Paris Academy of Science in early February 1896. Less than a month later he was back before the Academy, this time reporting the astonishing fact that the blackening of the photographic plate occurred even when the mineral had not been exposed to sunlight. Moreover, the effect was quantitatively greater because of the longer exposure time of the photographic plate to the mineral. Thus, on the second day of March, 1896, Becquerel reported his conclusion to the Academy: the uranium salt was spontaneously and continuously emitting rays that penetrated the opaque wrapping and exposed the photographic plate.

Certain parallels with x-rays lie in the discovery of radioactivity. Long before Becquerel, another French scientist, the obscure de St. Victor, had noted the blackening of photographic plates by uranium nitrate that had previously been exposed to sunlight. However de St. Victor failed to pursue his studies and thus did not recognize that the uranium itself, independently of the sunlight, was responsible for the effect. And, across the channel in England, Silvanus P. Thompson was carrying out studies similar to those of Becquerel at the same time, but apparently failed to realize that sunlight was not a necessary condition for the effect. Although Thompson thus came close to making the discovery—and indeed there is some evidence to suggest that he may have done so independently—he nonetheless graciously deferred to Becquerel, who was awarded the 1903 Nobel Prize in Physics for the discovery.

Uranium and its strange radiations aroused the curiosity of many scientists, including Pierre and Marie Curie, who began their studies of radioactivity in 1897. The first entry in their laboratory notebook relating to uranium was made by Marie on December 17 of that year and discussed measurement of the radiation from uranium by means of piezoelectrometer. Their first publication, authored by Marie, appeared in *Comptes Rendus* dated April 12, 1898, and reported that all uranium compounds were active, with the amount of activity proportional to the amount of uranium present. More significantly, she noted that certain ores of uranium, specifically pitchblende and chalcolite, exhibited more activity that could be accounted for by the uranium, thus suggesting the existence of a more active element than uranium. In July of the following year, the new element was isolated in a precipitate of bismuth and given the name polonium by Marie in honor of Poland, the country of her birth. In that same paper by Marie and Pierre appeared the first usage of the word 'radioactive'.

Late in 1898, the Curies succeeded in identifying another radioelement by its radiological properties. This they named radium, and were pleased to note its confirmation by the more conventional method of spectroscopy by their fellow countryman, M. Demarcay. However, before the two new elements could be accepted as such by the scientific community, it was necessary to establish their atomic weights, and to do this would require a much greater, albeit still tiny, quantity of the elements. For nearly four years, Marie labored in a small poorly lighted and ventilated shed, unheated in the winter and stifling in the summer, performing numerous fractional crystalizations of pitchblende ore from Czechoslovakia. In 1902, she realized her goal, isolating 100 mg of the pure chloride of radium, and unequivocally establishing its atomic weight. The following year, she shared the Nobel Prize in Physics along with her husband Pierre for this discovery.

As a member of the highly reactive Group IIA of the Periodic Chart, radium was not prepared in metallic form until 1910, also by Marie Curie, who obtained the purified metal by passing an electric current through a quantity of molton radium chloride. For this and related efforts, she shared the Nobel Prize in Chemistry in 1911, becoming the first person to ever receive two Nobel awards. Ironically, despite her extraordinary scientific achievements, and numerous other awards and honors, including election to membership in virtually every major scientific academy in the world, this shy and modest lady never was accorded such recognition by the country to which she had brought such fame, denied membership in the French Academy because she was a woman. She died in 1934 of leukemia, a disease quite possibly induced by her long exposure to the radioactive elements whose mysteries she did so much to unlock.

Environmental Radioactivity: Early Studies

In the first few years of the twentieth century (and especially prior to 1905) numerous studies and discoveries were made of the radioactivity and radioactive properties of the environment. Early studies of the radioactivity in air were carried out by Hans Geitel in Germany and C. T. R. Wilson in Scotland. In 1900, these two investigators independently discovered that an electroscope could be discharged simply from the air inside it, and not from insulator leakage as had previously been thought to be the case. From this rather simple discovery came the realization that the air contained radioactive components.

The following year Geitel, in collaboration with his colleague Julius Elster, performed what was for the time a bold experiment to extract the radioactivity from the air. Elster and Geitel strung a wire 20 meters long well above the ground and electrified it with a potential of 600 volts. After a period of hours, the wire was taken down and placed in an electroscope where it produced a rapid and large discharge. The effect was observed irrespective of whether a positive or negative discharge was applied to the wire. Further experiments showed that the discharge of the electrometer was attributable to radioactivity deposited on the wire, which Elster and Geitel were able to remove by rubbing with a piece of ammonia saturated leather. The material removed from the wire in this manner was decidedly radioactive: it would readily expose a photographic plate and would cause a barium platinocyanide screen to fluoresce. Further observations showed that the radioactivity had a half-life of about 30 minutes. These pioneering experiments were quickly confirmed by the great Ernest Rutherford and his colleague H. S. Allan at McGill University in Montreal, who also showed that the activity deposited on the wires emitted both alpha and beta particles.

In 1902, C. T. R. Wilson turned his attention to rain, evaporating freshly collected samples to dryness and examining the residue in an electroscope. As had been the case with air, radioactivity with a half-life of about 30 minutes was observed. As no such activity was seen in tap water, Wilson correctly concluded that radioactivity was removed from the air by the rain. In further work he was able to extract the radioactivity from the rain by precipitating it with barium following the addition of barium chloride and sulfuric acid, or by precipitation with aluminum by the addition of alum and ammonia. Independent studies were made of the radioactivity in snow by Wilson in Scotland and Allan and J. C. McLennan in Montreal. All three observed radioactivity with the characteristic 30 minute half-life. McLennan also observed that the radioactivity content of the air was reduced by a prolonged snowfall.

Elster and Geitel also initiated pioneering studies of terrestrial radioactivity, noting in 1902 that the air in caves had abnormally high concentrations of radioactivity. To test whether this was a self-induced effect—i.e. whether stagnant air itself might produce radioactivity—or was in fact emanating from the soil, they sealed a large boiler and examined the radioactivity content of air in it after several weeks. No increase—in fact a decrease—in the radioactivity content of the air was seen. To establish that the soil was producing the radioactivity observed in the air, Elster and Geitel sank a pipe into the ground and withdrew air from it, noting that this air contained more radioactivity than air collected above ground. They thus concluded that the earth itself was responsible for a radioactive emanation that gradually and continually diffused into the atmosphere.

In Munich, H. Ebert and P. Ewers confirmed of the terrestrial radioactivity emanation and also measured its half life as 3.2 days, very close to the modern value of 3.825 days for ^{222}Rn. Ebert was able to extract the radon emanation from soil by cooling it with liquid air and determined that it was identical to the radium emanation or radon gas that Rutherford had associated with the decay of radium. Radium and uranium were clearly a constituent of normal soils, and Rutherford and others, following the somewhat startling observation by Mme. Curie that radium was always warmer than its surrounding medium, busied themselves with calculations and observations of the amount of heat generated by radium. And, a few years later, British scientists N. R. Campbell and A. B. Wood showed that two elements heretofore considered stable—potassium and rubidium—were in fact radioactive and emitted beta particles. Numerous studies were conducted by these investigators and others of the radioactivity content of soil and rocks from all over the world. By 1910, so much data had been gathered on the radioactivity content of the earth and various kinds of rocks that Madame Curie found it necessary to devote several pages of her classic *Traite de Radioactivite* to tables summarizing the findings to that time.

The effects of weather on airborne radioactivity were also studied by

Elster and Geitel, who observed and characterized both seasonal and diurnal variations. These pioneers of the study of environmental radioactivity were able to establish an inverse relationship between airborne radioactivity and barometric pressure, and also observed that the radioactivity content of the air increased at temperatures below 0° C. Based on his own measurements and those of others, Rutherford was able to conclude as early as 1905 that variability in the radioactivity content of the air could be attributable to the radium content of the soil in the region in which the measurements were made.

Attempts to quantify the activity in the air were largely unsucessful. However as early as 1902, A. S. Eve, working with Rutherford in Montreal, measured the activity in the air in a large sealed tank, and determined that the quantity of radium emanation in one cubic kilo meter of air was approximately equal to the emanation from 0.56 g of radium bromide. This corresponds to 12.2 Bq/m^3 (3.3 × 10^{-10} μCi/cm^3), a typical value for air in that region. On the basis of this measurement, Rutherford computed the total quantity of atmospheric radon to be equivalent to the amount produced from 400 tons of radium bromide.

By 1902, it had been well established that radioactivity was ubiquitous in the environment, and several investigators turned their attention towards determination of the amount of radium and other naturally occurring radioactivity in common substances. In England, R. J. Strutt (later Lord Rayleigh) made measurements by lining an ordinary electroscope with various ordinary materials, publishing the tabulation below which demonstrated the considerable variability in the radioactivity content of ordinary matter:

Material	Electroscope Leakage
Tinfoil	3.3
Tinfoil (another sample)	2.3
Glass coated with phosphoric acid	1.3
Silver deposited on glass	1.6
Zinc	1.2
Lead	2.2
Copper	2.3
Copper (oxidized)	1.7
Platinum (3 samples)	2.0, 2.9, 3.9
Aluminum	1.4

Such studies were highly popular and carried out by many investigators.

The radioactivity in the oceans was examined prior to 1910 by Strutt who observed a concentration of 2.3×10^{-15} g of radium per gram of seawater, roughly equivalent to 850 μBq/m^3 (2 fCi/cm^3). Eve observed slightly lower concentrations in waters from the North Atlantic while Joly found significantly greater (and more in line with contemporary) values, particularly in the Indian and Mediterannean Oceans. Examination of waters in hot and mineral springs in general showed significantly greater concentrations of radioactivity than in other surface waters. The activity was almost invariably from radium and its radon daughter, leading to the speculation that radium was responsible for the curative or benefical effects of various spas. Helium was also found to be associated with the spring waters, lending support to the theory that this gas was produced as a result of the decay of radium.

Early Studies of Cosmic Rays

The early years of the twentieth century also saw the discovery of cosmic rays which had in origins in independent observations made by C. T. R. Wilson and Elster and Geitel in 1900. These experimenters observed small residual ionization in electroscopes, an effect originally thought to be the result of insulator leakage. The effect, however, could be lessened by shielding the electroscope with lead, suggesting that an external source of penetrating radiation was the cause. Later, Rutherford, in conjunction with H. L. Cooke at McGill University in Montreal, and independently J. C. McLennan in Ontario, observed a highly penetrating radiation inside buildings, even when using shielded electroscopes. The effect from this puzzling penetrating radiation was far greater than could be accounted for by the gamma rays associated with the radium in the earth and in the walls of the building, and the Canadians were unable to explain the cause what they had observed.

Similar penetrating radiations were observed in measurements made atop the Eiffel Tower in Paris by T. Wulf in 1909. As these measurements were made high above the earth, the influence of the radium and other naturally occurring radioactive substances should have been negligible, and hence such strongly positive results were questionable. But the following year, these findings were confirmed and enlarged upon by Swiss scientist Albert Gockel who sent electroscopes aloft in balloons to eliminate possible interferences from terrestrial radiations. His results were also not what might be expected if terrestrial sources were the cause of the mysterious penetrating radiations, for the intensity of the penetrating radiations increased with altitude. This work was further extended a short time later, first by Victor F. Hess in Austria who made measurements at altitudes to 16,000 feet (5 km) and then by W. Kolhorster in Germany who extended the altitude to 30,000 feet (9 km). It was Hess who apparently was the first to recognize that these highly penetrating radiations were not terrestrial, but were in fact extraterrestrial in origin.

After an interruption of some eight years (1914–22) caused by World War I, cosmic ray research continued, with the center of activity shifting from Europe to the United States. Largely throught the efforts of Robert A. Millikan, first at the University of Chicago and later the California Institute of Technology, numerous studies were made of the penetrating cosmic rays. Collaborating with I. S. Bowen in the spring of 1922, Millikan extended the high altitude measurements into the stratosphere, reaching a height of 10 miles (16 km) in a series of balloon experiments at Kelly Field, Texas. In other experiments performed at Muir Lake and Lake Arrowhead in California, he examined the penetrating power of the cosmic rays, noting that 68 feet (20.7 m) of water were required for 'complete' attenuation. On the basis of these and similar experiments, Millikan established conclusively the extraterrestrial nature of these penetrating radiations for which he coined the name 'cosmic rays' in 1925. The cosmic ray researches of Millikan and his chief collaborators, Carl Anderson, Seth Neddermeyer, Victor Neher, and William C. Pickering, led to the discovery of the positron and meson (then known as the mesotron) and also contributed greatly to the development of instrumentation and techniques for the measurement of low level radiation (Figure 1-2).

Two of the early investigators of the cosmic ray phenomenon—Wilson and Millikan—became Nobel laureates in Physics. For the Scottish-born Wilson, the award came in 1927, largely for his development of the cloud chamber, a device in which the paths of charged particles such as those produced in cosmic ray interactions could be made visible by the condensation of vapor on them. Millikan, who received his award in 1923, became the first American born scientist (but the third American after Michaelson and Einstein) to be so honored, receiving the award largely for work not directly related to cosmic rays, his measurement of the charge on the electron which he performed in conjunction with his graduate student Harvey Fletcher.

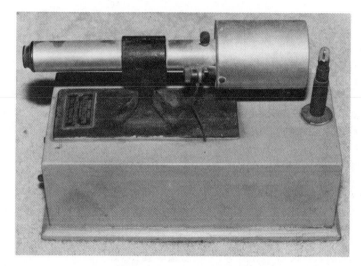

FIGURE 1–2   Electroscopes of this type were used in early cosmic ray studies as well as for studying radioactivity. The microscope eyepiece at left was used to observe an internal fiber whose deflection was governed by the amount of ionization in the chamber (top right). The bulb at the far right provided illuminaion for the fiber and calibrated scale through a window (not visible) in the end of the chamber. The box on which the unit is mounted was used to hold the battery power supply.

Applications of Radioactivity

Given the enormous interest generated by the discovery of x-rays and radioactivity, it is not at all surprising that practical applications were quickly made. The potential value of the x-rays in medical diagnosis was recognized immediately, and mentioned in the very same edition of the Vienna *Presse* that carried the announcement of the discovery. Within a month, several physicians had used the x-rays for diagnostic purposes, and within a half year, the x-ray was firmly entrenched as a prime diagnostic aid of the physician.

Numerous other applications were proposed for the x-rays, some quite reasonable and others bizarre. X-ray baths were advocated as beneficial for treatment of various diseases, including tuberculosis, and x-raying the head was suggested with perfect sobriety by more than one reputable therapist as a means of curing criminal behaviour. One serious experimenter even claimed to have induced a dog to salivate simply be projecting an x-ray image of a bone onto the poor beast's brain. A bill to ban the use of x-rays in opera glasses, thereby preserving the modesty of the ladies on the stage, came close to passage in the New Jersey legislature, and demonstrations of x-rays abounded at county and state fairs and similar gatherings at which people were given the opportunity to view their own bones with aid of a crude fluoroscope.

The x-ray craze died away within a few years and was replaced in the mind of the public by a fascination with radium, which was thought to have near magical curative properties. This gave rise to exploitation by quacks and charlatans who offered for sale to a gullible public a variety of radium based patent medicines and nostrums beginning about 1910, and lasting well into the 1930's. *Radol*, the creation of a charlatan with the adopted name of Dr. Rupert Wells was among the earliest of these products, and was billed as 'radium impregnated' and claimed as a cure for all forms of cancer. *Radol*, however, was not radium enriched, but merely an acidic solution of quinine sulfate with added alcohol.

Not all patent medicines were free from radioactivity in significant or harmful amounts. *Radithor* was a popular radium bearing water allegedly beneficial for the treatment of no less than 160 diseases and conditions. It was supplied in two ounce bottles which sold at the relatively high price for the times of more than a dollar, and was implicated in the deaths of several persons including a well known Pittsburgh industrialist, E. M. Byers, who died from radium poisoning in 1932. Byers habit was to drink several bottles of *Radithor* daily, and at the time of his death his body contained an estimated 36 μCi (13.3 MBq) of ^{226}Ra, a value which can be compared with the modern maximum permissible body burden for occupationally exposed individuals of 0.1 μCi (3.7 KBq).

Radium waters were highly popular as a tonic for many years and several patented devices were sold which would enable these waters to be made at home. The *Thomas Radium Cone* was one such device, and consisted of a simple cone of sintered ore about six inches tall from which radium would leach when placed in a container of water. A more sophisticated device was a ceramic crock lined with radium bearing ore known as the *Revigator* (Figure 1-3). This device was patented in 1912. It was equipped with a spring loaded spigot through which the radium bearing water could be drawn off. The directions called for the crock to filled with fresh water each night, with consumption of six to eight glasses the following day. Water treated in this manner would typically contain on the order of 50 pCi/l (1.85 KBq/m^3) of radium and 1300 pCi/l (481 KBq/m^3) of gross alpha-beta activity; a person following the directions would thus ingest about 0.026 μCi (1 KBq) annually. Over the years, *Revigator* crocks have become collectors' items, and are occasionally found for sale

FIGURE 1-3   A *Revigator* radium crock, popular about 1912. The capacity of this unit is about 6 liters. Directions for use were provided on the sides, and called for filling with clear water nightly, with the water to be drunk the following day after having steeped in the ore lined crock overnight.

in antique stores and flea markets. They have also been put to use as flower pots and pickling crocks, with potential serious consequences, one instance of the latter having come to light in Oregon as late as the mid-1970's.

Radium and other naturally occurring radioactive materials appeared in a wide variety of health aids and cosmetics including belts, pads, tissue and facial creams, salves, hair tonics, mouthwashes, tooth pastes, and even candy bars (Figure 1-4). Radioactive belts enjoyed some popularity, and were designed to be wrapped around the body with the active portion centered over the afflicted organs. Among the more curious devices to capitalize on the supposed healing properties of radium was the *Radium Ear*, a gadget that was designed to be hung over the outer ear with a metal tube which was to be inserted in the ear canal. The tube contained a mysterious radioactive subtance known as 'Hearium' which originally

FIGURE 1–4   An advertisement for various preparations of radium, ca. 1915. Belief in the general broad therapeutic and tonic properties of radium was widespread during the firt half of the twentieth century.

was not radioactive but was later replaced with a low grade radium bearing ore. Another such device was a pad impregnated with 95% radium sulfate (said by the manufacturer to be superior to a competitor's radium bromide) sold by the Los Angeles based International Radium Company. The pad was affixed over the nose by a strap that encircled the head, and, if worn, allegedly would cure catarrh, which the company claimed caused every disease in the body.

Waters in spas and mineral springs were, upon analysis, frequently found to have elevated concentrations of natural radioactivity and were bottled and sold to the public as tonics or for their supposed curative properties. At the spas themselves, the practices were not limited to bathing and drinking the water, but were extended to exposure to the radon which was collected in a specially enclosed room known as emanatorium or inhalatorium. At Badgastein, Austria, the underground chambers are fitted with bunk beds and attendents to care for the clientele. In the United States, The Free Enterprise Mine in Montana began operations in 1951, and was greatly aided by a favorable article in the popular magazine *LIFE* the following summer. According to a book written by one of its founders, exposure to the mine is beneficial for the treatment of arthritis, asthma, sinusitis, and similar ailments.

Radioactive toothpastes made their debut about 1920 with the appearance of a German product named *Radiogen*. This radium bearing product would suppposedly release a constant quantity of radon while the teeth were being brushed, promoting not only dental hygiene but aiding the digestive processes as well. During World War II the top secret Alsos project was created by the Allies to follow the German progress towards development of an atomic bomb. In 1944, the Alsos team determined that a German chemist known to have an interest in radium had obtained all the available thorium in occupied France and ordered it shipped to Germany. As this amount of thorium was more than adequate for ordinary industrial uses for all of Europe for the next twenty years, the suspicions of the team were aroused. The mystery was unravelled in the best cloak-and-dagger tradition by dint of good intelligence and a little luck. The Alsos mission determined that there was no sinister purpose earmarked for the thorium; officials of a large German chemical company, preparing for the postwar years, planned to take advantage of some patents they owned for thoriated toothpaste which the company hoped to market with the aid of American mass merchandising techniques.

Radium captivated and mesmerized the American public for decades after its discovery. The great American 'serpentine lady', dancer Loie Fuller, created a series of dances using fluorescent salts and pitchblende residues. Radium was essential to the plot of the popular Broadway musical *Piff*, *Paff*, *Pouff* and to the mystery novel *The Radium Terrors*, both written before 1910. As with the x-rays, some suggested applications were humorous, even though made with complete seriousness. A chicken farmer, noting the internal heat generated by radioactive decay, suggested that radium be mixed with chicken feed to produce self-incubating or even self-cooking eggs. And, in the 1920's, at least two radium bearing fertilizers were successfully marketed. On a more practical level, radium was successfully applied to industrial radiography because its high energy gamma rays provided much greater penetrating power than could be achieved with x-rays.

The most widespread and well known application of radium was the manufacture of self-luminous compounds and paints which enjoyed enormous popularity. The idea of self-luminous compounds was not new, predating the discovery of radium by more than a century. Naturally phosphorescent biological materials were used by American patriot David Bushnell to illuminate the interior of the one man submarine he invented for use in the Revolutionary War. In the 1870's, a luminous compound based on calcium sulfide was patented but was not widely used as it produced phosphorescence with only a limited lifetime, and required previous exposure to light to activate it.

The commercial potential for a continuous permanent or long-lived self-luminizing material using radium as the activating agent were realized quickly. As early as 1903, George Kunz, a jeweller at Tiffany's in New York City, prepared a self-luminous radium compound which was painted on the face of a watch. The material was less than satisfactory, however, and never found widespread usage. Radioluminescent paints and compounds were not perfected until World War I, largely through the efforts of Austrian born physician Sabin A. von Sohocky, who pioneered the used of radium dials in German U-boats. After the war, von Sohocky emigrated to the United States where he became technical director of the U.S. Radium Company until his untimely death at the age of 44 in 1926 from aplastic anemia (leukemia?) undoubtedly induced by radium poisoning.

Radium dial painting (Figure 1-5) was a procedure almost exclusively carried out by young women who would use their lips to moisten their brushes to produce a pointed tip. In so doing, they ingested significant amounts of radium; they were also exposed to the airborne radioactivity and external gamma radiation associated with use of radium. In 1924, New York City dentist Theodore Blum mentioned a new disease he called 'radium jaw', prevalent among dial painters, in a footnote to an article he published in the *Journal of the American Dental Association*. This was read by Harrison Martland, then the medical examiner for

FIGURE 1-5 Radium dial painters at work in a watch factory in 1922. An extensive study of the radium dial painters is being carried out in Argonne National Laboratory; many of the women shown in this picture have been identified and have participated in this study.

18

Essex County, New Jersey, who started on a personal study of the radium dial painters which was to continue for more than 20 years. Ironically, it was Martland who signed von Sohocky's death certificate in 1926. Despite vastly improved industrial hygiene measures, many radium dial painters were to suffer illness and tragically early and painful deaths from their exposure to this naturally occurring radioelement, whose use in the United States for radioluminescent paints ceased in the United States during the mid-1970's.

References

1. Badash, L. 1979. *Radioactivity in America*, The Johns Hopkins University Press, Baltimore.
2. Cajori, F. 1962. *A History of Physics*, Dover, New York.
3. Cramp, A. J. 1936. *Nostrums and Quackery and Pseudomedicine*, Volume III, American Medical Association, Chicago.
4. Curie, E. 1937. *Madame Curie*, Doubleday Doran, New York.
5. Curie, Mme. P. 1910. *Traite de Radioactivite*, Gauthier-Villiers, Paris.
6. Dampier, W. C. 1966. *A History of Science*, Fourth Edition, Cambridge University Press, Cambridge.
7. Glasser, O. 1943. *Wilhelm Conrad Roentgen and the Early History of the Roentgen Rays*, Charles C.Thomas, Springfield.
8. Glasstone, S. 1968. *Sourcebook on Atomic Energy*, D. Van Nostrand, New York.
9. Goudsmit, S. 1947. *Alsos*, Henry Schuman, New York.
10. Grigg, E. R. N. 1965. *The Trail of the Invisible Light*, Charles C. Thomas, Springfield.
11. Heisenburg, W. 1946. *Cosmic Rays*, Translated by T. H. Johnson, Dover, New York.
12. Jones, H. C. 1906. *Electrical Nature of Matter and Radioactivity*, D. Van Nostrand, New York.
13. Kallet, A. and F. J. Schlink. 1933. *100,000,000 Guniea Pigs*, Grosset and Dunlap, New York.
14. Kathren, R. L. 1979. "Historical Development of Radiation Protection and Measurement" in *Handbook of Radiation Protection and Measurement*, A. Brodsky, Ed., CRC Press, Boca Raton.
15. Lewis, W. V. 1955. *Arthritis and Radioactivity*, Christopher Publishing House, Boston.
16. Millikan, R. A. 1935. *Electrons (+ and -), Protons, Photons, Neutrons and Cosmic Rays*, University of Chicago Press, Chicago.
17. Rutherford, E. 1905. *Radioactivity*, Second Edition, Cambridge University Press, Cambridge.
18. Tilden, Sir W. A. 1926. *Chemical Discovery and Invention in the Twentieth Century*, Fith Edition, Dutton, New York.
19. Whetham, W. C. D. 1904. *The Recent Development of Physical Science*, John Murray, London.
20. Young, J. H. 1967. *The Medical Messiahs*, Princeton University Press, Princeton.

CHAPTER 2

Radioactivity and Radiations of Cosmic Origin

Cosmic Rays

The term cosmic rays or cosmic radiation refers to primary energetic particles of extraterrestrial origin that enter the earth's atmosphere, and to the secondary radiations produced by their interactions with the atmosphere. Cosmic rays can be conveniently divided into galactic rays, which originate outside our solar system, and solar radiations which are emitted by the sun. Primary galactic radiation impinges isotropically upon the earth's atmosphere and is largely composed of protons (87%) and helium nuclei (11%), with a small fraction of nuclei with atomic numbers greater than three and electrons (Table 2–1). In general, the elemental constituents of the primary galactic cosmic rays are basically the same as those of the known universe (Table 2–2), although the relative abundances may differ slightly but are still within experimental error (Tobias and Wallace 1961).

Primary cosmic radiations are highly energetic and penetrating, having energies ranging to 10^{20} eV, and possibly even higher (UNSCEAR 1977, 1982). Typically, most have energies in the range of 10^8 to 10^{11} eV, with a mean of about 10^{10} eV. Since the earth's magnetic field makes it impossible for particles below a given energy to reach the top of the atmosphere at certain latitudes, it is possible unfold the original energy spectrum of the primary galactic radiation based on high altitude measurements at various locations. For protons, which constitute most of the primary galactic cosmic ray intensity, the integral energy spectrum, N_p, given as the number of protons/cm^2-sec-ster with energies $>E$ expressed in MeV, is (Tobias and Wallace 1961)

$$N_p \, (> E) = \frac{0.3}{(1 + E)^{1.5}} \qquad (2\text{-}1)$$

TABLE 2-1   Composition of Galactic Radiation.

Particle	Range of Abundance	Mean Abundance
Protons	75–89%	87%
He Nuclei	10–18%	11%
Nuclei with $Z > 2$	1–7%	1%
Electrons	—	1%

This empirical equation is valid for protons in the energy range of 5×10^6 to 2×10^{10} MeV and yields an expectedly low mean energy of about 4×10^9 MeV as it does not take into account the higher end of the energy spectrum. Similar data have been developed for other constituents of cosmic rays and are shown in Figure 2-1. The average cosmic ray energy fluence rate at the top of the atmosphere is 2×10^3 MeV/cm^2 - sec, which, assuming a mean particle energy of 10^{10} eV, indicates a particle fluence rate incident on the earth of 0.2 per square centimeter per second.

In general, the cosmic radiation field is constant with time outside the solar system, but within the solar system is modified by disordered mag-

TABLE 2-2   Atomic Abundance in the Universe and in Cosmic Rays.

Z	Element	Atoms per 10^5 H atoms	Relative Cosmic Ray Abundance
1	Hydrogen	100,000	100,000
2	Helium	7,700	15,500
3–5	Li, Be, B	0.1	240
6	Carbon	23	260
7	Nitrogen	46	?
8	Oxygen	63	260
10	Neon	2.6–70	30
12	Magnesium	2.5	40
14	Silicon	2.9	30
26	Iron	5	30
>10		30	400
<30, not listed		2.7	100
$30 < Z < 92$		0.1	<10

Taken from Tobias and Wallace, p. 422.

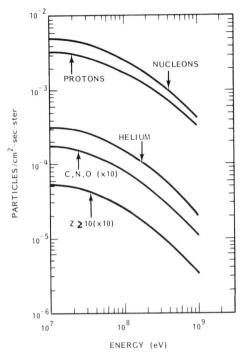

FIGURE 2-1   Primary Cosmic Ray Flux as a Function of Energy. (Tobias and Wallace 1961).

netic fields in the solar wind, or the extension of the solar corona out into space. Thus, the intensity of galactic cosmic rays with energies greater than 1 GeV varies by a factor of 3 following an 11 year cycle along with sunspots. The magnetic field of the earth also affects cosmic ray intensity and in fact provides the lower energy cutoff which ranges from 0 at the magnetic poles to 16 GeV at the magnetic equator for particles penetrating the atmosphere.

Based on track etch studies of meteorites, the cosmic ray fluence rate has remained more or less constant for at least 2,000 years. Studies based on terrrestrial cosmic ray induced and meteoritic radionuclides suggest that the fluence rate has not changed by more than a factor of 2 over the past 10^9 years (UNSCEAR 1977). Maximum levels occurred 700,000 years ago as a result of magnetic field reversals, but only represented a 10 per cent increase in fluence rate. The origins of the galactic radiations are unknown, although the subject of much speculation, as an enormous,

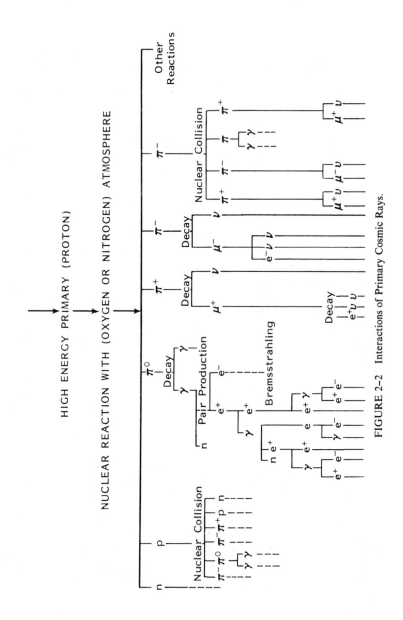

FIGURE 2-2 Interactions of Primary Cosmic Rays.

almost incomprehensible energy source is suggested by the knowledge gained to date (Adams 1952, Scarsi 1960).

Primary cosmic rays undergo various and complex interactions in the atmosphere, and a simplified scheme giving the principal reactions is shown in Figure 2-2 (NCRP 1976). Principal reactions include the following:

$$p + air \rightarrow p + n + \pi^+ + \pi^0 \tag{1}$$
$$n + air \rightarrow p + n + \pi^+ + \gamma^0 \tag{2}$$
$$\pi^+ \rightarrow \mu^+ + \gamma \tag{3}$$
$$\pi^0 \rightarrow 2\gamma + 4e^+ \tag{4}$$
$$\mu^+ \rightarrow e^+ + 2\gamma \text{ (etc.)} \tag{5}$$

Equations (1) and (2) are nucleonic cascades that illustrate the production of mesons from the interactions of protons and neutrons with the atmosphere. Equation (3) shows the production of muons and Equation (4), pion decay. The final equation shows muon decay to both negatrons and positrons and an electromagnetic cascade.

The probability of specific reactions taking place is highly energy dependent, and basically can be characterized as shown in Table 2-3. The highest energy events occur at the top of the atmosphere, produced largely by heavy charged particles—protons, alpha particles and other heavy nuclei—colliding with air nuclei. These inelastic nuclear collisions produce a number of relativistic pions, which travel roughly in the same direction (i.e. towards the surface of the earth) as the incident cosmic particle. Elastic collisions involving smaller energy transfers also take place, producing protons, neutrons, alpha particles, and occasionally

TABLE 2-3 Energy Relationship of Cosmic Ray Interactions.

Energy Range	Typical Interactions
<10 MeV	Electromagnetic radiation, neutron production, etc.
10–300 MeV	Spallation, fission, nuclear excitation followed by neutron, proton emission, etc.
300 Mev–1 BeV	Meson production
>1 Bev	Mesons, electron-photon cascades, heavy primary thindown events
>10 BeV	Build-up of secondaries

Adapted from Pickering et al., p. 383.

larger fragments of the interacting nucleus. Deeper in the atmosphere, the interactions are less energetic, and involve these secondary particles to a great extent. At sea level, nearly all of the ionization from cosmic rays is from secondary electromagnetic radiations, produced largely from incident primary radiations with energies greater than 10 GeV. About 965 MeV/cm^2-sec-ster are expended in particle production by cosmic rays in the atmosphere of which about two-thirds (615 MeV/cm^2-sec-ster) goes into ionization, and about a fourth (232 MeV/cm^2-sec-ster) into neutrinos (Tobias and Wallace 1961).

Solar radiation consists almost exclusively of protons and helium nuclei with energies to several GeV. Energies are usually in the range of 1 to 100 MeV. Bursts or flares of solar radiation tend to follow sunspot activity. The solar wind is a continuous flux outward from the sun of very low (<1 keV) particles which cannot penetrate the earth's magnetic field to reach the atmosphere, and hence is of no significance from an environmental point of view. Solar radiation also generates few secondaries, and thus contributes very little to the radiation intensity at ground level.

In addition to the radiations discussed above, two radiation belts known as the Van Allen belts after their discoverer, extend outward from the earth. These belts consist mainly of large numbers of protons and electons with energies ranging from a few kilovolts to a few BeV that have been trapped by the earth's magnetic field. Thus they spiral along the lines of magnetic force, with reflection occurring as the magnetic field converges near the poles, producing a gap in the belts. The Van Allen belts do not produce measurable radiation levels on earth, being situated at closest approach about eight thousand miles above the equator. Although neither the solar wind nor geomagnetically trapped radiations contribute significantly to the radiation fields or doses at the surface of the earth, and hence of of no radiobiological consequence to ordinary surface dwellers, these radiations can be important sources of exposure in manned space flight or even high altitude flights by the supersonic transport (Langham 1967).

Variations in Cosmic Rays

Variations in cosmic ray intensity at the surface of the earth are primarily the result of variations in the geomagnetic field of the earth, which varies with time, latitude, and altitude. Temporal variations have been observed directly for about a half century, and occur in cycles of 11 years, (seemingly associated with or following the sunspot cycle), 1 year, 27 days, and 1 day. The source of the latitude effect is basically geomagnetic, and is related to the location of the earth's magnetic poles. Solar radiation increases during periods of sunspot activity and solar flares, but inasmuch as the secondaries from galactic primary radiations constitute most of the ionizing cosmic radiations at sea level, the effect on the ionization density at the surface of the earth is small. At any given location on the earth, sea level ionization from cosmic rays will vary by less than 10 per cent, even during a large solar flare.

Related to temporal variation is the so-called Forbush effect, which was first observed in the late 1930's by E. M. Forbush. The Forbush effect is a characteristic sharp decrease in the primary cosmic ray fluence rate which occurs about twenty hours after a solar flare, and simultaneously with the sudden commencement of a geomagnetic storm or burst of geomagnetic activity of the sun. Forbush events occur at random intervals, gnerally on the order of a few times per year, and are of limited duration and effect, reducing the galactic cosmic ray intensity on earth by 10 per cent or less.

Variation of cosmic ray intensity with latitude at the earth's surface is small, and attributable to the magnetic field of the earth, which has the form of a dipole located 215 miles (310 km) from the center with the poles at 79° N, 69° W, in northwest Greenland, and at 76° S, 121° E in Antarctica. This produces the geomagnetic latitude effect illustrated in Figure 2–3.

From an environmental standpoint, the most important variation in cosmic ray intensity is altitude or distance above the surface of the earth (Figure 2–4). The cosmic ray dose rate approximately doubles with each 1800 meter rise in altitude, at least for the first 10 km or so above the surface of the earth. This variability is largely due to the change in attenuation brought about by the decreasing thickness and density of air, which results in reduced shielding as one ascends from the surface. The dose equivalent rate, however, shows a rather different variability, since it is determined not only by the energy deposition but also by the quality factor which is a function of the linear energy transfer of the exposing radiation. Thus, in any discussion of environmental radiation, care must be taken to distinguish the dose from the dose equivalent as these related but different quantities are frequently confused and erroneously used interchangeably.

Cosmic ray intensities are usually expressed in ionization density rate (I) or ion pairs per cm per second at standard temperature and pressure. At sea level, I is typically on the order of 2.22. Dose rates at sea level are

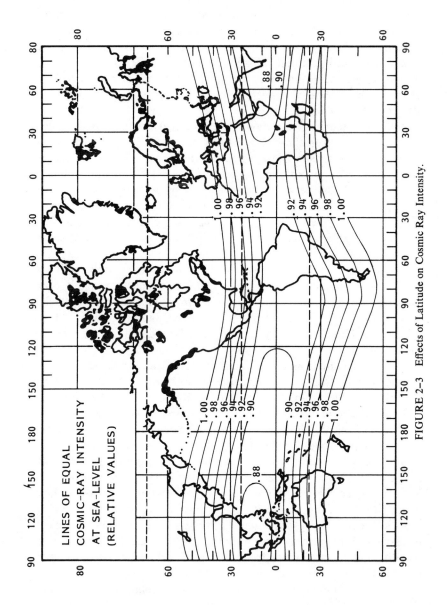

FIGURE 2–3   Effects of Latitude on Cosmic Ray Intensity.

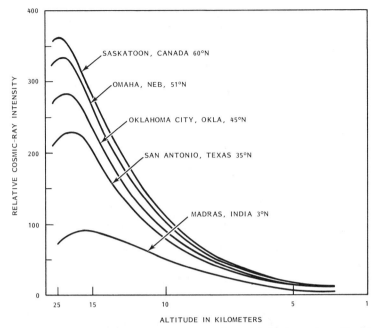

FIGURE 2-4   Effects of Altitude on Cosmic Ray Intensity. (Redrawn from Glasstone 1950).

on the order of 1.5 μrad/h (15 nGy/h) (NCRP 1975) and 1.65 μrem/h (16.5 nSv/h) for the dose equivalent rate (0akley 1972). The corresponds to a sea level dose rate of about 25 to 30 millirad (250–300 μGy) per year with the cosmic ray dose rate at higher altitudes propotionately greater. In Denver, for example, the dose rate is about twice that at sea level.

The neutron component of the total dose is quite small, generally on the order of about 0.7 mrad/y (7 μGy/y). However, the quality factor for neutrons associated with cosmic rays is relatively large, ranging from 3 to 8 depending on the investigator. The higher value of 8, however, appears to be more likely correct and is more commonly used, resulting in an annual dose equivalent rate of 5.6 mrem (56 μSv). Table 2–4 shows the estimated mean annual whole body dose equivalent for the United States and its territories. The high values for the Rocky Mountain States are related to the relatively high altitude at which the population in these

TABLE 2-4　Annual Cosmic Ray Dose Equivalents in the United States.

Political Unit	mSv/y	mrem/y	Political Unit	mSv/y	mrem/y
Alabama	0.40	40	New Jersey	0.40	40
Alaska	0.45	45	New Mexico	1.05	105
Arizona	0.60	60	New York	0.45	45
Arkansas	0.40	40	North Carolina	0.45	45
California	0.40	40	North Dakota	0.60	60
Colorado	1.20	120	Ohio	0.50	50
Connecticut	0.40	40	Oklahoma	0.50	50
Delaware	0.40	40	Oregon	0.50	50
Florida	0.35	35	Pennsylvania	0.45	45
Georgia	0.40	40	Rhode Island	0.40	40
Hawaii	0.30	30	South Carolina	0.40	40
Idaho	0.85	85	South Dakota	0.70	70
Illinois	0.45	45	Tennessee	0.45	45
Indiana	0.45	45	Texas	0.45	45
Iowa	0.50	50	Utah	1.15	115
Kansas	0.50	50	Vermont	0.50	50
Kentucky	0.45	45	Virginia	0.45	45
Louisiana	0.35	35	Washington	0.50	50
Maine	0.50	50	West Virginia	0.50	50
Maryland	0.40	40	Wisconsin	0.50	50
Massachusetts	0.40	40	Wyoming	1.30	130
Michigan	0.50	50	District of		
Minnesota	0.55	55	Columbia	0.40	40
Mississippi	0.40	40	Puerto Rico	0.30	30
Missouri	0.45	45	Guam	0.35	35
Montana	0.90	90	Samoa	0.30	30
Nebraska	0.75	75	Virgin Islands	0.30	30
Nevada	0.85	85			
New Hampshire	0.45	45	U.S. Mean	0.45	45

From Klement et al., 1972, p. 10

differences among the sea level states and territories (compare, for example, Hawaii, Florida, District of Columbia, Guam, and Samoa) is attributable to the geomagnetic latitude location.

The altitude variations of cosmic rays are of special significance not only to people resident at high altitudes, but also to air travellers and in space exploration. Modern commercial jet aircraft fly at an altitude of approximately 35,000 feet (11 km), and a cross country traveller might incur a dose of 2–5 mrad (20–50 μGy) from a round trip flight on a typical subsonic jet aircraft flight; transatlantic travellers in the SST would incur similar dose equivalents, although their exposure would be at a greater rate because of the greater altitude of flight (19 km) but exposure would be for a shorter time period because of the greater rate of speed (UNSCEAR 1982). Typically, the quality factor for galactic exposures is about 1.7; hence the dose equivalents incurred by air travellers are about 5–8 mrem (50–80 μSv) on a round trip cross-country flight (O'Brien 1975).

Incurred dose equivalents could be considerably greater for supersonic travellers if flight was undertaken during times of solar flares, perhaps increasing by an order of magnitude. Similarly, astronauts incur higher doses during their orbital flights; on the earth orbital American Apollo VII (August 1968) and IX (February 1969) missions, the astronauts incurred doses of 120 and 200 mrad (1.2 and 2 mGy) respectively. Doses incurred on the circumlunar Apollo VIII (December 1968) and X (May 1969) missions were 185 and 410 mrad (1.85 and 4.1 mGy). The four Apollo lunar landings in 1969 and 1971 resulted in doses in the range of 200–500 mrad (2–5 mGy), the greatest occurring on the Apollo XIV mission in January 1971. Similar doses were incurred by astronauts on the Soviet Vostok, Voskhad and Soyuz missions (UNSCEAR 1982).

Cosmogenic Radioactivity

Several radionuclides are produced by the reactions of cosmic rays with air, largely from spallation and neutron capture. Some radioactivity may also be added to the earth's environment from extraterrestrial dust and meteorites, although this source is relatively small. Approximately 1×10^7 kg of dust bombards the earth from outer space each year, containing radioactivity at concentrations of up to 450 pCi/kg (17 Bq/kg). Assuming the maximum concentration of 450 pCi/kg (17 Bq/kg), the upper limit for radioactivity from this source is only $(450 \times 10^{-12}$ Ci/kg$)(10^7$ kg$) = 4.5 \times 10^{-3}$ Ci $= 4.5$ mCi $(1.7 \times 10^8$ Bq). This activity is mostly from lighter nuclides—^{7}Be, ^{22}Na, ^{26}Al, ^{46}Sc, ^{48}V, ^{51}Cr, ^{53,54}Mn, ^{56,57,58,60}Co, and ^{59}Ni, although thorium, uranium, and other heavy elements have been detected in meteoritic materials (Klement 1982).

Most cosmogenic radioactivity is produced from spallation, a process in which a nucleus is split into several lighter nuclei by collision with a high energy particle, usually a neutron. The term "spallation", derived from the verb "to spall", which means to break up by chipping off small fragments, was coined by W. H. Sullivan and first used by Glen Seaborg in 1947 to describe the breakup of intermediate mass particles bombarded by 400 MeV alpha particles or 200 MeV deuterons in the 184 inch cycloton at the University of California (now Lawrence) Radiation Laboratory (Glasstone 1950). A similar process, observed with cosmic ray interactions, had been previously known as star formation or simply stars because of the multipronged tracks left behind on photographic emulsions. Spallation reactions are typically have thresholds around 50 MeV, although in a few cases the thresholds may rise to a few hundred MeV.

Thermal neutron reactions are involved in the production of at least two cosmogenic radionuclides, ^{14}C by the (n,p) reaction on ^{14}N, and ^{81}Kr by the (n,γ) reaction on ^{80}Kr. Although by far the bulk of the cosmogenic radionuclides are produced by interactions with air, a wide variety of nuclear species is potentially possible from cosmic ray reactions with the earth and from cosmic ray induced fissions (Klement 1982).

Cosmogenic radionuclides (Table 2–5) are generally light (low Z) elements with half-lives ranging from 32 min (^{34m}Cl) to 2.5 million years (^{10}Be). The shorter lived ones usually decay before settling to earth and entering the ecosphere. The decay and settling processes may vary considerably with altitude; variation in ^{7}Be concentrations over two orders

TABLE 2–5   Principal Cosmogenic Radionuclides.

Nuclide	Half-life	Production Rate (atoms/ cm²-sec)	Global Inventory (kg)	Air Activity Bq/m³
H-3	12.26 y	0.25	3.5	0.167
Be-7	53 d	0.08	0.0032	0.017
Be-10	2,700,000 y	0.05	3.9×10^5	10^{-7}
C-14	5760 y	2.5	6.8×10^4	0.067
Na-22	2.6 y	8.6×10^{-5}	0.0019	1.7×10^{-6}
Al-25	740,000 y	1.4×10^{-5}	1000	—
Si-32	280 y	1.6×10^{-4}	1.4	3.3×10^{-8}
P-32	14.3 d	8.1×10^{-4}	0.0004	0.00033
P-33	24.4 d	5.8×10^{-4}	0.0006	0.00025
S-35	87.9 d	0.0014	0.0045	0.00025
Cl-36	380,000 y	0.0011	1.4×10^4	5×10^{-10}
Ar-39	270 y	0.0056	23	—
Kr-81	210,000 y	10^{-6}	16.2	—

of magnitude have been recorded with even greater variation noted for ^{24}Na (NCRP 1975). Smaller but still significant variations may occur with latitude; for example, in the case of ^{24}Na, roughly a 5-fold variation has been noted over the latitude 0 to 60° N, with the minumum concentrations occurring at the equator.

Production may vary considerably with altitude and latitude, although overall has been essentially constant with time for at least the past 1000 years. Approximately 70% of the ^{7}Be is produced in the stratosphere, with the remaining 30% produced in the troposphere. Activity produced in the stratosphere has a residence time of about a year and is transferred to the troposphere; the residence time in the troposphere is about six weeks. Transfer to the earth's surface is largely accomplished by gravitational settling and precipitation processes. Concentrations of cosmogenic nuclides may thus vary significantly with altitude, and also with latitude, not only because of the location of production, but because of atmospheric mixing process and half-life. However, with the exception of the normally gaseous elements (the two argon isotopes, tritium, and ^{81}Kr) and shorter lived nuclides and ^{22}Na, virtually all of the cosmogenic activity—99+%—is found in ocean sediments or the lithosphere.

Of the cosmogenic radionuclides, only ^{3}H and ^{14}C are of importance from a biological standpoint. While ^{7}Be, ^{22}Na, and other cosmogenic nuclides may also be present in living things, the rates of production and amounts and frequently the half-lives are so small that these nuclides are of little, if any biological significance. Tritium, however, is produced in relatively large amounts by cosmogenic processes. The world inventory of cosmic ray produced tritium is 70 MCi (2.59×10^{18} Bq), which corresponds to an annual rate of production of 4 MCi (1.48×10^{17}) per year (ICRP 79). This is but a small fraction of the current world inventory,which has been greatly swelled by nuclear weapons tests (Chapter 5). Tritium is oxidized or exchanges with ordinary hydrogen to form tritiated water which is transferred from the atmosphere to the surface of the earth by precipitation. Surface deposition is greatest in the latitudes where tritium production is higest and also over the oceans, with the observed deposition velocity ranging from 0.4–0.8 cm/s.

Once in the ocean, tritium mixes and exchanges horizontally within the upper warm mixing layer, which supports most of the marine life and has a depth of 50 to 100 meters. The residence time here is on the order of 22 years. Mixing in lakes and small bodies of waters may be more vertical, with the tritium ultimately reaching the ocean via streams, rivers, and other surface runoff, or through the groundwater. Typically, cosmogenic tritium concentrations in ground waters are on the order of 19 pCi/l (700 Bq/m^3). As tritium is an isotope of hydrogen, it finds its way into all living things, which derive most of their tritium content from tritiated water. For man, the annual dose from cosmogenic tritium is about 1 μrad (10 nGy) (NCRP 1975).

Of even greater bioenvironmental importance than tritium is ^{14}C, which has been extensively studied in relation to the biosphere. Carbon-14 is produced by the (n,p) reaction on ^{14}N. Most of the production takes place in the stratosphere, and the ^{14}C thus produced is oxidized to $^{14}CO_2$ by unknown processes. Atmospheric CO_2 has always contained ^{14}C from cosmogenic sources, containing about 13.5 disintegrations per minute (0.23 Bq) per gram of carbon.

Carbon is fundamental to life processes and is thus found throughout the biosphere. The radiocarbon cycle (Figure 2–5) describes the dynam-

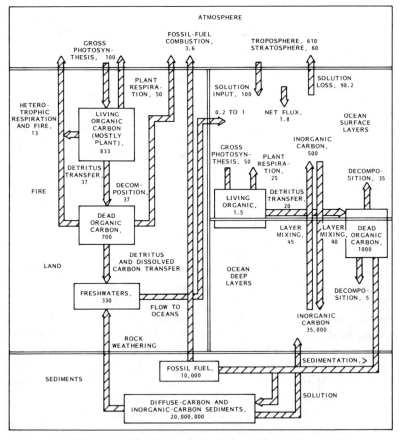

FIGURE 2-5 The Carbon Cycle. (Adapted from Reiners 1973).

ics of environmental ^{14}C (Reiners 1973). Cosmic ray produced ^{14}C is oxidized to carbon dioxide, taken up photosynthetically by plants, which in turn are eaten by animals. The death and excretory processes of plants and animals put organic ^{14}C into both the aquatic and terrestrial environment, where it remains in the active reservoir subject to weathering and other natural chemical and physical processes which may convert it to inorganic carbonates or geologically buried carbonaceous deposits such as coal or oil which constitute the dormant reservoir. Most of the carbon ultimately reaches the ocean, and is subject to evaporation and rainfall as well as biological processes. Carbon is returned to the atmosphere not only by evaporation, but also by combustion of fossil fuels, vulcanism, and weathering.

The above description is greatly simplified, for the carbon cycle (or CO_2 cycle as it is sometimes called) is highly complicated, and affected by many complex and as yet incompletely understood factors, both physical and biological (Livingstone 1973, Odum 1971, Reiners 1973). However, the distribution and amounts of ^{14}C are well known, at least on a worldwide basis (NCRP 1975). The worldwide inventory of natural ^{14}C is estimated at 310 MCi (1.15×10^{19} Bq), corresponding to 68,000 kg, distributed as shown (NCRP 1975):

Deep ocean	91.7%
Ocean sediments	0.4%
Upper mixed ocean layer	2.2%
Land surface	4 %
Troposphere	1.6%
Stratosphere	0.3%

The amount of radiocarbon, as well as its environmental dynamics, is affected by many factors (Livingstone 1973). The most important of these are weapons tests (Chapter 5), which have significantly increased the quantity in the environment, and the burning of fossil fuels with the concomittant release of carbon dioxide (Bacastrow and Keeling 1973). Although fossil fuel burning increases the overall quantity of ^{14}C in the environment, it contains a greatly reduced concentration of ^{14}C because of its long underground storage time and decreases the relative concentration of ^{14}C in the carbon at the surface of the earth. This is known as the Suess effect after Professor Hans Suess of the University of California at San Diego who first described it. In the nineteenth century, carbon at the earth's surface contained 6.1 pCi (0.22 Bq) of ^{14}C per gram. Fossil fuel burning reduced this by 3.2% by 1950, and 7% by 1970. By the year 2000, it is estimated that the total reduction from the Suess effect will be about 23% (Stuvier 1973).

The annual dose to man of about 1 mrad (10 μGy) from cosmogenic ^{14}C is the greatest by far of any dose delivered by the cosmogenic radionuclides, accounting for more than 99% of the total (O'Brien and McLaughlin 1972, NCRP 1975). However, the dose from cosmogenic radionuclides is small when compared to that from cosmic radiations, accounting for only about 1–3% of the total dose delivered to man by radiations and radioactivity of cosmic origin.

The short lived cosmogenic nuclides can have application as tracers in the study of various atmospheric processes (Young et al. 1970). Thus, ^{39}Cl (55 min), ^{38}Cl (37 min) and ^{7}Be (53 days) have been applied to the study of precipitation scavenging. As the ratios of the production rates of the short lived cosmogenic nuclides vary little as a function of altitude and latitude, these ratios can thus be utilized to determine the relative efficiencies of nucleation and post-nucleation scavenging.

References

1. Adams, R. V. 1952. "Nuclear Interactions of Cosmic Rays", in *Annual Review of Nuclear Science*, Vol. 1, Annual Reviews, Stanford, pp. 107–136.
2. Bacastrow, R., and C. D. Keeling. 1973. "Changes from A.D. 1700 to 2070 as Deduced from a Geochemical Model", in *Carbon and the Biosphere*, G. M. Woodwell and E. V. Pecan, Eds., U.S. Atomic Energy Commission Symposium Series 30, CONF-720510, pp. 86–135.
3. Glasstone, S. 1950. *Sourcebook on Atomic Energy*, D. Van Nostrand, New York, p. 259.
4. Klement, A. W. Jr. 1982. "Natural Sources of Environmental Radiation", in *Handbook of Environmental Radiation*, A. W. Klement, Jr., Ed., CRC Press, Boca Raton, pp. 5–21.
5. Klement, A. W., Jr., C. R. Miller, R. P. Minx, and B. Schleien. 1972. "Estimates of Ionizing Radiation Doses in the United States 1960–2000", U.S. Environmental Protection Agency, Publication ORP/CSD 72-1.
6. Korff, S. A. 1964. "Production of Neutrons by Cosmic Radiation", in *The Natural Radiation Environment*, J. A. S. Adams and W. A. Louder, Eds., University of Chicago Press, Chicago.
7. Langham, W. H. 1967. *Radiobiological Factors in Manned Space Flight*, National Academy of Sciences, Washington, D. C.
8. Livingstone, D. A. 1973. "The Biosphere", in *Carbon and the Biosphere*, G. M. Woodwell and E. V. Pecan, Eds., U.S. Atomic Energy Symposium Series 30, CONF-72051, pp. 1–5.

9. National Council on Radiation Protection and Measurements (NCRP). 1975. "Natural Background Radiation in the United States", NCRP Report No 45, Washington.

10. National Council on Radiation Protection and Measurements (NCRP). 1976. "Environmental Radiation Measurements", NCRP Report No. 50, Washington.

11. National Council on Radiation Protection and Measurements (NCRP). 1979. "Tritium in the Environment, NCRP Report No. 62, Washington.

12. O'Brien, K. 1972. "The Cosmic Ray Field at Ground Level", in *The Natural Radiation Environment II"*, Vol. 1, J. A. S. Adams, W. M. Lowder and T. F. Gesell, Eds., CONF–720805-P1, pp. 15–54.

13. O'Brien, K., and J. E. McLaughlin. 1972. "The Radiation Dose to Man From Galactic Cosmic Rays", *Health Phys.* 22:225.

14. Oakley, D. T. 1972. "Natural Radiation Exposure in the United States", U.S. Environmental Protection Agency, Report ORP/SID 71–1.

15. Odum, E. P. 1971. *Fundamentals of Ecology*, Third Edition, W. B. Saunders, Philadelphia.

16. Pickering, J. E., R. G. Allen, Jr., and O. L. Ritter. 1960. "Problems in Shielding", in *Medical and Biological Aspects of the Energies of Space*, P. A. Campbell, Ed.,Columbia University Press, New York (1960), pp 381–390.

17. Reiners, W. A. 1973. "Summary of World Carbon Cycle and Recommendations for Critical Research", in *Carbon and the Biosphere*, G. M. Woodwell and E. V. Pecan, Eds., U. S. Atomic Energy Commission Symposium Series 30, CONF–720510, pp. 368–382.

18. Scarsi, L. 1960. "Cosmic Radiation", *Am. J. Phys.* 28:213 (1960).

19. Stuiver, M. 1973. The C-14 Cycle and Its Implications for Mixing Rates in the Ocean-Atmosphere System", in *Carbon and the Biosphere*, G. M. Woodwell and E. V. Pecan, Eds., U. S. Atomic Energy Commission Symposium Series 30, CONF–72510, pp. 6–20.

20. Tobias, C. A., and R. Wallace. 1961. "Particulate Radiation: Electrons and Protons Up to Carbon", in *Medical and Biological Aspects of the Energies of Space*, P. A. Campbell, Ed., Columbia University Press, New York, pp. 421.

21. United Nations Scientific Committee on the Effects of Atomic Radiation (UNSCEAR). 1977. *Report to the General Assembly, Sources and Effects of Ionizing Radiation*, Publication U.N.E.77.IX.1, United Nations, New York.

22. United Nations Scientific Committee on the Effects of Atomic Radiation (UNSCEAR). 1982. *Ionizing Radiation: Sources and Biological Effects*, Publication E.82.IX.8, United Nations, New York.

23. Young, J. A., et al. 1970. "Short Lived Cosmic Ray-Produced Radionuclides as Tracers of Atmospheric Processes", in *Radionuclides in the Environment, Advances in Chemistry Series 93*, American Chemical Society, Washington, pp. 506–521.

CHAPTER 3

Terrestrial Radiations and Radioactivity

By far, the majority of the naturally occurring radioactivity found on earth is from internal or terrestrial sources. Primordial radionuclides are those with half-lives great enough relative to the age of the earth, estimated as 4.5×10^9 years, to remain in detectable amounts. Such nuclides may occur singly or as part of series or chains of radioactive species decaying sequentially. Other terrestrial radionuclides are short-lived and could not have been present when the earth was formed; these are constantly being produced from the decay of the long lived members of the series.

Terrestrial radionuclides contribute to the external radiation field (Table 3–1), as well as to the internal doses incurred by people. Virtually everything in our world contains trace amounts of radioactivity of terrestrial origin, in varying degrees. Thus, terrestrial radionuclides are indeed environmentally ubiquitous.

Singly Occurring Nuclides

At least twenty-two naturally occuring single or non-series primordial radionuclides have been identified (Table 3–2). Most of these have such long half-lives, small isotopic and elemental abundances and small biological uptake and concentration that they are of little significance in terms of environmental dose. Only two are important from a biological standpoint: ^{40}K and ^{87}Rb. Potassium-40 is a primordial radioisotope of potassium that is quite possibly the single most important radionuclide of terrestrial origin, at least from a biological standpoint. Potassium, a member of the highly reactive Group 1A alkali metal family, has three isotopes with mass numbers 39, 40, and 41; only ^{40}K is radioactive and has a half-life of 1.3×10^9 years. As potassium is essential to life, ^{40}K is found in all living and formerly living things.

TABLE 3–1 Range of Ambient External Radiation
Fields from Terrestrial Sources.

Location	mrad/y	mGy/y
Clallam Bay, WA	24	0.24
Typical U.S.A.	60	0.60
Denver, CO	114	1.14
Kerala, India	1,600	16
Black Forest	1,800	18
Central City, CO	2,200	22
Guarapari, Brazil	17,000	170
Portion of U.S.S.R.	70,000	700

Adapted from Whicker and Schultz (1982), p. 80.

TABLE 3–2 Primordial Singly Occurring Radionuclides.

Nuclide	Half-Life (years)	% Isotopic Abundance	Decay Mode	Energy (Mev)
K-40	1.3×10^9	0.0118	Beta	1.32
V-50	6×10^{14}	0.25	Beta	—
Rb-87	4.7×10^{10}	27.83	Beta	0.273
Cd-113	9×10^{15}	12.3	Beta	—
In-115	5×10^{14}	95.7	Beta	0.49
Te-123	1.2×10^{13}	0.87	EC	—
La-138	1.1×10^{11}	0.09	Beta	0.27
Ce-142	$>5 \times 10^{16}$	11.1	Alpha	1.5
Nd-144	2.1×10^{15}	23.9	Alpha	1.83
Sm-147	1.1×10^{11}	15.0	Alpha	2.23
Sm-148	8×10^{15}	11.2	Alpha	1.95
Sm-149	$>10^{16}$	13.8	Alpha	<2.0
Gd-152	1.1×10^{14}	0.20	Alpha	2.14
Dy-156	2×10^{14}	0.06	Alpha	3 (?)
Lu-176	2.7×10^{10}	2.6	Beta	0.57, 0.31
Hf-174	2×10^{15}	0.17	Alpha	2.50
Ta-180	$>1.6 \times 10^{13}$	0.012	Beta	—
Re-187	5×10^{10}	62.5	Beta	0.0026
Pt-190	7×10^{11}	0.013	Alpha	3.16
Pb-204	1.4×10^{17}	1.48	Alpha	2.6

Data from Lederer et al. (1977), *Table of Isotopes*.

The isotopic abundance of ^{40}K is small, only 0.012% of naturally occurring potassium, which gives a specific activity of 855 pCi/g (31.6 Bq/g) of natural potassium. The decay scheme for ^{40}K is shown in Figure 3–1; ^{40}K, with a half-life of 1.26 × 10^9 years, undergoes decay to stable ^{40}Ca 89% of the time, emitting a 1.314 MeV$_{max}$ beta particle in the process. With the exception of a tiny fraction of decays (1 × 10^{-3}%) by electron capture, ^{40}K undergoes decay by positron emission the remaining 11% of the time, emitting a characteristic photon with an energy of 1.460 MeV. This photon is highly useful for identification and quantification of ^{40}K by gamma spectrometry, and makes an excellent calibration point because of the presence of potassium in essentially all environmental samples.

Potassium is widely distributed in nature with concentrations ranging from about 0.1% for limestones to as much as 3.5% for some granites. Sandstones are intermediate, having a potassium content on the order 1% or less. Making adjustment for the natural abundance of ^{40}K, this translates a few parts per million or tens of picocuries per gram for most rocks and soils. For crustal rock, the mean ^{40}K activity is 17 pCi/g (0.63 Bq/ g), while some granites, particularly those low in calcium, and syenites may have concentrations exceeding 50 pCi/g (1.85 Bq/g) (Table 3–3). Soils are somewhat lower, with a mean of around 12 pCi (0.44 Bq)/g. However, soil concentrations can be appreciably altered by agricultural activities, particularly the application of potassium bearing fertilizers. Klement (1982) noted a greater than ten fold increase in the potassium content of soils under cultivation for 20 years when potassium fertilizers were used. Concentrations in seawater are approximately 300 pCi (11 Bq)/m^3.

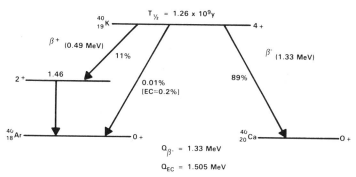

FIGURE 3–1   Decay Scheme of ^{40}K. (Lederer et al. 1977)

TABLE 3-3   Potassium-40 in the Earth's Crust.

Rock Type	Typical Concentration (μg/g)
Ultrabasic	0.0047
Basaltic	0.98
High Ca Granite	2.97
Low Ca Granite	4.96
Syenites	5.66
Shales	3.14
Sandstones	1.26
Carbonates	0.32
Deep Sea Sediments	
Carbonates	0.34
Clay	2.95

Adapted from Klement, 1982, p. 9.

From a biological standpoint, potassium is ubiquitous. A 70 kg (154 lb) man contains about 140 g of potassium, mostly in muscle (ICRP 76). Hence, given 855 pCi/g (31.6 Bq/g) as the specific activity of ^{40}K in potassium, this corresponds to 0.12 μCi (4440 Bq) of ^{40}K in the body. The quantity of potassium in the body is variable according to age and sex, generally decreasing linearly with age (Figure 3-2). The potassium-40 concentration (i.e. g of K per kg of body weight) in men is about 25–30% greater than in women of the same age, larely due to the larger muscle mass and lower fat content in men.

While the mean overall concentration of potassium in the soft tissues of the body is about 0.2%, there is considerable variability among the various tissues and organs. The lowest concentrations are found in the bones and teeth and range from 0.05 to 0.1% wet weight (0.4–0.9 pCi/g or 15-34 Bq/kg) and the fat (0.06% or 0.5 pCi/g [18.5 Bq/kg]). The highest concentrations are found in the muscles, which contain approximately 0.36% potassium by wet weight, which corresponds to about 3 pCi/l (111 Bq/kg). Slightly lower concentrations are found in the brain.

These differences have been put to practical use. Since lean muscle contains a higher percentage of potassium (and hence ^{40}K) than fat, *in vivo* whole body counting techniques can be used to obtain the body potassium content. When taken together with the body weight, the lean to fat ratio can be computed. This information may be of value medically

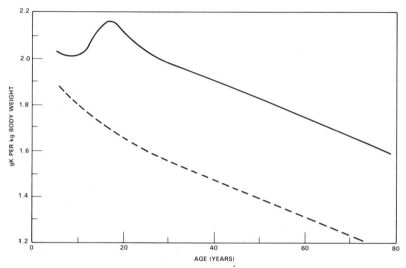

FIGURE 3–2   Potassium Content as a Function of Age and Sex. (UNSCEAR 1977)

and to some athletes; it has also been used with meat animals to determine the lean to fat ratio from certain diets.

Another significant aspect of the variability relates to the dose delivered to the various tissues. Potassium-40 is the major naturally occuring source of internal radiation dose, contributing up to a quarter of the naturally occuring background radiation dose. Obviously, those tissues with smaller concentrations of potassium will receive a lower dose from the ^{40}K incorporated in them. Typically, the soft tissues—those with the greatest concentrations of potassium such as the muscles and gonads—will incur a dose of about 19 mrad (190 μGy)/y from ^{40}K, about 85% of which is from the beta particles associated with the decay of the nuclide, and the remainder from photons. Because of its ubiquity in the environment and the 1.460 MeV photon associated with its decay, ^{40}K also delivers a small external radiation dose as well, which has been calculated as 0.18 μrad/h per pCi/g (0.05 μGy/h per Bq/g) of soil at a height of 1 m above the surface of the ground (Beck 1972).

The other singly occurring nuclide of significance is ^{87}Rb. Like potassium, rubidium is also an alkaline earth metal, and may replace potassium chemically within the body. Two isotopes of rubidium—85 and 87—have been found in nature. The radioactive one, ^{87}Rb, has an abundance of 27.8% and undergoes beta decay to stable Sr-87 with a half-life of 4.8×10^{10} years. The maximum energy of the ^{87}Rb beta is 0.274 Mev; no gamma rays are emitted.

As is the case with potassium, rubidium is widespread in trace amounts in soil and rock. Typically most rocks contain from 10 –200 ppm (0.2–0.5 pCi/g or 0.007–0.019 Bq/g), with granites having the highest concentrations. In most soils, ^{87}Rb concentrations are on the order of 10 ppm (0.2 pCi/g or 0.007 Bq/g). Seawater typically contains about 2.8 pCi/l (104 Bq/m^3), and fish and other marine life a few hundredths of a pico-curie (several ten thousandths of a Bq) per gram.

In humans, the concentration of ^{87}Rb is about 17 mg/kg, which corresponds to 400 pCi/kg (14.7 Bq/kg) or about 0.03 μCi (1 KBq) for a 70 kg man. This is about one-fourth the activity from ^{40}K, and delivers a dose of 0.3 mrad (3 μGy) annually to the soft tissues, 0.4 mrad (4 μGy)/y to small tissue inclusions in the bone, and 0.6 mrad (6 μGy)/y to the bone marrow (UNSCEAR 1977). The entire dose is from internally deposited rubidium, since, as has already been noted, there is no penetrating component.

As seen in Table 3–2, the remainder of the primordial radionuclides are largely isotopes of elements lying in the middle of the periodic chart, and have half-lives $> 10^{10}$ and usually around 10^{15} years. Some such as ^{147}Sm and ^{152}Gd are alpha emitters, but the majority undergo beta decay. In general, the alpha emitters are very long lived and emit alpha particles with relatively low energies. Several have appreciable abundances; in the case of bismuth, all of the naturally occurring element consists of Bi-209, which is weakly radioactive, having a half-life of $> 2 \times 10^{18}$ years. This long half-life, of course, results in a very low specific activity; the specific activity of naturally occurring bismuth (i.e. Bi-209) is 8.6×10^{-16} Ci/g (3.2×10^{-5} Bq/g), or, considering the inverse case, 1.17×10^{15} g of pure bismuth equal 1 Ci (31.4 kg = 1 Bq). Measurement of radionuclides with long half-lives is difficult and is usually accomplished by statistical determination of the time interval between decays; this interval can be sufficiently long and decays so infrequent that accuracy is poor. Indeed for many years the half-life of ^{209}Bi was thought to be an order of magnitude lower than the currently accepted value. And, there is a theory that all nuclides are radioactive, with those that are considered stable simply having half-lives that are too long to measure with the present state of the art. Since 2×10^{18} or thereabouts is the upper limit to measurable half-life, it is unlikely that very many more, if any, additional singly occuring primordial long-lived radionuclides will be discovered, or experimental verification of the theory obtained without advancements in techniques of measuring low specific activity radionuclides.

Chain or Series Decaying Radionuclides

There are four radioactive series, or chains of radionuclides. Any one series can be characterized in terms of the mass numbers of its constituents by the simple expression

$$A = 4n + m$$

in which A is the mass number, n is the largest whole integer divisible into A, and m is the remainder. Thus, the four possible series are 4n, 4n + 1, 4n + 2, and 4n + 3. Since radioactive decay of heavy nuclei is predominantly by alpha emission, which results in a stepwise reduction in mass number of four, or by beta emission which does not affect the mass number, a series can thus only have nuclides of one of the four possible types.

Three of these series are found in nature and account for most of the natural radioactivity of terrestrial origin. These are the 4n series, headed by ^{232}Th and also known as the thorium series; the 4n + 2 series, headed by ^{238}U and also known as the uranium series; and the 4n + 3 series, headed by ^{235}U but also known as the actinium series. Each of these series is headed by a radionuclide with a half-life that is long relative to the age of the earth. The fourth series, the 4n + 1 or neptunium series, has not been found in nature, except for its essentially stable end product ^{209}Bi. The 4n + 1 series is headed by plutonium-241, and with the exception of ^{209}Bi already noted, consists entirely of radionuclides with half-lives at least three orders of magnitude smaller than the age of the earth. Tables 3–4 through 3–7 show the individual nuclides in the four series, along with their half-lives and other salient radiological features, including the original historic names given to each at the time of discovery in the early twentieth century before it was known that these were isotopes of various elements.

The three series found in nature all have as their final product a stable isotope of lead. Thus, the 4n series ends up as ^{208}Pb, which has an abundance of 51.55%; the 4n + 2 series ends up as ^{206}Pb, which is 26.26% abundant; and the 4n + 3 series as ^{207}Pb, 20.8% abundant. These three radiogenic isotopes make up 98.6% of the lead found in nature. A fourth lead isotope, ^{204}Pb, with an abundance of 1.4% makes up essentially all the remainder. The origins of ^{204}Pb are uncertain; this nuclide is weakly radioactive, emitting an alpha particle and decaying to ^{200}Hg with a half-life of 1.4×10^{17} years, eight orders of magnitude greater than the age

TABLE 3-4 The Thorium (4n) Series

Nuclide	Historical name	Half-life	Major radiation energies (MeV) and intensities		
			α	β	γ
$^{232}_{90}$Th	Thorium	1.41×10^{10}y	3.95 (24%) 4.01 (76%)	---	---
$\longrightarrow$ $^{228}_{88}$Ra	Mesothorium I	6.7y	---	0.055 (100%)	---
$\longrightarrow$ $^{228}_{89}$Ac	Mesothorium II	6.13h	---	1.18 (35%) 1.75 (12%) 2.09 (12%)	0.34c (15%) 0.908 (25%) 0.96c (20%)
$\longrightarrow$ $^{228}_{90}$Th	Radiothorium	1.910y	5.34 (28%) 5.43 (71%)	---	0.084 (1.6%) 0.214 (0.3%)
$\longrightarrow$ $^{224}_{88}$Ra	Thorium X	3.64d	5.45 (6%) 5.68 (94%)	---	0.241 (3.7%)

	Half-life	α (MeV)	β (MeV)	γ (MeV)
Emanation Thoron (Tn)	55s	6.29 (100%)	---	0.55 (0.07%)
Thorium A	0.15s	6.78 (100%)	---	---
Thorium B	10.64h	---	0.346 (81%) 0.586 (14%)	0.239 (47%) 0.300 (3.2%)
Thorium C	60.6m	6.05 (25%) 6.09 (10%)	1.55 (5%) 2.26 (55%)	0.040 (2%) 0.727 (7%) 1.620 (1.8%)
Thorium C′	304ns	8.78 (100%)	---	---
Thorium C″	3.10m	---	1.28 (25%) 1.52 (21%) 1.80 (50%)	0.511 (23%) 0.583 (86%) 0.860 (12%) 2.614 (100%)
Thorium D	Stable	---	---	---

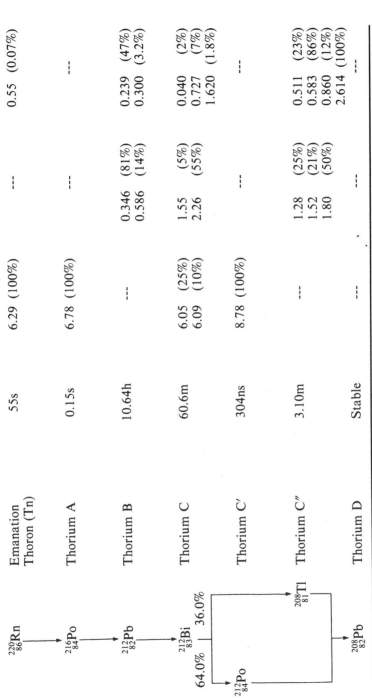

$^{220}_{86}$Rn → $^{216}_{84}$Po → $^{212}_{82}$Pb → $^{212}_{83}$Bi

64.0% → $^{212}_{84}$Po

36.0% → $^{208}_{81}$Tl

→ $^{208}_{82}$Pb

From *Radiological Health Handbook*, United States Department of Health, Education, and Welfare, Public Health Service, 1970.

47

TABLE 3-5 The Neptunium (4n + 1) Series

Nuclide	Element name	Half-life	Major radiation energies (MeV) and intensities		
			α	β	γ
$^{241}_{94}\text{Pu}$	Plutonium	13.2y	4.85 (0.0003%) 4.90 (0.0019%)	0.021 (~100%)	0.145 (.00016%)
$^{241}_{95}\text{Am}$	Americium	458y	5.44 (13%) 5.49 (85%)	--	0.060 (36%) 0.101c (0.04%)
$^{237}_{92}\text{U}$	Uranium	6.75d	--	0.248 (96%)	0.060 (36%) 0.208 (23%)
$^{237}_{93}\text{Np}$	Neptunium	2.14×10^6y	4.65c (12%) 4.78c (75%)	--	0.030 (14%) 0.086 (14%) 0.145 (1%)
$^{233}_{91}\text{Pa}$	Protactinium	27.0d	--	0.145 (37%) 0.257 (58%) 0.568 (5%)	0.31c (44%)
$^{233}_{92}\text{U}$	Uranium	1.62×10^5y	4.78 (15%) 4.82 (83%)	--	0.042 (?) 0.097 (?)
$^{229}_{90}\text{Th}$	Thorium	7340y	4.84 (58%) 4.90 (11%) 5.05 (7%)	--	0.137c (~3%) 0.20c (~10%)
$^{225}_{88}\text{Ra}$	Radium	14.8d	--	0.32 (100%)	0.040 (33%)

Decay chain:

$^{241}_{94}\text{Pu}$ — ~100% → $^{241}_{95}\text{Am}$; 0.0023% → $^{237}_{92}\text{U}$ → $^{237}_{93}\text{Np}$ → $^{233}_{91}\text{Pa}$ → $^{233}_{92}\text{U}$ → $^{229}_{90}\text{Th}$ → $^{225}_{88}\text{Ra}$ →

Element	Half-life	Alpha (MeV)	Beta (MeV)	Gamma (MeV)
Actinium	10.0d	5.73c (10%) 5.79 (28%) 5.83 (54%)	--	0.099 (?) 0.150 (?) 0.187 (?)
Francium	4.8m	6.12 (15%) 6.34 (82%)	--	0.218 (14%)
Astatine	0.032s	7.07 (~100%)	--	--
Bismuth	47m	5.87 (~2.2%)	1.39 (~97.8%)	0.437 (?)
Polonium	4.2μs	8.38 (~100%)	--	--
Thallium	2.2m	--	1.99 (100%)	0.12 (50%) 0.45 (100%) 1.56 (100%)
Lead	3.30h	--	0.637 (100%)	--
Bismuth	Stable (>2 × 10^{18}y)	--	--	--

$^{225}_{89}$Ac →
$^{221}_{87}$Fr →
$^{217}_{85}$At →
$^{213}_{83}$Bi
2.2% → $^{209}_{81}$Tl
97.8% → $^{213}_{84}$Po
→ $^{209}_{82}$Pb →
$^{209}_{83}$Bi

From *Radiological Health Handbook*, United States Department of Health, Education, and Welfare, Public Health Service, 1970.

TABLE 3-6 The Uranium (4n + 2) Series

Nuclide	Historical name	Half-life	Major radiation energies (MeV) and intensities		
			α	β	γ
$^{238}_{92}\text{U}$	Uranium I	$4.51 \times 10^9\text{y}$	4.15 (25%) 4.20 (75%)	---	---
$^{234}_{90}\text{Th}$	Uranium X$_1$	24.1d	---	0.103 (21%) 0.193 (79%)	0.063c (3.5%) 0.093c (4%)
$^{234}_{91}\text{Pa}^{\text{m}}$	Uranium X$_2$	1.17m	---	2.29 (98%)	0.765 (0.30%) 1.001 (0.60%)
$^{234}_{91}\text{Pa}$	Uranium Z	6.75h	---	0.53 (66%) 1.13 (13%)	0.100 (50%) 0.70 (24%) 0.90 (70%)
$^{234}_{92}\text{U}$	Uranium II	$2.47 \times 10^5\text{y}$	4.72 (28%) 4.77 (72%)	---	0.053 (0.2%)
$^{230}_{90}\text{Th}$	Ionium	$8.0 \times 10^4\text{y}$	4.62 (24%) 4.68 (76%)	---	0.068 (0.6%) 0.142 (0.07%)
$^{226}_{88}\text{Ra}$	Radium	1602y	4.60 (6%) 4.78 (95%)	---	0.186 (4%)
$^{222}_{86}\text{Rn}$	Emanation Radon (Rn)	3.823d	5.49 (100%)	---	0.510 (0.07%)

99.87%   0.13%

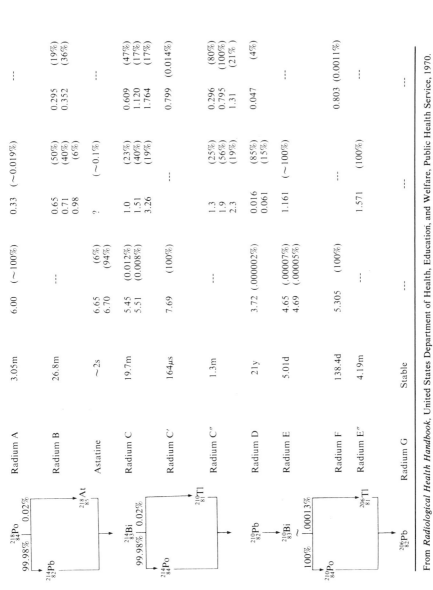

	Half-life	Alpha (MeV)	Beta (MeV)	Gamma (MeV)
Radium A	3.05m	6.00 (~100%)	0.33 (~0.019%)	---
Radium B	26.8m	---	0.65 (50%), 0.71 (40%), 0.98 (6%)	0.295 (19%), 0.352 (36%)
Astatine	~2s	6.65 (6%), 6.70 (94%)	? (~0.1%)	---
Radium C	19.7m	5.45 (0.012%), 5.51 (0.008%)	1.0 (23%), 1.51 (40%), 3.26 (19%)	0.609 (47%), 1.120 (17%), 1.764 (17%)
Radium C'	164μs	7.69 (100%)	---	0.799 (0.014%)
Radium C"	1.3m	---	1.3 (25%), 1.9 (56%), 2.3 (19%)	0.296 (80%), 0.795 (100%), 1.31 (21%)
Radium D	21y	3.72 (.000002%)	0.016 (85%), 0.061 (15%)	0.047 (4%)
Radium E	5.01d	4.65 (.00007%), 4.69 (.00005%)	1.161 (~100%)	
Radium F	138.4d	5.305 (100%)	---	0.803 (0.0011%)
Radium E"	4.19m	---	1.571 (100%)	---
Radium G	Stable	---	---	---

From *Radiological Health Handbook*, United States Department of Health, Education, and Welfare, Public Health Service, 1970.

TABLE 3-7 The Actinium (4n + 3) Series.

Nuclide	Historical name	Half-life	Major radiation energies (MeV) and intensities		
			α	β	γ
$^{235}_{92}$U	Actinouranium	7.1×10^8y	4.37 (18%) 4.40 (57%) 4.58c (8%)	---	0.143 (11%) 0.185 (54%) 0.204 (5%)
$^{231}_{90}$Th	Uranium Y	25.5h	---	0.140 (45%) 0.220 (15%) 0.305 (40%)	0.026 (2%) 0.084c (10%)
$^{231}_{91}$Pa	Protoactinium	3.25×10^4y	4.95 (22%) 5.01 (24%) 5.02 (23%)	---	0.027 (6%) 0.29c (6%)
$^{227}_{89}$Ac	Actinium	21.6y	4.86c (0.18%) 4.95c (1.2%)	0.043 (~99%)	0.070 (0.08%)
$^{227}_{90}$Th	Radioactinium	18.2d	5.76 (21%) 5.98 (24%) 6.04 (23%)	---	0.050 (8%) 0.237c (15%) 0.31c (8%)
$^{223}_{87}$Fr	Actinium K	22m	5.44 (~0.005%)	1.15 (~100%)	0.050 (40%) 0.080 (13%) 0.234 (4%)
$^{223}_{88}$Ra	Actinium X	11.43d	5.61 (26%) 5.71 (54%) 5.75 (9%)	---	0.149c (10%) 0.270 (10%) 0.33c (6%)

Decay chain:
$^{235}_{92}$U → $^{231}_{90}$Th → $^{231}_{91}$Pa → $^{227}_{89}$Ac ; 1.4% → $^{223}_{87}$Fr ; 98.6% → $^{227}_{90}$Th → $^{223}_{88}$Ra →

52

Name	Half-life	Alpha	Beta	Gamma
Emanation Actinon (An)	4.0s	6.42 (8%) 6.55 (11%) 6.82 (81%)	---	0.272 (9%) 0.401 (5%)
Actinium A	1.78ms	7.38 (~100%)	0.74 (~.00023%)	---
Actinium B	36.1m	---	0.29 (1.4%) 0.56 (9.4%) 1.39 (87.5%)	0.405 (3.4%) 0.427 (1.8%) 0.832 (3.4%)
Astatine	~0.1ms	8.01 (~100%)	---	---
Actinium C	2.15m	6.28 (16%) 6.62 (84%)	0.60 (0.28%)	0.351 (14%)
Actinium C'	0.52s	7.45 (99%)	---	0.570 (0.5%) 0.90 (0.5%)
Actinium C"	4.79m	---	1.44 (99.8%)	0.897 (0.16%)
Actinium D	Stable	---	---	---

$^{219}_{86}Rn \xrightarrow{\sim 100\%} {}^{216}_{84}Po \xrightarrow{.00023\%} {}^{215}_{85}At$

$^{216}_{84}Po \longrightarrow {}^{211}_{82}Pb \longrightarrow {}^{211}_{83}Bi \xrightarrow{99.7\%} {}^{207}_{81}Tl$

$^{211}_{83}Bi \xrightarrow{0.28\%} {}^{211}_{84}Po \longrightarrow {}^{207}_{82}Pb$

From *Radiological Health Handbook*, United States Department of Health, Education, and Welfare, Public Health Service, 1970.

53

TABLE 3–8   Typical Uranium Concentrations in Crustal Rock.

Rock Type	Uranium Concentration (ppm)
Ultrabasic	0.001
Basaltic	1
Granites	3
Syenites	3
Shales	3.7
Sandstones	0.45
Carbonates	2.2
Deep Sea Clays	1.3

Adapted from Klement, 1982, p. 9.

of the earth. Thus, the ^{204}Pb may be primordial in origin, having been present at the formation of the earth. Another possibility is suggested by noting that ^{204}Pb has a 4n configuration, and could be an extension of the 4n series beyond ^{208}Pb by the following additional chain:

$$\text{Pb-208} \xrightarrow{\beta} \text{Bi-208} \xrightarrow{\alpha} \text{Pb-204}$$

This sequence would fit in very nicely with the theory that all nuclides are in fact radioactive, mentioned above. Certainly such a decay chain is pociated with all three naturally occurring series (Table 3–8). Three isotopes of uranium are found in nature: ^{238}U, which has an abundance of 99.27%; ^{235}U, 0.72% abundant; and ^{234}U, 0.0057% abundant. Typically, most crustal rocks contain a few parts possible, for ^{208}Pb may simply have a half-life that is too great to be measured. However, in that case, ^{208}Bi would normally be found along with ^{208}Pb and ^{204}Pb in lead ores, but this has not been observed.

Series Decay Relationships

The radioactivity decay or activity-time relationship of any radionuclide can be relatively easily derived. The activity, A, or number of atoms decaying in a time period can also be written as N/t, in which N is the number of atoms decaying and t is the period of time. This is, of course, simply the rate of decay. If there are initially N atoms, the initial rate of decay will be N/t, but inasmuch as the radioactive atoms are constantly being removed from the system by decay, the rate of decay, or the number of atoms decaying in a unit time interval, continually becomes smaller. This change in the decay rate can be expressed in differential notation as -dN/dt, being negative as it is decreasing with time, and is mathematically equal to the number of atoms, N, times the probability of an atom undergoing decay in a time interval, which is represented by λ as shown in Equation 3–1

$$-dN/dt = \lambda N \qquad (3-1)$$

This equation can be rearranged

$$\frac{dN}{N} = -\lambda \, dt$$

and both sides integrated with respect to time from zero to infinity

$$\int_0^\infty \frac{dN}{N} = \int_0^\infty -\lambda \, dt$$
$$\ln\left(\frac{N}{N_0}\right) = -\lambda t$$

and the terms rearranged to give

$$N = N_0 e^{-\lambda t} \qquad (3-2)$$

The final form, Equation 3–2, is known as the fundamental decay law; in this equation, N is the number of radioactive atoms remaining at any time t, N_0 is the original number of atoms (i.e., the number of atoms at zero time), and λ is the decay constant for the particular radionuclide, expressed in units of reciprocal time.

The half-life is related to the decay constant, or probability of decay, by Equation 3–3, which is derived by setting $N/N_0 = 0.5$ as shown:

$$\frac{N}{N_0} = 0.5 = e^{-\lambda t_{1/2}}$$

taking the logarithm of both sides, to give

$$\ln 0.5 = -\lambda t_{1/2}$$

and solving for $t_{1/2}$:

$$t_{1/2} = \frac{\ln 0.5}{-\lambda} = \frac{0.693}{\lambda} \tag{3-3}$$

It should be noted that the half-life, $T_{1/2}$, is the time required for half of the atoms of a specific radionuclide to decay; it is not, however, the mean lifetime of an individual atom. The mean or average life, τ, is inversely proportional to the decay constant, as shown in Equation 3-4 which is derived below:

$$\tau = -\frac{1}{N_0} \int_0^\infty t \, dN = \frac{1}{N_0} \int_0^\infty t\lambda N \, dt$$

$$\tau = \lambda \int_0^\infty te^{-\lambda t} = -\left[\frac{\lambda t + 1}{\lambda} e^{-\lambda t}\right]_0^\infty \tag{3-4}$$

$$\tau = \frac{1}{\lambda}$$

The form of the fundamental decay law given in Equation 3-2 is applicable with either activity, A, or number of atoms, N, since one is a constant multiple of the other $(A = \lambda N)$ and applies only in the case of radioactive parent decaying to a stable daughter. The activity relationships among members of a chain may be very complex, and are related to the decay constants of the individual members of the series, which, as can be seen from Tables 3-4 to 3-7, are highly variable. If only a two member radioactive chain is considered—that is, a chain with a radioactive parent giving rise to a radioactive daughter, three possibilities can be seen to exist:

1) the half-life of the parent can be greater than that of the daughter (in which case the decay constant of the parent is smaller than that of the daughter).

2) the half-life of the parent and daughter are equal (or nearly so).

3) the half-life of the parent is less than that of the daughter.

With these in mind, the concept of radioactive equilibrium between radioactive parent and daughter can be examined.

There are basically three conditions of equilibrium between a radioactive parent and radioactive daughter: secular, transient, and no equilibrium. In the first case, secular equilibrium (Figure 3-3), the half-life of the parent is very much greater than that of the daughter, and, given

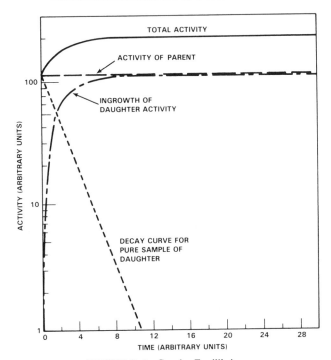

FIGURE 3-3   Secular Equilibrium.

an intially pure parent, the activity of the daughter will gradually increase or grow in until it is exactly equal to that of the parent, or

$$\lambda_1 N_1 = \lambda_2 N_2 \qquad\qquad (3-5)$$

in which the subscript 1 refers to the parent and the subscript 2 to the daughter.

At equilibrium, then, the total activity is twice what was originally present from the initially pure parent; reduction in activity of the parent may be neglected inasmuch as its half-life is so much longer than that of the daughter. Essentially what happens at equilibrium is that every time the parent decays to produce a daughter atom, a daughter atom also decays. Thus, the decay of the mixture of parent and daughter follows the decay curve of the parent. Note that although the activities of parent and daughter are equal, the numbers of atoms present are not, because of the higher specific activity of the daughter, and are related to the ratios of their decay constants as shown in Equation 3–6 below, which is simply a rearrangement of the terms in Equation 3–5.

$$\frac{N_1}{N_2} = \frac{\lambda_2}{\lambda_1} \qquad (3-6)$$

The relationship of parent to daughter activity can be expressed by Equation 3–7,

$$N_2 = \frac{\lambda_1 N_1}{N_2} (1 - e^{-\lambda_2 t}) \qquad (3-7)$$

Nin which N_1 and N_2 refer to the number of atoms of parent and daughter, respectively, present at any time t, λ_1 and λ_2 are the respective decay constants. The term in parentheses is simply a buildup term, and represents the fraction of equilibrium achieved; as $\lambda_2 t \rightarrow \infty$, $e^{-\lambda t} \rightarrow 0$ and therefore

$$N_1 \lambda_1 = \lambda_2 N_2$$

For practical purposes, 100% equilibrium is reached after about 7 daughter half-lives, assuming an intitally pure parent at zero time. There are many examples of potential secular equilibrium in nature, as can be seen by reference to Tables 3–4 to 3–7; a good example is given by the first two members of the 4n chain, ^{232}Th with a half-life of 1.4×10^{10} years, and its daughter ^{228}Ra, with a half-life of only 6.7 years. If one were to start with an initially pure sample containing 1 Bq (1 disintegration per second) of ^{232}Th, about 50 years later, the activity from the thorium would not have changed, since 50 years is small relative to its half-life, but there would then be a comparable activity from the ^{228}Ra daughter.

The condition of transient equilibrium occurs when the half life of the parent is greater (but not manyfold so) than that of the daughter. In this situation, illustrated graphically in Figure 3–4, the parent undergoes measurable decay while the buildup of the daughter is occurring. As is the case with secular equilibrium, the decay of the combined mixture of parent and daughter follows the decay of the parent after equilibrium is reached. However, because the half-lves of the two are not widely different, the combined activity from the parent and daughter never reach a value of twice the initial activity of the initially pure parent.

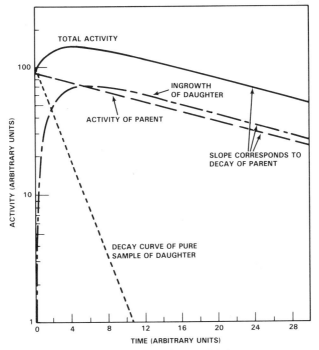

FIGURE 3-4 Transient Equilibrium.

Mathematically, the general equation for the number of daughter atoms is expressed by Equation 3-8:

$$N_2 = \frac{\lambda_1}{\lambda_2 - \lambda_1} N_{0_1} e^{-\lambda_1 t} \qquad (3-8)$$

and, since

$$N_1 = N_0 e^{-\lambda t}$$

by substitution and rearrangement, Equation 3-8 reduces to:

$$\frac{N_1}{N_2} = \frac{\lambda_2 - \lambda_1}{\lambda_1}$$

The condition of no equilibrium occurs when the parent is shorter lived than the daughter, and is graphically illustrated in Figure 3-5. The lim-

RONALD L. KATHREN

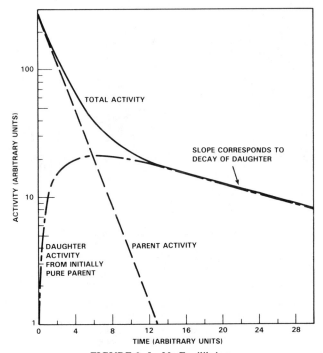

FIGURE 3-5   No Equilibrium.

iting case for no equilibrium is the situation where a radioactive parent decays to a stable daughter; in this case the number of daughter atoms will increase exponentially with time, and will be exactly equal to the number of parent atoms that have decayed. Both the transient and no equilibrium cases are sometimes analyzed in terms of t_m, the time requited for the daughter to reach maximum activity when growing in from an initially pure quantity of the parent. The value of t_m can be readily calculated from the general case by differentiating Equation 3–8 and setting the differential equal to zero when $t = t_m$, to yield:

$$\frac{\lambda_2}{\lambda_1} = e^{(\lambda_2 - \lambda_1)t_m}$$

which can then be solved for t_m:

$$t_m = \frac{\ln \lambda_2 - \ln \lambda_1}{\lambda_2 - \lambda_1}$$

The growth of a radioactive daughter from a radioactive parent can be described mathematically by recalling that the activity of the parent can be expressed by the simple differential equation

$$-\frac{dN}{dt} = \lambda N_1$$

and that the daughter thus grows in at that same rate. However, since the daughter is also radioactive, it undergoes decay itself, which is described by the term $-\lambda_2 N_2$.

Thus, the rate of decay of the daughter is expressed by Equation 3–9:

$$\frac{dN_2}{dt} = -\lambda_1 N_1 - \lambda_2 N_2 \qquad (3\text{--}9)$$

Assuming an initially pure parent with N atoms at zero time, the number of atoms of the parent at any time t can be calculated by Equation 3–10:

$$N_1 = N_{0_1} e^{-\lambda t} \qquad (3\text{--}10)$$

which is simply a restatement of the fundamental decay law. Combining Equations 3–9 and 3–10 and rearranging terms gives:

$$\frac{dN_2}{dt} + \lambda_2 N_2 - \lambda_1 N_0 e^{-\lambda_1 t} = 0 \qquad (3\text{--}11)$$

which is a first order differential equation and can be solved by assuming a solution of the form.

$$N_2 = uv \qquad (3\text{--}12)$$

in which u and v are functions of t. Differentiating Equation 3–12 and substituting in Equation 3–11 gives

$$u\frac{dv}{dt} + v\frac{du}{dt} + \lambda_2 uv - \lambda_1 N_{0_1} e^{-\lambda_1 t} = 0 \qquad (3\text{--}12)$$

which can be rearranged to yield

$$u\left(\frac{dv}{dt} + \lambda_2 v\right) + v\frac{du}{dt} - \lambda_1 N_{0_1} e^{-\lambda_1 t} = 0 \qquad (3\text{--}13)$$

If the function v is selected such that the term in parentheses in Equation 3–13 is 0, then

$$\frac{dv}{dt} + \lambda_2 v = 0$$

which upon integration gives

$$v = e^{-\lambda_2 t} \tag{3–14}$$

Equation 3–14 can be substituted in Equation 3–13 to yield a differential equation in terms of u, which can then be solved for u as shown

$$e^{-\lambda_2 t} \frac{du}{dt} - \lambda_1 N_{0_1} e^{-\lambda_1 t} = 0$$

rearranging

$$du = \lambda_1 N_{0_1} e^{(\lambda_2 - \lambda_1)t} \, dt$$

and integrating to give

$$u = \frac{\lambda_1}{\lambda_2 - \lambda_1} N_0 e^{(\lambda_2 - \lambda_1)t} + C$$

which by substitution for N_2 yields.

$$N_2 = uv = \frac{\lambda_1}{\lambda_2 - \lambda_1} N_{0_1} e^{-\lambda_1 t} + C e^{-\lambda_2 t} \tag{3-15}$$

Equation 3–15 can be resolved by evaluating the constant C at zero time, when $N_{0_2} = N_2$

$$C = N_{0_2} - \frac{\lambda_1}{\lambda_2 - \lambda_1} N_0 \tag{3–16}$$

(Equation 3–16)and then substituting Equation 3–16 in Equation 3–15 to obtain Equation 3–17:

$$N_2 - \frac{\lambda_1}{\lambda_2 - \lambda_1} N_0 \left(e^{-\lambda_1 t} - e^{-\lambda_2 t} \right) + N_{0_2} e^{-\lambda_2 t} \tag{3–17}$$

The result is the general equation for ingrowth of a radioactive daughter from a radioactive parent, as shown by Equation 3–17. Inspection of the right hand side of Equation 3–17 reveals the ingrowth of the daughter represented by the first group of terms, followed by the decay of these atoms, and finally a term which gives the contribution from any daughter atoms that were present initially. If the parent was initially pure—i.e. no daughter was present—this last term will be zero. The equations for the special cases of transient and secular equilibrium presented above are derivable from the general equation 3–17.

The above equations apply only to a two member chain, but may be extended to a chain or series of any length. The result is a very complex series of equations which is generally solved today with the aid of a computer. These are known as the Bateman equations, named after H. Bateman who first put forth the solution to this system of equations in 1910 for the special case of an initally pure parent (Bateman 1910). The general Bateman solution for n successive decays is shown below:

$$N_n(t) = C_1 e^{-\lambda_1 t} + C_2 e^{-\lambda_2 t} \ldots C_n e^{-\lambda_n t} \qquad (3-18)$$

in which

$$C_1 = \frac{\lambda_1 \lambda_2 \lambda_3 \ldots \lambda_{n-1}}{(\lambda_2 - \lambda_1)(\lambda_3 - \lambda_1) \ldots (\lambda_n - \lambda_1)} N_{0_1}$$

and

$$C_2 = \frac{\lambda_1 \lambda_2 \lambda_3 \ldots \lambda_{n-1}}{(\lambda_1 - \lambda_2)(\lambda_3 - \lambda_2) \ldots (\lambda_n - \lambda_2)} N_{0_1}$$

$$\vdots$$

$$C_n = \frac{\lambda_1 \lambda_2 - \lambda_{n-1}}{(\lambda_1 - \lambda_n)(\lambda_2 - \lambda_n) \ldots (\lambda_{n-1} - \lambda_n)} N_{0_1}$$

Important Nuclides from Chains

The three naturally occurring radioactive series, as has been noted, account for most of the natural terrestrial radioactivity, and include some environmentally important radionuclides. The uranium (4n + 2) series

(Table 3–6) includes ^{230}Th, which has a half-life of 80,000 years, and which may be the most significant contributor to the lung dose from inhaled uranium-bearing dusts (Harley and Pasternack 1979). Radium-226—the radium of Mme Curie—is another important member of this series as is its daughter ^{222}Rn, a radioactive gas that diffuses out of the ground and into the atmosphere. The radium decay sequence is an important subseries of the 4n + 2 series; indeed, it should be noted that there are several important subseries among the naturally occurring series. What determines a subseries is largely pragmatic and includes such factors as half-life relationships and the biological significance of the nuclides.

The thorium or 4n series also contains environmentally important radionuclides. Like the uranium series, it also includes a radioactive isotope of the inert gas radon, with mass number 220. This radon isotope is usually referred to by its historic name, thoron. Other important members of the thorium chain from the standpoint of dose include Ra-228, Th-228, and Ra-224. As the radionuclides in the thorium series, other than the parent ^{232}Th are relatively shortlived, freshly separated thorium will achieve equilibrium in about 60 years. During this period, however, the radioactivity composition is constantly changing, along with the radiotoxicity, external dose rate, and related chemical and physicial characteristics.

Uranium is an environmentally ubiquitous element that is associated with all three naturally occurring series (Table 3–8). Three isotopes of uranium are found in nature: U-238, which has an abundance of 99.27%; U-235, 0.72% abundant; and U-234, 0.0057% abundant. Typically, most crustal rocks contain a few parts per million of uranium, averaging about 2.8 ppm (UNSCEAR 1977,1982); phosphate rocks, often used as fertilizers, may contain significantly more—120+ ppm. Igneous rocks and granites also generally contain higher amounts of uranium, while lower concentrations are generally found in sedimentary rocks such as limestone. Uranium is also found in coal and hence in coal ash and other coal plant effluents. Natural fresh waters typically contain on the order of 0.024 to 200 μg/l, with sea water showing a more or less uniform concentration of 2 to 3.7 μg/l.

A typical man will contain about 100–125 μg of uranium. The daily intake and excretion in man is about 1 μg; intake is largely from food and excretion is in the feces (ICRP 1977). Soft tissue concentrations are relatively low, being on the order of a few hundred picograms per gram of wet tissue, while bone contains a few thousand pg of uranium per gram of bone ash (Hamilton 1972). Doses are small, less than 1 mrad (10 μ Gy) being delivered annually to the bone.

Note that the concentrations of uranium are expressed in units of mass rather than activity. This is commonly done in the environmental literature, and in part is due to the difficulties encountered by early investigators in determining the specific activity of uranium, which is, of course, affected by the state of the equilibrium. Thus, freshly separated uranium will have a lower specific activity than uranium that has been removed from the ground, or that has been allowed to sit for several years since it has been separated from the daughter products. One gram of freshly separated natural uranium contains 0.33 μCi (1.2 $\times$ 10^4 Bq) of ^{238}U, 0.015 uCi (555 Bq) of ^{235}U, and a negligible amount—about 12 pCi (1584 Bq)—of ^{234}U. The specific activity of natural uranium is 3.5 $\times$ 10^{-7} Ci (1.3 $\times$ 10^4 Bq)/g; this value refers only to the uranium, and does not include any contribution from daughters. Another source of confusion has been the definition of the curie as applied to uranium, which in the past has been taken as 3.7 $\times$ 10^{10} disintegrations per second from ^{238}U, plus an identical amount from U-234, plus 9 $\times$ 10^8 from U-235, rather than the standard definition of 3.7 $\times$ 10^{10} disintegrations per second. This special definition has been the basis for internal dosimetry calculations in the past (NCRP 1959).

Like uranium, thorium is widely found in crustal rocks, being more abundant in acidic than in alkaline materials. Typically, crustal rocks contain a few parts per million of thorium, perhaps averaging about 12 ppm (Eisenbud 1973). This is about a fourfold greater concentration than uranium, but because of the lower specific activity of Th-232, the radioactivity content is about the same (Table 3–9). There are, however, several locations in the world where very high thorium and thorium daughter concentrations have been noted, giving rise to very high external natural radiation fields. These include the monazite sand areas of the Brazilian coast and the State of Kerala in India, and an area in the Soviet Union. Monazite is an insoluble rare earth mineral composed largely of the phosphates of cerium, lanthanum, and thorium. In beach sands in

TABLE 3–9 Typical Activity Concentrations of Uranium and Thorium in Crustal Rock.

Rock Type	U (pCi/g)	Th (pCi/g)
Igneous	1.3	1.3
Sandstones	0.4	0.65
Shales	0.4	1.1
Limestones	0.4	0.14

Adapted from UNSCEAR 1977.

Brazil, the monazite is usually mixed with the mineral ilmenite, which produces the famous black sands or 'aerea preta' of the Brazilian coast. Monazite sands contain varying amounts of thorium, usually on the order of 10 per cent. The external dose rates associated with these sands are manyfold higher than normal; measurements in the monazite sand regions reveal levels up to several mrad/h (several tens of μGy/h), in sharp contrast to the approximately ten microrad per hour usually associated with natural terrestrial background. People living in these regions thus typically receive a background dose of several rads (several tens of milligrays) per year. Despite this high dose, which in many cases exceeds the maximum permissible levels for occupational exposure, no untoward effects have been documented in residents of the monazite sand areas.

The human body contains about 2.2 pCi (0.08 Bq) of ^{232}Th, about three-fifths of which is found in the bone, with another 20% in the lungs. Somewhat greater activities of ^{228}Th and ^{230}Th are found: 15.2 and 3.1 pCi (0.56 and 0.11 Bq), respectively. On a mass basis, the quantity of these latter two isotopes is far less than that of ^{232}Th because of their very much shorter half-lives, and hence greater specific activity. Thorium-230 is distributed in the tissues similarly to ^{232}Th, but about 95% of the Th-228 is found in the bone, probably reflecting its origin from its parent ^{228}Ra (Wrenn et al. 1980).

Radium is an extremely important member of the uranium decay series, having been extracted and concentrated for various commercial, medical, and educational applications. Unlike uranium and thorium, radium is soluble and readily forms compounds that can be taken up by plants. Radium belongs to Group 2a of the Periodic Table along with calcium and thus is metabolized by plants and animals in a manner similar to calcium. Both the thorium and uranium series produce isotopes of radium, and since on the average the activity concentrations of U and Th in soil and rock are about equal, the same is true for the Ra-226 and Ra-228 members of these chains (Table 3–9)(UNSCEAR 1977, 1982). The activity of these two radium isotopes in soils and rocks is usually in the range of a few tenths of a picocurie (several tens of becquerels) per gram.

In surface waters, radium concentrations are highly variable. In fresh waters, typical concentrations may range from 0.01 to 1 pCi/l (0.37–37 Bq/m^3) with concentrations to 100 pCi/l (3.7×10^3 Bq/m^3) or greater found in certain mineral waters. Drinking water supplies drawn from wells may have relatively high concentrations of radium, particularly if untreated; tap water in Joliet, Illinois has been reported to have a Ra-226 concentration of greater than 50 pCi/l (185 Bq/m^3) (Hursh 1953). In the ocean, the concentration of Ra-226 is roughly constant at the rel-

TABLE 3-10   Alpha Activity in Various Foodstuffs.

Foodstuff	Maximum observed alpha activity (pCi/g)
Brazil nuts	14
Cereals	0.06
Teas	0.04
Organ meats	0.015
Flours	0.014
Peanuts, peanut butter	0.012
Chocolate	0.008
Cookies	0.002
Milk (evaporated)	0.002
Fish	0.002
Cheeses	0.0009
Eggs	0.0009
Vegetables	0.0007
Meat (muscle)	0.0005
Fruits	0.0001

Adapted from Mayneord et al. (1958).

atively low value of about 0.05 pCi/l (1.85 Bq/m^3). Concentrations of Ra-228, which is from the 4n +2 series, are even lower.

Radium is relatively readily taken up by plants in comparison with uranium and thorium, although generally not as well as calcium. An exception to this is the Brazil nut, which selectively accumulates radium and barium over calcium, and has been reported to contain up to 7.1 pCi/ g (0.26 Bq/g) (Penna Franca et al. 1968). The alpha radioactivity content of various foodstuffs is shown in Table 3-10.

Radium-226 is also found in small concentrations in the atmosphere. Natural sources of atmospheric radium are volcanic ash and fumes, resuspension and wind transfer of soils and water, forest fires, and meteorites. Coal burning may also contribute to the atmospheric radium burden, for coals typically contain on the order of one pCi (0.037 Bq) per gram. Levels of atmospheric radium have been estimated to range from 3×10^{-21} μCi/cm^3, which corresponds to 3×10^{-15} pCi/l or 1.1×10^{-10} Bq/m^3 (Jaworowski et al. 1972), to 7×10^{-18} μCi/cm^3 (2.6×10^{-7} Bq/m^3) (UNSCEAR 1977, 1982).

For man, the mean daily intake of each of the two radium isotopes is about 1.4 pCi (0.05 Bq) with a range from 0.7-2.4 pCi (0.03-0.09 Bq) throughout much of the world, although in some locations such as upstate New York and parts of the USSR the levels may by significantly higher (Holtzman 1980). In portions of the USSR, an average daily intake of 17 pCi (0.63 Bq) of Ra-226 has been reported; in Kerala, India, daily intake of 160 pCi (5.9 Bq) of Ra-228 has been reported (UNSCEAR 1977). Intake of radium is largely through grains, green vegetables, and fruits, except in the areas of high intake where animal products and water become the primary sources. Pacific salmon, which forms a large portion of the diet among Eskimos and North American Indians, is rich in radium, containing about 20 pCi (0.74 Bq) per gram and contributing to their higher dietary intake (Jenkins 1969). In humans, about 80–85% of the radium is found in the skeleton, which typically contains about 40 pCi (1.5 Bq) of Ra-226 and about 20 pCi (0.74 Bq) of Ra-228, delivering a dose to the osteocytes of 3.5 mrad/y (35 μGy/y), approximately equally divided between the two nuclides (UNSCEAR 1977, 1982).

Airborne Radioactivity

Most of the radioactivity in the atmosphere at sea level is attributable to two isotopes of radon and their daughters. These are Rn-222, which is the direct daughter of Ra-226 and is thus part of the uranium series, and Rn-220, which is part of the thorium or 4n series. Although these nuclides are both isotopes of the element radon, they are frequently referred to by their historic or generic names. Thus, Rn-222 is known as radon (Rn) and Rn-220 as thoron (Tn), nomenclature which can sometimes produce confusion.

Radon-220 and 222 diffuse from the earth into the atmosphere producing a number of short-lived decay products (Tables 3–4, 3–6). The concentration of the two radon isotopes in air is highly variable, and determined by the following:

1. Concentration of precursers in soil.
2. Altitude
3. Soil porosity and grain size.
4. Temperature.
5. Pressure.
6. Soil moisture, rainfall and snow cover.

7. Atmospheric conditions.

8. Season.

As might be expected, atmospheric radioactivity concentrations of radon and daughters are greatest over areas in which the soil is richest in thorium and radium-226, and at low altitudes. Radon emanation from the ground depends to a large extent on the specific characteristics of the soil matrix, particularly its porosity and grain size. Rn-222, with its 3.825 day half-life has greater opportunity to escape from the soil prior to decay than does its sister Rn-220, which has a 54 second half-life. As a result, atmospheric concentrations of Rn-222 are generally greater than those of radon-220. Soil porosity is greatly affected by soil moisture, and diffusion is considerably less from wet soil. There may also be a scavenging effect inasmuch as radon is soluble in water. Snow or ice cover on the ground will act as a blanket and restrict the emanation from the soil. Airborne radioactivity concentrations over the oceans or other large bodies of water are generally considerably less than over the land masses, in part because of the lower concentrations of the radon precursers in water and also because the gases are absorbed or mechanically prevented from evolving. Typically, the mean diffusion rate for Rn-222 from the ground is about 1.4 pCi(0.05 Bq)/m^2-sec; the diffusion rate for ^{220}Rn is somewhat less.

Increased temperature enhances the evolution of radon from the soil, as does decreased atmospheric pressure. Other atmospheric conditions also affect the buildup and retention of the radon and daughter products in the air. A low lying inversion will result in a buildup of airborne activity, while winds and unstable atmospheric conditions will tend to reduce the air concentrations of radon and daughters. Winds from the ocean or other large bodies of water are particularly effective at dilution. Seasonal and diurnal variations can be marked. Radon concentrations are lowest during winter and greater by perhaps as much as a factor of ten on the average during the summer months, largely as a result of ground moisture. Diurnally, radon levels are greatest during the early morning hours, and least during the afternoon during characteristic unstable conditions, typically varying by a factor of 3–5. There may be a second peak in activity, usually smaller than that in the early morning hours, in the late afternoon.

In the atmosphere, the radon undergoes decay into a variety of short-lived daughter products, depending on which chain or isotope is involved (Tables 3–4, 3–7). Since decay involves a sudden nuclear transformation, the decay products are electrically charged when they are formed and readily attach to dust particles present in the atmosphere. Most of the daughter activity is associated with particles with diameters <0.035 μm, with the mean diameter of the carrier aerosol being 0.025 μm, the range 0.006–0.2 μm, and the concentration in air about 3×10^4 particles per cubic centimeter (Haque and Collinson 1967). In general, the dustier the atmosphere, the greater the concentration of radon and daughters. This is not only from the daughters attached to the dust particles but also from the dust itself which contains the progenitors of radon and so acts as a source of emanation. The dust particles with the attached radon daughters settles to earth by natural gravitational processes. This occurs rather slowly because of the small size of the radioactivity bearing aerosols. Removal of the dust from the air is greatly expedited by rain or snow which serve to scrub or wash the particles out of the air. In addition, precipitation removes some of the gaseous radon and also reduces the emanation from the soil. Radioactive equilibrium between radon and its daughter products is seldom achieved in the atmosphere because of the deposition processes and the constantly changing rate of evolution and atmospheric mixing.

Air concentrations of radon and daughters indoors are a function of the concentration in the outside air, the rate of emanation from inside the building, and the ventilation. Brick, concrete, and aggregate buildings have a greater emanation rate of radon than wood fram buildings because the former contain more of the radon precursers. Some paints and wall treatments also are a good source of radon evolution. In general, concentrations indoors exceed those outside, particularly in basements and other locations where ventilation is poor. In poorly ventilated rooms, radioactive equilibrium of radon and its decay products may be achieved.

In the continental United States, Rn-222 concentrations in air at 1–3 meters above the ground range over about two orders of magnitude, from about 10^{-11} to 10^{-9} μCi/cm^3 (0.01–1 pCi/l or 0.37–37 Bq/m^3), depending on seasonal and temporal considerations, as well as the amount of radium-226 in the soil. Concentrations of radon-220 typically are three- to tenfold lower. Similar values are found elsewhere in the world over the large land masses; howver, the Rn-222:Rn-220 ratio is smaller in Europe because of the generally higher levels of thorium in the soil. Over the oceans and at island locations such as Hawaii, American Samoa, and the Philippines, mean concentrations of Rn-222 in air are on the order of 1–4 $\times$ 10^{-12} μCi/cm^3 (equivalent to 0.037–0.15 Bq/m^3 or 0.001–0.004 pCi/l), which can be compared with the mean concentration of 1.2 $\times$

10^{-10} uCi/cm^3 (4.4 Bq/m^3 or 0.12 pCi/l) reported for Washington, D.C. (UNSCEAR 1977, 1982). The dose from radon and daughters is largely to the epithelial tissue lining the lung, with the major contribution being from the daughters. Rn-222 delivers an annual dose equivalent of only about 2 mrem (20 μSv); the annual dose equivalent from the daughters to the alveolar region is about 100 mrem (1 mSv). The largest dose equivalent of 550 mrem/y (5.5 mSv/y) is incurred by the segmental bronchioles. The principal contributor to the lung dose from alpha emitting isotopes of polonium. Tobacco smoke is enriched in these isotopes, and the annual dose equivalent to the lungs is increased by 75–150 mrem (0.75–1.5 mSv) among cigarette smokers. (NCRP 1975, UNSCEAR 1982).

References

1. Bateman, H. 1910. "The Solution of a System of Differential Equations Occurring in the Theory of the Radio-Active Transformations",*Proc. Cambridge Phil. Soc.*, 16:423.
2. Beck, H. L. 1972. "The Physics of Environmental Gamma Radiation Fields", in *The Natural Radiation Environment II*, Vol 1, J. A. S. Adams, W. M. Lowder, and T. F. Gesell, Eds., CONF–720805-P1, pp. 101–133.
3. Eisenbud, M. 1973. *Environmental Radioactivity*, Second Edition, Academic Press, Inc., New York, p. 177.
4. Hamilton, E. I. 1972. "The Concentration of Uranium in Man and His Diet", *Health Phys.* 22:149 (1972).
5. Harley, N. H. and B. S. Pasternack. 1979. "Potential Carcinogenic Effects of Actinides in the Environment", *Health Phys.* 37:291.
6. Haque, A. K. M. and A. J. L. Collinson. 1967. "Radiation Dose to the Respiratory System due to Radon and Its Daughter Products", *Health Phys.* 13:431.
7. Holtzman, R. B. 1977. "Comments on "Estimate of Natural and Internal Radiation Dose to Man" ", *Health Phys.* 32:324.
8. Hursh, J. B. 1953. "The Radium Content of Public Water Supplies", U.S. Atomic Energy Commission Report UR-257.
9. International Commission on Radiological Protection. 1976. *Reference Man: Anatomical, Physiological, and Metabolic Characteristics*, ICRP Publication No. 23, Pergamon Press, Oxford.
10. Jaworowski, Z., J. Bilkiewicz, L. Kownacka, and S. Wlodek. 1972. "Artificial Sources of Natural Radionuclides in Environment", in *The Natural Radiation Environment II*, Vol. 2, J. A. S. Adams, W. M. Lowder,and T. F. Gesell, Eds., CONF–720805-P2, pp. 809–818.

11. Jenkins, C. E. 1969. "Radionuclide Distribution in Pacific Salmon," *Health Phys.* 17:507.
12. Klement, A. W. 1982. *Handbook of Environmental Radiation*, CRC Press, Boca Raton.
13. Lederer, C. M., et al. 1977. *Table of Isotopes*, Seventh Edition, Wiley, New York.
14. Mayneord, W. V., R. C. Turner, and J. M. Radley. 1958. "The Alpha-Ray Activity of Humans and Their Environment, in *Proc. Second International Conf. Peaceful Uses of Atomic Energy*, United Nations, New York.
15. National Council on Radiation Protection and Measurements. 1959. "Maximum Permissible Body Burdens and Maximum Permissible Concentrations of Radionuclides in Air and Water for Occupational Exposure", U. S. Department of Commerce, National Bureau of Standards Handbook 69, Washington, D. C.
16. National Council on Radiation Protection and Measurements. 1975. "Natural Background Radiation in the United States", NCRP Report No. 45, Washington, D. C.
17. Penna Franca, E., et al. 1968. "Radioactivity of Brazil Nuts", *Health Phys.* 14:95.
18. United Nations Scientific Committee on the Effects of Atomic Radiation (UNSCEAR). 1977. *Sources and Effects of Ionizing Radiation*, United Nations Publication E.77.IX.1, New York.
19. United Nations Scientific Committee on the Effects of Atomic Radiation (UNSCEAR). 1982. *Ionizing Radiation: Sources and Biological Effects*, United Nations Publication E.82.IX.8, New York.
20. United States Department of Health, Education and Welfare, Public Health Service. 1970. *Radiological Health Handbook*, Revised Edition, Rockville, MD.
21. Whicker, F. W. and V. Schultz. 1982. *Radioecology: Nuclear Energy and the Environment*, CRC Press, Boca Raton.
22. Wrenn, M. E., et al., "Thorium in Human Tissues", in The Natural Radiation Environment III, T. F. Gesell and W. M. Lowder, Eds., DOE Symposium Series 51, CONF-780422 (1980), pp. 783–799.

CHAPTER 4

Dose from Environmental Radiations

In addition to the exposure received from naturally occurring cosmic and terrestrial radiation sources, environmental radiation dose may also be incurred from artificially produced or enhanced natural radiation sources. There are many such anthropogenic sources that contribute to human radiation doses. For convenience, these can be categorized as:

1. Technologically enhanced natural radiations.
2. Consumer products.
3. Fallout from weapons tests.
4. Nuclear power.
5. Medical radiation.
6. Occupational exposure.

The contribution of each of these sources to the aggregate is quite variable, ranging from insignificant or very slight fractions to one or more orders of magnitude of the dose incurred from normal levels of background radiation. Thus, quantification is difficult and can only be considered in a generic rather than a specific sense.

Technologically Enhanced Natural Radiation

Human activities frequently cause significant disruption of the natural radiation environment, resulting in increased or altered patterns of environmental radiation exposure. With the exception of radium, which is sought and concentrated for its radiological properties, most of these activities are not carried out because of the radioactivity, but rather because of the other properties of the material that contains the radioactivity. The term "technological enhanced natural radiation" (TENR) was introduced in 1975 (Gesell and Pritchard 1975) to describe this modification of the natural environment and the possible increased human radiation exposure associated with it. This term should not be confused with the acronyms NORM, which is sometimes used for "naturally occurring radioactive material(s)" and NARM, which is used in place of "natural and accelerator produced radioactive material(s)".

A major source of TENR results from mining and milling operations, in which ores with relatively high radioactivity content are extracted from their natural locations (which may be deep underground), possibly concentrated, and redistributed throughout the environment. The presence of radium or thorium in these ores may result in increased concentrations of radon or thoron and daughters in the general location of the mining or milling operation. In some cases, the radioactive metal is what is being sought; this is true for radium, which finds its way into a variety of consumer products in concentrated form, and uranium, which is used for nuclear power production as well as industrially in pigments and glazes. Thorium metal is mined not only for its value in certain industrial processes, such as the production of certain phosphors and in welding, but also because of its ability to produce a strong and lightweight metal when alloyed with magnesium. Magnesium-thorium alloys generally contain a few per cent ($<5\%$) thorium and have been used largely for aircraft and in the space program.

The manufacture of Welsbach gas mantles is an important use of thorium. These mantles consist largely of thoria, ThO_2, which is obtained from monazite sands from Brazil as well as locations in North Carolina and West Virginia. Although electricity has largely replaced gas for lighting purposes, these mantles still are widely used for camping lanterns and in locations bereft of electricity. An interesting consumer application of thorium bearing material has been its use in paperweights and as ballast for cellophane tape dispensers and similar products.

Mining of phosphate rock for use as fertilizer is a major industry. In the United States, about 150 million tons are mined annually, largely in Florida, (91%) and Tennessee (3%), with lesser amounts mined in Idaho, Missouri, Montana, Utah, Wyoming, Georgia, and the Carolinas. Phosphate rock contains relatively high concentrations of radium, thorium, and uranium; typical concentrations of ^{232}Th are on the order of 0.5–2 pCi/g (19–74 Bq/kg), while those of ^{238}U and ^{226}Ra range from a few pCi/g to 130 pCi/g (4800 Bq/kg), this latter value having been reported for material from South Carolina (UNSCEAR 1977, 1982; Boothe 1977). Potassium bearing phosphate fertilizers such as PK and NPK type may contain appreciable quantities of ^{40}K, on the order of several tens to hundreds of pCi/g (thousands of Bq/kg) (UNSCEAR 1982). There is both a wet and dry mining process, with the former being favored in the southeastern U.S. and the latter in the west. Treatment involves benefication, milling, and manufacture of ammonium phosphate bearing fertilizers for both commercial and general agricultural usage. About 12 to 15 million tons of phosphate fertilizers are produced in this manner annually in the United States, resulting in the distribution of about 120 Ci (4.4 GBq) of Ra-226 over agricultural lands.

Radon-222 figures significantly as a source of TENR. This nuclide is a gaseous daughter of ^{226}Ra, and has been used medically at least as early as 1914 (Jennings and Russ 1948). The medical application generally involves allowing the radon gas to evolve from a solution or compound of radium and collecting it in glass capillaries which can then be sealed in platinum. The source can thus be made virtually any strength desired. Radon generators or "cows" as they are often called are devices in which the radon gas is generated; removal of the radon from the generator is a process known as "milking", and, although carefully done, may result in the leakage of radon to the environment. The use of radon sources in medicine is on the decline, and this source of TENR is not of great significance.

Enhanced concentrations of radon may occur in homes and buildings as a result of

1. Use of water containing high concentrations of radon.

2. Release from building materials used for construction.

High radon concentrations are not uncommon in potable waters. Typically for the United States, about a fourth of the potable waters show ^{222}Rn concentrations in excess of 2000 pCi/l (7.4 $\times$ 10^4 Bq/m^3), with about 5% >10,000 pCi/l (3.7 $\times$ 10^5 Bq/m^3); in general, the ^{222}Rn is not in equilibrium with the parent ^{226}Ra (Grune et al. 1964). Radon, a noble gas, is released from the water by heating or by gentle agitiation. Thus, simple domestic uses such as bathing or showering or even running a tap, as well as industrial processes, can readily release the radon dissolved in water into the atmosphere. Water containing 1000 pCi/l (37 KBq/m^3) of ^{222}Rn could produce air concentrations indoors of 1pCi/l (37 Bq/m^3) from ordinary domestic uses (Duncan, Gesell, and Johnson 1976). Similarly, domestic water containing 500 pCi/l (18,500 Bq/m^3) could produce 20 health effects annually in a population of 1,000,000 from inhalation of the radon released.

Experiments have been performed in Texas with water used for both tub and shower bathing (Gesell and Pritchard 1980) and suggest that in a typical household with ^{222}Rn concentrations in water of 500 pCi/l

(18,500 Bq/m^3), the annual dose equivalent to the bronchial epithelium could be as great as 500 mrem/y (5 mSv/y), although more likely on the order of 100–200 mrem (1–2 mSv) annually. This was based on a house size of only 8000 cubic feet (2264 cubic meters) containing a bathroom measuring $5 \times 5 \times 8$ feet, a ventilation rate of one air change per hour, and a Rn:daughter equilibrium of 50%. An estimated 10,000 Ci/y (3.7 $\times 10^{14}$ Bq/y) are released to the environment from potable water (Travis et al. 1979). The potential problem of radon released from domestic waters has as yet received scant serious consideration.

Numerous building materials may release radon or produce direct radiation exposure because of their content of radium, thorium, uranium, and potassium. Wood construction houses and buildings in general have a lower external dose rate than do buildings constructed from concrete or concrete block, brick, or stone. The latter may produce significant external dose rates from the radium and other naturally occurring gamma emitters present in trace amounts. Granites used for construction can result in exposure rates of even a few tenths of millirad (few μGy) per hour; a notable example of high dose rates from granite used for construction is found in the Grand Central Station in New York City.

Radium is usually found in association with other alkaline earth metals and thus is present in plasters and gypsum bearing materials, as well as in concrete and rock. The radium not only produces an external radiation field, but more importantly gives rise to radon gas which can then evolve out into the building and buildup in the enclosed spaces. Typical radon evolution values, determined in England, ranged from 6 to 100 pCi/m^2 per hour (0.22–37 Bq/m^2) (Haque et al. 1965). Typical room concentrations depend not only on the rate of generation but also on the ventilation rate, or rate of removal. A calculation for a room with a volume of approximately 800 cubic feet (22.6 m^3) in which the walls and ceiling were emanating 30 pCi/m^2-h (1.1 Bq/m^2-h) for a total of 2500 pCi/h (92.5 Bq/h) would produce an equilibrium concentration of 4.4×10^{-9} uCi/cm^3 (163 Bq/m^3) if unventilated. With 4 air changes per hour and an outdoor concentration of 7×10^{-11} $\mu Ci/cm^3$ (2.6 Bq/m^3), the room air concentration would be $<8 \times 10^{-11}$ $\mu Ci/cm^3$ (3.0 Bq/m^3) (NCRP 1975). Typically, air concentrations of radon and daughters are greatest in enclosed rooms such as basements (Yeates, Goldin and Moeller 1972). External dose equivalent rates from radon in building materials is on the order of 10–20 mrem (100–200 μSv) per year, and about 5600 Ci (2.1 $\times 10^{14}$ Bq) of radon are released annually from this source (Travis et al 1979).

In some cases, houses have been constructed on land that has been filled with waste materials from the mining and milling of minerals that contain significant amounts of natural radioactivity. Clearly, this practice will produce both elevated ambient external levels as well as increased concentrations of radon and daughters. The problem has been extensively studied and is discussed more fully in Chapter 6.

Radon is also present in natural gas and natural gas products which may be combusted in homes and other buildings. While concentrations to 1450 pCi/l (5.4×10^4 Bq/m^3) have been observed, a value of 20–50 pCi/l (740–1850 Bq/m^3) is more realistic at the point of consumption (NCRP 1975). Combustion of natural gas at these levels has been estimate to deliver an average annual tracheobronchial dose equivalent of 54 mrem (540 μSv), with a maximum of 4.25 rem (42.5 mSv), depending on concentration and combustion conditions (Johnson et al. 1973). Burning of natural gas is estimated to release about 11,000 Ci (4.1×10^{14} Bq) of radon annually, largely from industrial sources (Travis et al. 1979).

Enhanced levels of radon are also found in caves and mineral hot springs and in association with geothermal energy production. Spelunkers and other cave visitors may incur significant doses as a result of high levels of radon and daughters in the relatively undisturbed atmospheres of caves; this source of radiation, however, does not affect a large number of people and is relatively contained. Geothermal energy production may release significant amounts of ^{222}Rn and daughters to the atmosphere (Church 1975); the Geysers plant in California has been estimated to release several hundred curies (tens of gigabequerels) annually (Travis et al. 1979). Radon-222 concentrations in the geothermal brine concentrate at the Geysers is about one to two orders of magnitude greater than in ordinary domestic waters, averaging 16.7 nCi/l (6.2×10^5 Bq/m^3) (Kruger, Stoker and Umana 1977).

Combustion of coal is a significant source of technologically enhanced exposure to naturally occurring radioactivity. Coal contains significant quantities of ^{14}C, in addition to varying amounts of thorium, uranium and their daughters, and ^{40}K. Th-230 and ^{226}Ra, members of the uranium chain, are particularly significant nuclides in coal because of their relatively high radiotoxicity. The uranium content of coals from the Western United States (Wyoming, Idaho, Montana) is significantly greater than that of coals from the Eastern part of the country. Western coals may contain several hundred parts per million of uranium and daughters, although ordinarily the concentrations are of the order of 10 ppm or less, with 1.8 ppm suggested as a good average value (Swanson et al. 1976).

When coal is burned, the radioactivity is released in the flue gases and fly ash directly to the atmosphere; a significant fraction is also retained in the bottom ash. The amount of radioactivity retained in the ash is a function of many variables including the combustion temperature, additives used, the type of coal, and the fraction of the ash. In general, most of the activity from the heavy elements remains in the ash, while virtually all the ^{14}C is released in the gaseous effluents. The enormous quantity of coal consumed makes this source highly significant, even for those elements largely retained in the ash. Enhanced concentrations of various naturally occurring radionuclides have been found around coal fired power plants, both in the atmosphere and as deposits on the ground (UNSCEAR 1982). Whole body dose equivalents to people residing in the vicinity of large coal fired plants could run to as much as 100 mrem (1 mSv) annually.

Aircraft crews and passengers incur an enhanced radiation exposure from enhanced cosmic radiation fields at high altitudes. The dose received will be not only a function of the duration of the flight and the altitude, but also the latitude flown and the sunspot activity. Typically, a cross-continental (North America) round trip flight results in a dose of about 3 mrad (30 μSv) under average solar conditions (UNSCEAR 1977); a transatlantic trip will deliver about the same dose if the flight is in a subsonic aircraft and about 2 mrad (20 μGy) in an SST under average solar conditions (UNSCEAR 1982). Doses to space travellers have been significantly greater, generally on the order of a few hundred millirad (few mGy), and depend to a great extent on solar conditions and passage through the various radiation belts around the earth (UNSCEAR 1982).

Radioactivity in Consumer Products

Many different consumer products contain radioactive materials or emit ionizing radiations incidentally to their operation. These can be conveniently divided into six categories (UNSCEAR 1982):

1. Radioluminous products.
2. Electronic and electrical devices.
3. Antistatic devices.

4. Smoke and fire detectors.

5. Ceramics and U-Th alloys.

6. Other products, including scientific apparatus.

Perhaps the earliest and the most well known of all consumer applications of radioactivity is for radioluminescent dials and markers of various kinds. ^{226}Ra, ^{90}Sr, ^{147}Pm, and ^{3}H have all been used for this purpose. Use of ^{226}Ra has largely been replaced by ^{147}Pm and tritium, which are pure beta emitters with considerably lower radiotoxicity. However, many radium bearing timepieces, aircraft and marine instrument dials, and similar devices are still in active or inactive use. A typical radium dial watch might contain 0.1 to 3 μCi (3700–11,000 Bq) of ^{226}Ra; dials of aircraft and marine instruments contain considerably greater quantities, perhaps up to 20 μCi (7.4 $\times$ 10^5 Bq).

In addition to the direct radiation field produced by the photons associated with the decay of radium and its daughters, radon gas is continuously being produced, and released into the atmosphere. Leakage rates are highly variable, but the value of 1 nCi/h-μCi (1 mBq/h-Bq) of ^{226}Ra appears reasonable. At this rate, the air concentration of radon and daughters in a typical home with a volume of about 3000 cubic feet (85 cubic meters) would be 3 $\times$ 10^{-12} μCi/cm^3 (0.11 Bq/m^3), delivering an annual absorbed dose to the lungs of about 0.1 mrad (1μGy) (UNSCEAR 1977).

Many studies have been made of the radiological hazards associated with the use of radium bearing timepieces and instrument dials. The dose equivalent rate to the wrist from a radium dial watch has been estimated as 0.275 mrem/uCi-h (7.4 $\times$ 10^{-4} μSv/Bq-h), which would provide an annual dose equivalent of 2.4 rem (24 mSv) if the watch were worn continuously (Moghissi and Carter 1975). An estimated 8.4 million radium bearing timepieces were in use in the United States in 1977, with a mean activity of 0.5 uCi (18.5 KBq) per unit. These devices produced an estimated population dose-equivalent of 2500 person-rem (25 person-seivert) per year (Moghissi et al. 1978).

Somewhat smaller individual doses would be incurred from pocket watches, which have not been manufactured with radium dials since 1966 in most countries (UNSCEAR 1977). The concern is the gonadal dose received by the wearer, which has been estimated at about 8 mrad (80 μGy) annually, based on a radium content of 0.1 μCi (3700 Bq). Radium bearing alarm clocks, typically containing about 0.15 μCi (5550 Sv), would deliver an estimated annual gonadal dose of only about 0.1 mrad (10 μGy).

Data regarding the use of radium in aircraft and marine instruments is scanty. However, radium content was typically on the order of few microcuries, applied in unsealed forms, allowing leakage of radon gas plus flaking and chipping of the radioluminescent compound. Many instruments bearing radium were produced for military uses during and shortly after World War II; these were in many cases released to the general public and the environment through war surplus sales. Equipment may have also contained radium illumination buttons and switches in addition to dial markings; the external radiation fields associated with some of these instruments has been measured at several rad/h (several tens of mGy/h) a few inches from the surface.

Radium is particularly poor from a hazard standpoint, for not only does it emit penetrating radiations and the radioactive gas radon, but also provides a heavy dose to the radioluminescent compound matrix from the alpha particles associated with its decay. This results in radiation damage and degradation of the radioluminescent compound, and can lead to flaking and chipping of the material, which can then readily get into the environment. Neither timepieces nor instrument dials are hermetically or tightly sealed, and leakage of their radium content may be significant as these products age.

The use of ^{90}Sr for radioluminescent compounds was transitory and relatively small. However, in the late 1950's, ^{90}Sr was used as a substitute for radium in some expensive Swiss wrist watches. ^{90}Sr is a pure beta emitter, and is usually found in equilibrium with its daughter ^{90}Y, also a pure beta emitter. Like radium, ^{90}Sr has a fair degree of radiotoxicity, and in addition gives rise to bremsstrahlung because of the energetic betas associated with the decay of its daughter ^{90}Y. Relatively large amounts of ^{90}Sr were used in some watches, resulting in a recall of these products. Thus as a practical matter, this source of radiation in consumer products can be ignored.

The relatively low energy beta particles from ^{147}Pm (E_{max} = 224 keV) and tritium (E_{max} = 18.6 keV) are now widely utilized for radioluminescent materials. These two nuclides, and tritium in particular, have relatively low radiotoxicity, and the relatively weak betas they emit are largely absorbed in the plastic or glass coverings over the dial face. The weak betas are not as effective at producing radioluminescence. ^{147}Pm is only 1/170 as effective at producing luminescence as is ^{226}Ra; thus, 170 times as much activity must be used. For tritium, the comparable value is 5,000. In addition, unlike ^{226}Ra and ^{90}Sr, whose half-lives of 1620 and 28 years are highly appealing from a commercial standpoint, ^{147}Pm and tritium have half-lives of only 2.62 and 12.3 years. To obtain the 10 year useful lifetime for timpieces put forth as a standard by the International Atomic Energy Agency (IAEA 1967), more additional activity must be used for a total of 880 times as much for ^{147}Pm and 23,000 times as much for tritium as compared with radium (Moghissi et al. 1978). Thus, time-pieces may contain from 1–25 mCi (37–925 MBq) of tritium and from 65–200 μCi (2.4–7.4 MBq) of ^{147}Pm. Doses from these sources are, how-ever, relatively small. For ^{147}Pm, Moghissi and his coworkers (1978) esti-mate a total population dose equivalent of 500 person-rem (5 person-seivert) per year in the United States, based on 2.6 million timepieces containing a mean activity of 41 μCi (1.5 MBq). For tritium, the esti-mates are 800 person-rem (8 person-seivert) annually, based on 28 mil-lion devices containing an average of 1 mCi (37 MBq) each. Individual doses have been estimated at much less than 1 mrad for persons possess-ing such timepieces.

Both ^{147}Pm and tritium are now found in a wide variety of commercial consumer products, including instrument dials and markers, automobile key lock illuminators, gun sights, door bell pushes, automobile shift quad-rants, telephone dials, and exit marker signs. The latter almost exclu-sively use tritium, and may contain 20 to 30 Ci (7.4 $\times$ 10^{10}–1.1 $\times$ 10^{11} Bq). A few such devices have been made with ^{85}Kr and ^{14}C. In general, the radioactivity is contained in sealed tubes within the device. Despite the relatively high activity in these devices, the resultant collective pop-ulation dose is insignificant in comparison to that from radioluminous timepieces (UNSCEAR 1977).

Many electronic and electrical devices make use of radioactive mate-rials, most commonly tritium, ^{147}Pm, ^{85}Kr, ^{60}Co, and ^{63}Ni. Radium, sometimes in amounts approaching 100 μCi (3.7 MBq) , has been used in electronic tubes of early vintage. Typical applications are for cold cath-ode, voltage discharge, and glow discharge tubes. Natural thorium has been applied to various high pressure lamps, including mercury vapor lamps, sunlamps, and germicidal lights, ^{226}Ra and other nuclides to flu-orescent lamp starters. The total activity and dose from these uses is not known with certainty, but is relatively small relative to that from radiol-uminous timepieces. Hazards would primarily be the result of breakage of the radioactivity containing component with subsequent release to the environment.

Electronic products may emit stray levels of ionizing radiation inci-dental to their operation. This is particularly true of cathode ray devices—television sets, primarily—which operate at voltages in excess of 10–15 kV. Standards have been established for limitation of stray emissions from electronic products by the federal government in the

United States, as well as on an international level; the ICRP recommends a limit of 0.5 mR/h measured at a distance of 5 cm from the surface of the receiver, while the U.S. regulatory standard is somewhat lower at 0.1 mR/h. Measurements of color television receivers used in households has been made worldwide, with exposure rates generally falling below the 0.1 mR/h limit.

Because of its ionizing nature, radioactivity has found application in a variety of antistatic devices. Thus various sizes and shapes of static elimination devices have been manufactured, ranging from the tiny units designed to remove static and dusts from records to lightning rods and large industrial devices. Both ^{210}Po and ^{241}Am are found in anti-static brushes, the former being far more common. Typically, quantities are in the range of 0.05–0.5 mCi (1.85–18.5 MBq) for ^{210}Po and about an order of magnitude lower for ^{241}Am. Alpha emitting substances, largely ^{226}Ra and ^{241}Am. have found application in lightning rods; such devices may contain a few tens of microcuries (approximately 1 MBq). The number of radioactive lightening rods was conservatively estimated at 200,000 world wide in 1978 (Fornes and Ortiz 1978).

Fire detectors are devices making use of an ionization chamber to detect the presence of smoke. Air in the chamber is contiually ionized by an alpha emitting radioactive source, thus producing a small current flow. The presence of smoke particles blocks the alpha particles producing an alarm. Early ionization chamber smoke detectors used a few microcuries (about 100,000 MBq) of ^{226}Ra as the ionization source, but current models utilize ^{241}Am as an oxide mixed with gold in a foil. Some of the earlier models contained as much as 100 μCi (3.7 MBq) of ^{241}Am; more recent models may contain only 1 μCi (37,000 Bq), or even less. In some models, millicurie (tens of millions of becquerels) amounts of ^{85}Kr have been used; ^{238}Pu has also been used (20 μCi) as has natural or depleted uranium. This source of environmental radioactivity is expected to grow as fire detectors are required by many building codes in all new construction.

In the United States, ^{241}Am bearing smoke detectors are issued under the provisions of a general license with the proviso that the active portion of the unit be returned to the manufacturer at the end of the useful life of the device. This is probably highly unlikely, and thus these devices should be considered as environmentally dispersed. The average population dose from use of these devices has been estimated to be in the region of 1 to 1.4 μrad (10–14 μGy) annually with release of activity during a fire of low probability and little consequence (Johnson 1978, UNSCEAR 1977).

In addition to the industrial uses of thorium and uranium, these two naturally occurring radioactive elements are found in several consumer products. The use of thorium in gas lantern mantles and alloys has already been discussed supra. Thorium is also used in some pigments, and in certain optical glasses, some of which may contain up to 30 per cent by weight of thorium. Similar applications have been made of uranium. Uranium has also been rather extensively used in dental porcelain to simulate the fluorescence of the natural teeth. More than ten per cent of the false teeth in the United States are estimated to contain uranium compounds added for this purpose (UNSCEAR 1977). In the U.S., the quantity of uranium has been standardized at 300 ppm, and the mean dose equivalent rate to the gums and buccal region from dental porcelain has been estimated at 0.7 rem (7 mSv) annually.

An interesting consumer use of uranium and thorium has been as a glaze for ceramics, thus providing a unique orange or yellow color. Dishes and crockery as well as ceramic tile has been made utilizing these glazes. In general, such products are no longer commonly found on the market. Studies of pottery made with uranium glazes have shown a propensity of the uranium to leach in food acids, and doses to users could be several mrad (tens of μGy) annually (Simpson and Schuman 1978).

There have been over the years other consumer products making use of radioactivity, either because of the radioactive properties of the substance or its great density (e.g. U and Th). Thus, occasionally paperweights may included uranium or thorium bearing materials. C-14 has been used to a limited extent in paper products such as bank checks, and in vending machine coins. Certain scientific apparatus, notably gas chromatographs, may contain small sources of ^{63}Ni or tritium, and school chemical laboratories are repositories of thorium and uranium compounds. Chemical and other products containing uranium will vary in their isotopic composition, for most post-war uranium is depleted, the ^{235}U having been removed for use as a fissile material.

Other Sources of Environmental Exposure

In addition to the technologically enhanced radiation sources and radioactivity in consumer products, there are several other significant sources of environmental radiation exposure resulting from human activities. These include medical radiations, which are more prevalent in the more affluent and well developed countries, and therefore provide greater aver-

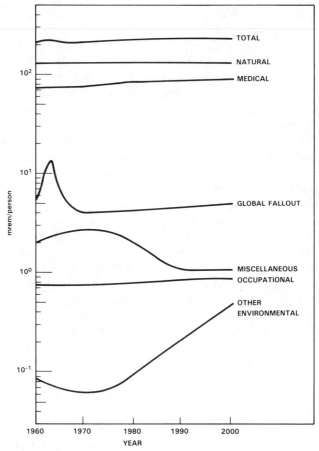

FIGURE 4–1 Estimated Mean Annual External Whole Body Dose Equivalents in the United States (Klement 1972, p. 167).

age exposure to people in those countries. Medical exposure in the advanced nations is a significant source of exposure, with relative magnitude approaching that of exposure from natural radiation sources (Figure 4–1).

A more ubiquitous source of environmental radioactivity and radiation is fallout from nuclear explosives and weapons tests. Fallout levels are temporally variable as well as affected by locale, and in past years have been a significant source of exposure to some populations. Fallout is both a source of external and internal exposure, and generally delivers on the order of five per cent or less of the dose from naturally occurring sources. Fallout as a source of exposure will be more fully addressed in the next chapter.

A growing and important source of ionizing radiation exposure as well as environmental radioactivity is from the generation of electricity by nuclear reactors. The nuclear fuel cycle begins with the mining and milling of uranium, and continues through the manufacture of fuel and its use in the reactor to generate electricity, culminating in the reprocessing of the fuel and the generation of radioactive wastes. These important sources of environmental exposure will be more fully adressed in subsequent chapters. For now, it will suffice to state that in general this source of exposure is trivial compared to natural background, although it will grow over the next several decades.

Miscellaneous exposure can come from various industrial uses of radioactive materials and radiation. Exposure from these sources is not limited to those involved occupationally but rather includes the general population as well. Small quantities of radioactivity may be produced and treated as low level radioactive wastes. Low level radioactive effluents may be directly released to the environment. Industrial and commercial processes involving radioactivity and radiation may produce non-occupational exposures to nearby residents, visitors, or to personnel who may be incidentally in the area of usage, but who are not radiation workers. In general, the average exposure from these uses is very small.

Summary of Environmental Radiation Exposure

Environmental radiation exposure comes from a number of sources as summarized below:

1. Natural radiations of cosmic origin.
2. Natural radiations of terrestrial origin.
3. Technologically enhanced natural radiations.
4. Consumer products utilizing radioactivity or emitting radiation.
5. Fallout from nuclear explosions.
6. Generation of electricity by nuclear fission.
7. Medical applications of radioactivity and radiation.
8. Occupational radiation exposure.

The importance and magnitude of any one of these sources is variable, and frequently difficult to assess. Thus, many different values for like or similar exposures have been reported in the literature, creating considerable confusion in some cases. Exposure estimates must be carefully examined in order to avoid misinterpretation and error. In addition to the error inherent in any measurements or calculations, a major source of error and confusion lies in the units used for reporting and interpretation of dose. Specification should be made as to whether the quantity is in fact dose (i.e. units of rad or gray), rather than dose equivalent (i.e. rem or seivert) or exposure in the special sense (roentgen units). It is also necessary to specify to what groups or portion of the anatomy the exposure is delivered, and whether it is from external or internal sources, and over what time period. Other conditions should also be specified, such as whether the values represent means or maxima, or involve special situations such as referring to certain seasons or specific locations, or, in the case of external exposures, whether credit is taken for shielding from dwellings and other buildings. The special conditions and assumptions made in the calculation or with regard to the measurement are also very important. Unfortunately, all too frequently these and other qualifications are unstated, particularly in summaries or in comparisons which attempt to 'prove' such points as the safety or lack of same of nuclear power plants.

In the United States, the average annual whole body dose equivalent from cosmic rays is about 45 mrem (450 μSv), ranging from a low of 30 in Hawaii and Puerto Rico to a high of 130 (1.3 mSv) in Wyoming (Figure 2–5). The annual whole body external dose equivalents from terrestrial sources are somewhat higher, averaging 60 mrem (600 μSv) for the U.S. as a whole, with statewide averages ranging from 30 in Texas to 115 in Colorado (Table 4–1). Thus, in the United States, the total average external whole body dose equivalent from natural sources is 105 mrem/y, (1.05 mSv/y) being the smallest in Louisiana (75 mrem/y or 750 μSv/y) and the greatest in Colorado (225 mrem/y or 2.25 mSv/y) (Klement et al. 1972). Internal exposure from natural radiation sources is largely from ^{40}K, with considerably smaller contributions from from ^{3}H, ^{14}C, and the members of the series decaying nuclides, largely ^{226}Ra and ^{228}Ra and their daughters. Doses from ^{40}K, ^{3}H, and ^{14}C are relatively uniform across the United States, but concentrations of radium and other members of the chains is highly variable across the country. On the average, the annual whole body dose equivalent in the United States is 25 mrem (250 μSv) from internal sources (although more recent estimates are 35 mrem (350 μSv), with about 16 mrem/y (160 μSv/y) from the

TABLE 4-1 Estimated Annual Dose Equivalents to the Whole Body from Natural Sources of External Gamma Radiation in the United States.

Political Unit	Mean Annual Whole Body Dose Equivalent			
	Terrestrial		Total	
	mrem	mSv	mrem	mSv
Alabama	70	0.7	110	1.1
Alaska	60	0.6	105	1.05
Arizona	60	0.6	120	1.2
Arkansas	75	0.75	115	1.15
California	50	0.5	90	0.9
Colorado	105	1.05	225	2.25
Connecticut	60	0.6	100	1.0
Delaware	60	0.6	100	1.0
District of Columbia	55	0.55	95	0.95
Florida	60	0.6	95	0.95
Georgia	60	0.6	100	1.0
Hawaii	60	0.6	90	0.9
Idaho	60	0.6	145	1.45
Illinois	65	0.65	110	1.1
Indiana	55	0.55	100	1.0
Iowa	60	0.6	110	1.1
Kansas	60	0.6	110	1.1
Kentucky	60	0.6	105	1.05
Louisiana	40	0.4	75	0.75
Maine	75	0.75	125	1.25
Maryland	55	0.55	95	0.95
Massachusetts	75	0.75	115	1.15
Michigan	60	0.6	110	1.1
Minnesota	70	0.7	125	1.25
Mississippi	65	0.65	105	1.05
Missouri	60	0.6	105	1.05
Montana	60	0.6	150	1.5
Nebraska	55	0.55	130	1.3
Nevada	40	0.4	125	1.25
New Hampshire	65	0.65	110	1.1
New Jersey	60	0.6	100	1.0
New Mexico	70	0.7	170	1.7
New York	65	0.65	110	1.1
North Carolina	75	0.75	120	1.2
North Dakota	60	0.6	120	1.2

TABLE 4–1 (*Continued*)

Political Unit	Terrestrial		Total	
	mrem	mSv	mrem	mSv
Ohio	65	0.65	115	1.15
Oklahoma	60	0.6	110	1.1
Oregon	60	0.6	110	1.1
Pennsylvania	55	0.55	100	1.0
Rhode Island	65	0.65	105	1.05
South Carolina	70	0.7	110	1.1
South Dakota	115	1.15	185	1.85
Tennessee	70	0.7	115	1.15
Texas	30	0.3	75	0.75
Utal	40	0.4	155	1.55
Vermont	45	0.45	95	0.95
Virginia	55	0.5	100	1.0
Washington	60	0.6	110	1.1
West Virginia	60	0.6	110	1.1
Wisconsin	55	0.55	105	1.05
Wyoming	90	0.9	220	2.2
Puerto Rico	60	0.6	90	0.9
Guam	60	0.6	95	0.95
Samoa	60	0.6	90	0.9
Virgin Islands	60	0.6	90	0.9
U.S. Mean	60	0.6	105	1.05

Mean Annual Whole Body Dose Equivalent

Adapted from Klement 1972.

thorium and uranium series (Holtzman 1977)) with about 68% of the total coming from ^{40}K (Table 4–2). Thus the total average whole body dose equivalent in the United States from natural sources is about 130 mrem/y (1.3 mSv).

The mean annual dose from natural sources in so-called "normal" areas has been estimated on a world wide basis by the United Nations Scientific Committee on the Effects of Atomic Radiation as 93 mrad (930 μGy) to the gonads, 92 mrad to the cells lining the bone, and 89 mrad (890 μGy) to the bone marrow (UNSCEAR 1977). Total body dose equivalents have been developed using the weighting scheme put forth by the International Commission on Radiological Protection (ICRP

TABLE 4–2 Estimated Annual Internal Dose Equivalents from Natural Radioactivity in the United States.

Nuclide	Annual Dose Equivalent to Whole Body	
	mrem	μSv
H-3	0.004	0.04
C-14	1.0	10
K-40	17	170
Rb-87	0.6	6
Po-210	3.0	30
Rn-222	3.0	30
Total	25	250

Adapted from Klement 1972.

TABLE 4–3 Estimated Annual Effective Dose Equivalents from Natural Sources in Normal Background Areas, Worldwide.

Source	Annual Effective Dose Equivalent					
	External		Internal		Total	
	mrem	μSv	mrem	μSv	mrem	μSv
Cosmic Rays						
Ionizing Component	28	280			28	280
Neutron Component	2.1	21			2.1	21
Cosmogenic Radionuclides			1.5	15	1.5	15
Primordial Radionuclides						
K-40	12	120	18	180	30	300
Rb-87			0.6	6	0.6	6
U-238 Series	9	90	95.4	954	104.4	1044
Th-232 Series	14	140	18.6	186	32.6	326
Total (Rounded)	65	650	134	1340	200	2000

Source: UNSCEAR 1982, p. 102

1977) and are 2 mSv, about half of which is contributed by the ^{238}U series (UNSCEAR 1982). Similar values for the United States are generally somewhat lower for gonads (80 mrem/y) and bone marrow (80 mrem/y) and somewhat higher for bone surfaces (120 mrem/y) (NCRP 1975). It should be noted that different things are being compared; the UNSCEAR numbers are in units of dose or the ICRP weighted dose equivalent, while those for the United States are in the older units of dose equivalent. The calculational methods differ somewhat, but in general the agreement is not unreasonable. The differences between the data provide an actual illustrative example of the caveat regarding comparability of data given above.

The dose to the lung is largely a function of the inhalation of radon and daughters. The local dose equivalent to the segmental bronchioles has been estimated at 450 mrem/y (4.5 mSv) from this source, with a mean lung dose equivalent of 100 mrem (1 mSv) (NCRP 1975). An appreciable additional dose to the lung is incurred from cigarette smoking and has been estimated as 600–700 mrem (6–7 mSv) per year (Walsh 1978).

To the exposure from natural sources must be added the exposure from human activities, which contribute a total body dose equivalent of about 225 mrem/y (2.25 mSv) per person (Klement et al. 1972). By far, the greatest single contribution is from medical uses of radiation, which contributes at least 90% of the total. Fallout from weapons tests is the next highest contributor, providing less than 10 mrem/y (100 μSv/y). Other sources—occupational, nuclear power, and miscellaneous—contribute on the order of 2 mrem/y (20 μSv/y) or less. Estimated mean whole body dose equivalents have been estimated for the United States for the forty year period ranging from 1960 to 2000, and are shown in Figure 4–1 (Klement et al 1972). Note the slight upward slope to the contribution from medical sources, and the peak contribution from global fallout that occurred during the heyday of atmospheric weapons testing in the early 1960's. The reduction in the dose equivalent from miscellaneous sources is in large measure attributable to control of consumer product radiation, and the steep increase in the contribution from other environmental sources is largely due to the projected growth in nuclear power.

References

1. Boothe, G. F. 1977. "The Need for Radiation Controls in the Phosphate and Related Industries", *Health Phys.* 32:285.

2. Church, L. B. 1975. "Geothermal Power Plants: Environmental Impacts", *Science* 189:328.
3. Duncan, D. L., T. F. Gesell, and R. H. Johnson. 1976. "Radon-222 in Potable Water", Proc. Health Physics Society Tenth Midyear Topical Symposium, Natural Radioactivity in Man's Environment, Saratoga, New York.
4. Fornes, E. and P. Ortiz. 1978. "Radioactive Lightening Rods, Static Eliminators, and Other Radioactive Devices", in *Radioactivity in Consumer Products*, A. A. Moghissi, et al., Eds., NUREG/CP-0001, Washington, D.C. pp. 462–479.
5. Gesell, T. F., and N. M. Pritchard. 1975. "The Technologically Enhanced Natural Radiation Environment", *Health Physics* 28:361.
6. Gesell, T. F., and H. M. Pritchard. 1980. "The Contribution of Radon in Tap Water to Indoor Radon Concentrations", in *The Natural Radiation Environment III*, T. F. Gesell and W. M. Lowder, Eds., CONF-780422, pp. 1347–63.
7. Grune, W. N. et al. 1964. "Feasibility Studies for the Establishment of Parameters for Background Radiation Measurements", in *The Natural Radiation Environment*, J. A. S. Adams and W. M.Lowder, Eds., University of Chicago Press, pp. 661–686.
8. Haque, A. K., A. J. Collinson, and C. O. Blyth. 1965. "Radon Concentration in Different Environments and Factors Influencing It", *Phys. Med. Biol.*, 10:505.
9. Holtzman, R. B., 1977. "Comments on "Estimate of Natural Internal Radiation Dose to Man", *Health Phys.*, 32:324.
10. International Atomic Energy Agency (IAEA). 1967. "Radiation Protection Standards for Radioluminous Timepieces", IAEA Safety Series No. 23, Vienna.
11. International Commission on Radiological Protection. 1977. "Recommendations of the International Commission on Radiological Protection", ICRP Publication 26, *Annals of the ICRP* 1(3):1.
12. Jennings, W. A., and S. Russ. 1948. *Radon: Its Technique and Use*, John Murray, London.
13. Johnson, J. E. 1978. "Smoke Detectors Containing Radioactive Materials", in *Radioactivity in Consumer Products*, A. A. Moghissi et al., Eds., U. S. Nuclear Regulatory Commission Report NUREG/CP-0001, Washington, D. C., pp. 434–440.
14. Johnson, R. H., et al. 1973. "Assessment of Potential Radiological Health Effects from Radon in Natural Gas", U. S. Environmental Protection Agency Report EPA-520/1-73-004, Washington, D.C. (1973).
15. Klement, A. W., et al. 1972. "Estimates of Ionizing Radiation Doses in the United States 1960-2000", U. S. Environmental Protection Agency Report ORP/CSD 72-1, Washington, D. C.
16. Kruger, P., A. Stoker, and A. Umana. 1977. "Radon in Geothermal Reservoir Engineering", *Geothermics* 5(1–4):13.
17. Moghissi, A. A., and M. W. Carter. 1975. "Public Health Implication of Radioluminous Material", U.S. Public Health Service Report FDA 76-8001, Washington, D. C.
18. Moghissi, A. A., et al. 1978. "Evaluation of Public Health Implications of Radioluminous Materials", in *Radioactivity in Consumer Products*, A. A. Moghissi et al., Eds., U. S. Nuclear Regulatory Commission Report NUREG/CP-0001, Washington, D. C., pp.256–276.
19. National Council on Radiation Protection and Measurements (NCRP). 1975. "Natural Background Radiation in the United States", NCRP Report No. 45, Washington, D. C.
20. Simpson, R. E., and F. D. G. Schuman. 1978. "The Use of Uranium in Ceramic Tableware", in *Radioactivity in Consumer Products* A. A. Moghissi et al, Eds., U.

S. Nuclear Regulatory Commission Report NUREG/CR-0001, Washington, D. C., pp. 470–474.

21. Swanson, V. E. et al. 1976. "Collection, Chemical Analysis, and Evaluation of Coal Samples in 1975", U. S. Geological Survey Open File Report 76–468.
22. Travis, E. F., et al. 1979. "Natural and Technologically Enhanced Sources of Radon-222" *Nuclear Safety*, 20:722 (1979).
23. United Nations Scientific Committee on the Effects of Atomic Radiation (UNSCEAR). 1977. *Sources and Effects of Ionizing Radiation*, United Nations Document E.77.IX.1, New York.
24. United Nations Scientific Committee on the Effects of Atomic Radiation (UNSCEAR). 1982. *Ionizing Radiation: Sources and Biological Effects*, United Nations Document E.82.IX.8, New York.
25. Walsh, P. J. 1978. "Radiation Dose to the Respiratory Tract Due to Inhalation of Cigarette Tobacco Smoke", in *Radioactivity in Consumer Products*, A. A. Moghissi et al., Eds., U.S. Nuclear Regulatory Commission Report NUREG/CP-0001, Washington, D.C.
26. Yeates, D. B., A. S. Goldin, and D. W. Moeller. 1972. "Natural Radiation in the Urban Environment," *Nuclear Safety* **13:275.**

CHAPTER 5

Fallout from Nuclear Explosives

Introduction

In 1932, Sir James Chadwick discovered an uncharged primary particle with approximately the same mass as that of the proton, thus fulfilling the prediction made a decade earlier by Sir Ernest Rutherford. Because of its lack of charge, Chadwick's particle—the neutron—was able to penetrate the nucleus itself without being deflected by the electric fields of the atom. By bombarding ordinary elements with neutrons, experimenters such as Frederic and Irene Joliot-Curie (she the daughter of the famous Mme. Curie) in France and Enrico Fermi and his coworkers in Italy were able to produce numerous radioactive isotopes as physicists continued the study of the nucleus of the atom.

In Germany in 1938, Otto Hahn and Fritz Strassmann identified radioactive barium and other lighter atomic weight radionuclides in a sample of uranium that had been irradiated with neutrons. Hahn and Strassmann communicated their results to the scientific journal *Nature*, and privately to their former coworker Lise Meitner who had been forced to flee from Nazi Germany. On Christmas Day, the exiled Meitner and her nephew Otto Frisch, while on a cross-country ski outing, worked out the detailed explanation of what Hahn and Strassmann had observed. The result was a clear and coldly logical explanation of fission: the breakup of the nucleus of a heavy atom of uranium following bombardment with neutrons, with the resultant production of two lighter weight nuclei or fission products and the release of a quantity of energy. By 1940, more than 100 papers had been published in the scientific literature on fission. Then the curtain of secrecy descended, and little more was published on the subject until after the end of World War II.

The possible military value of fission was realized almost immediately. As early as March 1939, only two months after Meitner and Frisch published their explanation of the fission process, H. van Halban, F. Joliot-Curie, and L. Kowarski in France, and Leo Szilard, a Hungarian physicist then in the United States, realized that a nuclear chain reaction was possible, and would result in the release of great quantities of energy from the fissioning of the uranium nucleus. Szilard, concerned about events in Europe and the possibility of the Nazis building and using an atomic bomb to conquer the world, convinced Albert Einstein to sign a letter that was sent to President Franklin Roosevelt in August, urging him to have the United States undertake a military nuclear explosive development program. Roosevelt, heeding Einstein and as well as many others, established the supersecret Manhattan Project, an intensive scientific development program with the single purpose of producing an atomic bomb.

The Manhattan Project was indeed sucessful; in the early morning hours of July 16, 1945, the first nuclear weapons test took place in the New Mexico desert near the town of Alamagordo. Code named Trinity, this first nuclear detonation utilized the fission of Pu-239 and achieved an explosive force equivalent to 19 kilotons (KT) of TNT, rattling windows and lighting up the sky for miles around. Less than a month later, on August 6, 1945, a U-235 device was exploded over the city of Hiroshima, followed by the detonation of another plutonium device over Nagasaki three days later. Thus, World War II was brought to a dramatic and rapid conclusion with the destruction of two major Japanese cities by explosion of the only two nuclear weapons ever actively used in the conduct of war.

The following summer, Operation Crossroads was undertaken by the U.S. Navy at Bikini Atoll to study the effects of nuclear explosions. Crossroads was a major effort involving 42,000 men, 242 ships, more than 150 aircraft, 25,000 radiation monitoring devices, and more than 5,000 test animals, mostly rats (Shurcliff 1947). Two weapons were detonated: the first, Code named Able, was detonated in the atmosphere at an altitude of 500 feet on June 30. Able was followed by the Baker shot on July 24, detonated 100 feet under water. The results were phenomenal. An enormous water spout was produced, nearly 2000 feet (610 m) across, rising to its maximum height in about a minute. The hollow spout had a mushroom top about a mile and a half across, in which most of the fission product radioactivity was contained. From the base of the huge column of water, a white mist rolled out at a speed of about 50 miles per hour, (80 km/h) producing an unexpected fallout containing rain. The explosive force was so great from this nominal ($<$ 20KT) explosion that warships near the point of detonation were tossed about like leaves in the wind.

The United States detonated three more test devices in 1948, followed by the first U.S.S.R. test in August of 1949. Great Britain detonated its first nuclear explosive in October 1952, followed a month later by the first American thermonuclear or H-bomb explosion, with a reported yield of 5 megatons (MT). The Russians followed with a thermonuclear detonation less than a year later, and the British in 1957. Through the end of 1958, when the moratorium on nuclear testing began, 250 known nuclear explosions with a cumulative yield of more than 32 MT had taken place, virtually all having been detonated in the atmosphere (Carter and Moghissi 1977). Despite the moratorium and limited test ban treaty, atmospheric testing was resumed by the U.S.S.R. in September 1961, and underground testing by the United States the same month. France, not a signatory to the treaties, conducted its first nuclear test in 1960, and the People's Republic of China its first in 1964. India, the sixth nation known to have tested nuclear weapons, conducted its first test in 1974.

Most tests have been conducted in the northern hemisphere; the United States testing has been largely carried out in Nevada and at the Pacific Proving Ground, which includes Bikini and Eniwetok Atolls, Johnston Island, and Christmas Island. Other American sites include Amchitka Island, Alaska, and locations in Colorado and Mississippi. British tests have been carried out at Australian locations and at Christmas Island. The U.S.S.R. is known to have carried out tests at several locations, primarily in the Ural mountains, and at Novaya Zemlya and Semipalatinsk. French testing has been in Algeria and the Tuamotu Islands in the Pacific, while the People's Republic of China has limited their testing to the Loop Nor area in northern China. India has conducted its test in the Rajasthan Desert on the Indian subcontinent. Thus, nuclear testing has been rather widespread throughout the world.

Nuclear Explosive Phenomenology

A nuclear explosion is simply the extremely rapid release of energy due to nuclear fission or fusion in a very small volume. The reaction generally occurs in a few milliseconds, releasing enormous quantities of energy. The fission reaction requires a supercritical mass of an appropriate fissionable nuclide and an initial source of neutrons. U-233, U-235 and Pu-239 are all suitable in that they are capable of undergoing fission with any energy of neutrons and have reasonably long half lives such that they do not decay to other non-fissionable materials. The basic fission reaction, as exemplified by U-235, is.

$$U\text{-}235 + n \rightarrow \text{Fission Fragments} + 2\text{-}3 \text{ neutrons} + \text{Energy}$$

In the equation, a uranium-235 nucleus is bombarded with a neutron and splits typically into two lighter nuclei or fission products along with the relaese of energy and two or three neutrons which then can fission other uranium nuclei. Splitting of one U-235 nucleus releases about 200 Mev of energy, distributed as shown in Table 5-1. About half of the energy goes into the explosive blast, a third to heat, and the remaining sixth or so to radioactivity (Glasstone 1977).

Fission products have characteristic yields, or percentage of the total of atoms fissioned. As approximately two fission products are produced per fission, the total fission product yield is about 200%, with that of each individual fission product varying according to its mass number, with highest yields occurring near the so-called "magic numbers" associated with closed nuclear shells. Examination of the fission yield curve for ^{235}U shown in Figure 5-1 shows two distinct groups of high yield fission products, one with mass numbers from 85 to 104, and the other with mass numbers from 130 to 149. Other fissionable nuclides have similar double humped fission product yield curves.

The energy released in the fission process produces extremely high temperatures which can bring about the fusion of certain light nuclei with the concomitant release of additional energy. Nuclear fusion is the basis of the so-called hydrogen bomb or thermonuclear explosives, in which the high temperatures generated by a fission explosion trigger are used to fuse various hydrogen isotopes with the concomitant release of

TABLE 5-1   Distribution of Fission Energy.

	MeV
Kinetic Energy of Fission Fragments	167
Prompt Gamma Radiation	7
Kinetic Energy of Fission Neutrons	5
Beta Particles from Fission Products	7 Delayed
Photons from Fission Products	6 Delayed
Neutrinos from Fission Products	11 Delayed
Total Fission Energy	203 MeV/fission

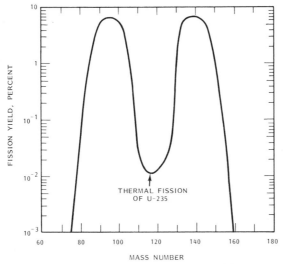

FIGURE 5-1   Fission Yield Curve for ^{235}U.

vast amounts of energy, neutrons, and the production of tritium. The neutrons produced in the thermonuclear reactions can be used to produce fissions also, and this is most efficiently and effectively accomplished by use of a nuclide such as ^{238}U which has a high cross-section for fission with fast neutrons such as those produced by the fusion reaction. Thus, there is also the "3F" bomb, or fission-fusion-fission device. In addition to the radioactive byproducts of the fission and fusion reactions, both fission and fusion explosives produce a variety of radioactive species from neutron activation of the bomb casing and other materials, including the air and ground.

   The nuclear explosion process is extremely rapid, taking place in less than a microsecond. The bomb materials are vaporized to hot gas with a pressure of several billion atmospheres by the tremendous temperatures (10^8 °C) produced by the explosion. A fireball begins to grow; 0.03 seconds after the detonation of a 1 MT device the fireball is 440 feet (134 m) in diameter, increasing to 7200 feet (2200 m) after 10 seconds. The fireball, which may also contain volatilized soil or rock if the explosion has occurred near the surface of the ground, rises rapidly, intially at a speed of perhaps a few hundred miles per hour, slowing as it cools. The altitude to which it rises is a function of the explosive yield of the device; the cloud from a 1 MT explosion would reach an altitude of about 14

TABLE 5-2   Rise of the Fireball from a 1 MT Nuclear Explosion.

Height			Rate of Rise	
(mi)	(km)	Time	(mph)	(km/h)
2	3	0.3	300	500
4	6	0.75	200	300
6	8	1.4	140	200
10	14	3.8	90	150
14	20	6.3	35	55

miles (22.4 km) after about 6.3 minutes (Table 5-2). When it reaches the top of the troposphere—an altitude of about 50,000 feet (15 km)— the fireball begins to spread out, expanding into the familiar mushroom shape. The cloud continues to rise into the stratosphere, and may achieve an altitude of 25 miles (40 km). Maximum height is achieved about 10 minutes after detonation, with the mushroom spreading laterally outward to about 100 miles (160 km) in less than an hour (Government of India 1958).

As the cloud rises, cool air is drawn up into the hot cloud, creating the so called afterwinds, which carry with them surface dust and debris to which the radioactive fission product nuclei can attach if the fireball touches the ground, thus producing additional fallout debris. The fireball develops a toroidal shape within a few seconds after the explosion, with updrafts in the center and downdrafts around the exterior. The fission products are largely confined to the torus, which has been described as a "smoke ring" (Kellogg et al. 1957).

Within a minute after the detonation, particles composed of the oxides of iron, aluminum and other refractory materials, with diameters in the range of 0.4–4 μm are formed (Sisefsky and Persson 1970). The refractory radionuclides tend to be incorporated in these particles while the lower melting point more volatile radionuclides condense into particles with diameters <0.4 μm. This division into two particle size ranges is known as fractionation; the larger particles tend to settle to earth rather quickly because of gravity, producing what is known as close-in or local fallout, depositing on the ground within a hundred miles or so of the point of detonation. The smaller particles, which contain most of the volatile fission products including the radiocesiums, radiostrontiums and radio-iodines, remain suspended for longer periods and fall out over a much larger area in the same hemisphere, perhaps many thousands of miles distant, and constitute the tropospheric fallout (Freiling 1961). For detonations with yields in the kiloton range, most of the radioactivity is injected into the troposphere, where it has a rate of diminution or residence half-time of about three weeks, with essentially all the debris falling to earth within 2–3 months (Stewart et al. 1957).

Particles carried into the stratosphere by the detonation produce a world-wide pattern of global or stratospheric fallout largely confined to the hemisphere in which the detonation occurred as there is very little mixing of air between the hemispheres (Figure 5–2). Once in the stratosphere, the radioactive particles tend to remain there for some length of time, generally averaging about a year. The actual residence time will vary, being determined by many factors including the explosive yield of device, and the altitude, latitude, and season at which the detonation occurred.

Movement through the stratosphere is described by the Brewer-Dobson model (Machta and List 1960), which in much simplified form, states that air from the troposphere rises into the stratosphere near the equatorial tropopause and moves towards the poles, where it sinks back into the troposphere. Thus, air flow in the stratosphere is generally south to north in the northern hemisphere and north to south in the southern hemisphere (Figure 5–2). Transfer to the troposphere occurs in the temperate and polar zones and is maximal during the late winter and early spring when the air in the high latitudes is coldest and hence densest;

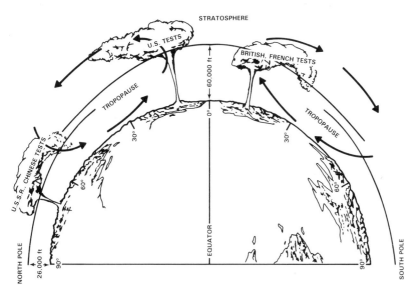

FIGURE 5–2   Hemispherical Air Circulation Patterns Illustrating Lack of Mixing of Bomb Debris Between Hemispheres.

this, in conjunction with rainfall and storm activity, gives rise to a characteristic increase in fallout activity during the spring that is known as the spring peak (Karol 1970). Removal of particulate radioactivity from the troposphere is fairly rapid, generally on the order of a month (Gavini, Beck, and Kuroda 1974), which limits mixing between the hemispheres. However, long-lived gaseous radionuclides such as Kr-85 (10.76 y) readily mix between hemispheres, having a mean exchange time of 1.5 years (UNSCEAR 1977, 1982).

Radioactivity from Nuclear Explosions

A nuclear explosion produces a great amount of radioactivity. An explosive force of 1 KT is produced by the complete fission of 56 grams of fissionable material, or about 1.4×10^{23} nuclei. This produces approximately twice as many fission product nuclei, most of which are radioactive. The total radioactivity produced by a fission explosion with an explosive yield of one kiloton is on the order of 1.5×10^{13} Ci (5.55×10^{23} Bq), mostly of shortlived fission gases. Fission of a heavy nucleus can occur in more than 80 different ways, producing more than 200 different radioactive species with half-lives ranging from a fractions of a second to 17 million years.

The decay rate for mixed fission products is the sum of the exponential decay rates for each of the radioactive species in the mixture. Thus, the decay rate for the complex mixture can be expressed as a power function. However, the mixture of fission products is constantly changing, and power thus varies. The rate of emission of beta and gamma rays can be empirically approximated by Equations 5–1 and 5–2:

$$\text{Betas/sec-fission} = 3.2t^{-1.2} \qquad (5\text{–}1)$$
$$\text{Photons/sec-fission} = 1.6t^{-1.2} \qquad (5\text{–}2)$$

in which t is the time in seconds after the instant of fission. Thus, the decay of the mixture of fission products can be approximated by the $t^{-1.2}$ rule, which indicates a tenfold drop in activity for every sevenfold increase in time. This is sometimes known as the Wigner-Way rule (Wigner and Way 1948), and can be expressed mathematically as

$$A = 1.03 \times 10^{-16}t^{-1.2} \qquad (5\text{–}3)$$

in which A is the activity in curies/fission and t is the time in days after fission. Alternatively this can be written

$$A = A_1 t^{-1.2} \qquad (5\text{-}4)$$

in which A is the activity at time t after fission, and A_1 is the activity at unit time after the detonation. Note that t can be in any units of time so long as the value of A is specified for the same time unit. Thus, if A_1 is the activity at 1 second after detonation, t must have the value of seconds, etc. Note that Equation (5-4) is compatible with the SI units in that A and A_1 can be in Bq. If Bq are used in Equation (5-3), the constant becomes 3.8×10^{-6}.

As has been shown in Figure 5-1, the yield of any specific fission product is a function of its mass number. The most probable mass numbers for thermal fission of ^{235}U are 95 and 139, each with yields of about six per cent. Production is both direct from the fission process itself and from ingrowth from chains. Important fission product chains are those which produce or include radionuclides of environmental significance. An example of such a chain is illustrated below:

$$^{90}Br \xrightarrow[3.2 \text{ min}]{15\%} {}^{89}Kr + \text{neutron}$$

$$\xrightarrow[1.6 \text{ sec.}]{85\% \ \beta} {}^{90}Kr \xrightarrow[33 \text{ sec}]{\beta} {}^{90}Rb \xrightarrow[2.7 \text{ min}]{\beta} {}^{90}Sr \xrightarrow[28 \text{ y}]{\beta} {}^{90}Y \xrightarrow[65 \text{ h}]{\beta} {}^{90}Zr \text{ (stable)}$$

The chain illustrated begins with the direct production of 1.6 second ^{90}Br, which undergoes a branching decay to ^{89}Kr 15% of the time and to 33 second ^{90}Kr 85% of the time. Because of its short half-life, the fission yield of ^{90}Br is not precisely known, but is thought to be about 4.8%. The total fission yield of the ^{90}Kr is 5.0%. Krypton-90 undergoes beta decay to 2.7 min ^{90}Rb which in turn beta decays to 28 year ^{90}Sr. The total fission yield of ^{90}Sr is 5.77%, 5.0% growing in from ^{90}Rb with the additional 0.77% attributable to direct production. Ingrowth of any species is governed by the Bateman equations described in Chapter 3, but as a practical matter the first members of the chain usually have very short half-lives such that the ingrowth of the significant long lived member of the chain is completed within minutes.

Because of the complexity of the numerous decay chains, and the wide variation in half-lives of the fission products, the relative contribution of activity from any given fission product to the total is highly variable with time. For example, ^{131}I, with a half-life of 8.05 days, contributes about two per cent of the total radioactivity three days after a nuclear detonation, growing to six per cent after one week and peaking at seven per cent of the total about 11 days after fission, and dropping to less than 1% after two months.. On the other hand, the 30 year half-life ^{137}Cs contributes very little to the total initially, rising to about one and a half per cent of the activity a year after the detonation, and to 18 % after ten years. Although the amount of ^{137}Cs is decreasing, its contribution or fraction of the total increases because of its long half-life, rising to 25% 30 years after detonation. A million years after detonation virtually no activity remains, but what does remain is almost wholly ^{129}I with a half-life of 1.7×10^7 years.

In addition to the activity from fission products, radioactive species will be produced from neutron activation of the materials of the explosive— the casing, the high explosive detonator, and various other systems—as well as from neutron activation of the air and the ground, assuming the detonation does not occur at a very high altitude, or under water or underground. The activation products include various iron and cobalt isotopes, ^{14}C, and tritium. Tritium is also produced directly by fission and in the fusion reaction along with numerous neutrons. Neutron activation of the soil produces, among others, ^{24}Na (15 h), ^{32}P (14 d), ^{45}Ca (152 d), ^{55}Fe (2.9 y), and ^{59}Fe (46 d), all of which may have bioenvironmental significance.

The amount of radioactivity introduced into the environment from weapons testing has been enormous. Through June 30, 1978, 905 detonations totalling 366 megatons of explosive force have been documented (Carter 1980). Major test programs were conducted during 1954–58 and 1961–2; 77 atmospheric detonations were carried out in 1962 alone—39 Russian totaling an estimated 180.3 MT and 38 American tests totaling 37.1 MT. Not all of these have been detonated in the atmosphere, nor has all the explosive yield been from fission. However, although the number of underground and hence presumably contained detonations exceeds that of atmospheric detonations, most of the explosive yield has taken place in the atmosphere. The United Nations Scientific Committee on the Effects of Atomic Radiation has documented 423 atmospheric tests through 1980 with an estimated total yield of 545.4 MT (UNSCEAR 1982). About 40% of this yield—217.2 MT—is attributable to fission, which, assuming 1.5×10^{13} Ci/KT, would indicate that about 3.3×10^{18} curies (1.2×10^{29} Bq) of fission products have been injected into the atmosphere from fission alone. To this must be added the activation products and fusion produced tritium. Granted most of the activity produced by fission is short-lived, but nonetheless the quantity is staggering and of great environmental significance.

The largest number of atmospheric tests—193 with an estimated total yield of 138.6 MT of which 72.1 MT were from fission—has been carried out by the United States. The USSR has conducted 142 atmospheric detonations which account for well over half of the total yield—357.5 MT of which 110.9 MT were from fission. These tests, plus the 21 British detonations and four French detonations, were all conducted prior to the Limited Test Ban Treaty which took effect in 1962. Together they account for fully 85% of the total number of atmospheric detonations but nearly 95% of the total yield and about 90% of the fission yield. Since 1962, only France and China have carried out atmospheric tests, totaling through 1980 64 in number, or 15% of the total, and accounting for 23.5 MT of fission yield.

Important Fission Products

Of the more than 200 fission products produced by nuclear explosions, only a relatively small number have half-lives sufficiently long to be important from a long term environmental point of view. These are ^{85}Kr, the radiostrontiums, radioiodines, radiocesiums, ^{106}Ru, ^{140}Ba, and ^{144}Ce with ^{89,90}Sr, ^{131}I, and ^{137}Cs of primary import. The basic physical properties of the environmentally important fission products are shown in Table 5–3.

TABLE 5–3 Physical Properties of Important Activation and Fission Products in Fallout.

Nuclide	% Yield	Half-life	Comments
H-3	—	12.33 y	Primarily activation product
C-14	—	5730 y	Activation product
Mn-54	—	313 d	Activation product
Fe-55	—	2.7 y	Activation product
Sr-89	4.8	50.5 d	Chemically
Sr-90	5.9	28 y	similar to calcium
Ru-106	0.4	369 d	
I-131	3.1	8.85 d	Ingrowth from chain
Cs-134	7.19	2.06 y	
Cs-137	6.2	30 y	
Ba-140	6.4	12.8 d	Usually in equilibrium with daughter La-140
Ce-144	6.0	284 d	

Krypton-85 is the only noble gas isotope of significance because of its long half-life (10.76 years) and high fission yield (1.33%). This gaseous nuclide becomes fairly uniformly dispersed throughout the atmosphere within a few years after production by a nuclear explosion (Farges et al. 1974). As of 1983 an estimated 3.2 MCi (120 PBq) are resident in the atmosphere as a result of weapons tests, this being but a small fraction of the total atmospheric inventory, most of which has come from nuclear power production. Air concentrations of this nuclide world wide have been estimated as 1.67×10^{-11} μCi/cm^3 (0.62 Bq/m^3) (UNSCEAR 1982).

Krypton, an inert gas, does not concentrate in biota, and hence the primary radiological concern is the external exposure to the skin from immersion. Krypton-85 decays to stable ^{85}Rb by with the emission of a beta particle with an average energy of 250.5 keV ($E_{max} = 0.67$ MeV); a 0.514 Mev gamma photon is emitted in 0.41% of the decays (Lederer et al. 1977). The dose to the skin from this nuclide at the typical concentrations of a few pCi/l is essentially trivial, having been calculated as 21 μrad (0.21 μGy) over a lifetime (UNSCEAR 1982). Doses to other organs are typically two orders of magnitude lower.

Although at least eleven isotopes of strontium are produced by fission, only mass numbers 89 and 90 are environmentally significant. The higher mass number strontium isotopes all have relatively short half-lives, and ^{88}Sr, also produced by fission, is stable. Strontium-89 decays by emission of a beta particle with an average energy of 583 keV (1.463 MeV$_{max}$) to stable ^{89}Y (Lederer et al. 1977), and, with a half-life of 50.5 days and is a major component of fallout for a few months after the detonation. The direct fission yield of this nuclide is only about 0.2 per cent, but the total yield from the chain is about 4.8 per cent. In fresh fallout, the activity from ^{89}Sr may be 200 times that of ^{90}Sr, decreasing to about 10 after 1 year and to essentially zero after two years (Eisenbud 1973). The ingrowth of ^{89}Sr from the mass number 89 fission product chain shown below:

$$\text{Se-89} \xrightarrow[0.41 \text{ sec}]{\beta} \text{Br-89} \xrightarrow[4.5 \text{ sec}]{\beta} \text{Kr-89} \xrightarrow[3.16 \text{ min}]{\beta} \text{Rb-89} \xrightarrow[15.2 \text{ min}]{\beta} \text{Sr-89} \xrightarrow{\beta} \text{Y-89 (stable)}$$

Twenty kilocuries (740 MBq) are produced per kiloton of fission, producing an estimated injection of 2.9×10^9 curies (1.1×10^{20} Bq) into the atmosphere from nuclear weapons testing. Because of its relatively short half-life and the very short half-lives of its precursors, relatively little remains; about 100 days after fission, ^{89}Sr accounts for 10% of the total radioactivity from fission products, decaying to less than 1% within 15 months.

Strontium-89 is environmentally important largely because of foliar deposition, its short half-life making uptake from the soil of water of lesser import. The primary pathway to man is through milk and once within the body it behaves like calcium, concentrating in bone. Lifetime dose commitments to the bone marrow from this source have been estimated at less than 0.22 mrad (2.2 μGy) (UNSCEAR 1982). Dose commitments to the bone lining cells are somewhat greater, 0.34 mrad (3.4 μGy), with the largest commitments, 0.51 and 1.5 mrad (5.1 and 15 μGy) received by the upper and lower large intestine, respectively, from ingestion (UNSCEAR 1982).

Strontium-90 is a fallout radionuclide of great environmental significance. This nuclide is a pure beta emitter having an average energy of 195.8 keV and a maximum beta energy of 540 keV, decaying with a half-life of 28.6 years to ^{90}Y (Lederer et al. 1977). Yttrium-90 is also a pure beta emitter (E_{av} = 935 keV; E_{max} = 2.27 MeV) with a half-life of 64.1 hours. Thus, within a period of a few weeks, an equilibrium concentration of these two nuclides will exist. Production of ^{90}Sr is relatively small (approximately 1/185) in comparison to ^{89}Sr, about 100 Ci/KT (3.7 TBq/KT) for a total estimated historical production from weapons tests of about 18 MCi (660 PBq) (UNSCEAR 1982). The direct fission yield is about 0.9 per cent, and total yield about 5.9%.

Strontium-90 falling out on land is taken up by plants, ultimately finding its way into people through milk and through ingestion of various plant foods. Uptake by plants is more rapid when freshly added to the soil as opposed to aged deposits (Cline 1981). It is also concentrated by certain marine biota, particularly algae. The inventory of ^{90}Sr in the ocean was calculated for two years for which relatively complete data were available, 1961 and 1966 (Volchok et al. 1971); the levels were 24.5 and 31.4 MCi (907 and 1162 PBq), respectively. On land, most of the ^{90}Sr is resident in the northern hemisphere (Figure 5–3), with about half found in the first 4 cm depth of soil (UNSCEAR 1977).

Strontium-90 concentrates in bone, with concentrations varying according to residence location. Concentrations peaked during the mid-1960's when atmospheric weapons tests were in progress; for New York City, peak levels were about 2 pCi (0.074 Bq) per gram of Ca, with about half that value being observed in the southern hemisphere (Australia). The estimated dose commitments to bone tissues from fallout ^{90}Sr through 1975 have been computed by the United Nations Scientific Committee on the Effects of Atomic Radiation and are shown in Table

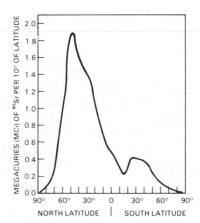

FIGURE 5-3   Latitudinal Distribution of Sr-90. (Adapted from UNSCEAR 1982).

5-4. Levels are expectedly greater in the northern hemisphere because of the greater amount of testing done there.

At least 14 different isotopes of iodine with mass numbers ranging from 127 to 141 have been identified as fission products. One—[127]I—is stable, and those with mass numbers greater than 135 have such short half-lives (<1.5 min in all cases, and ranging down to 0.4 s) are of little environmental consequence, even as precursors of other radionuclides, despite relatively high fission yields exceeding 6%. Thus, there are but six fission radioiodines of potential environmental significance (Table 5-3); [128]I and [130]I are not significant because they are only produced, if at all, directly, since each is blocked from chain production by beta decay by a stable precursor. I-129, with its long half-life (17 million years) and relatively low yield (0.9%) is also of little significance in fallout.

The potentially significant radioiodines have mass numbers 131–135

TABLE 5-4   Dose Commitment to Bone from Nuclear Testing through 1975.

Tissue	Northern Hemisphere		Southern Hemisphere	
	Temperate Zone	Mean	Temperate Zone	Mean
Bone marrow	85 mrad	56 mrad	24 mrad	17 mrad
Endosteal cells	116 mrad	77 mrad	33 mrad	24 mrad

Taken from UNSCEAR 1977, p. 137.

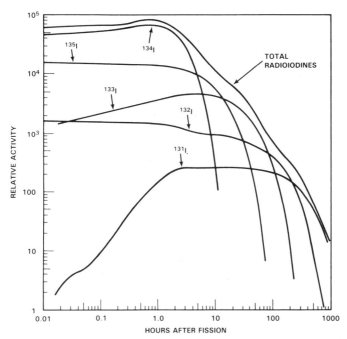

FIGURE 5-4   Radioiodines as a Function of Time After Fission. (Adapted from Kathren 1964).

and are all relatively short-lived. Their fission yields are relatively high (2.77, 4.13, 6.77, 6.7, and 6.12 per cent, respectively). All, with the exception of ^{135}I, grow in from chains in which at least one of the precursers has a relatively long half-life. Thus, the maximum activity from any given radioiodine may not occur until some time after fission with the fraction of the total iodine activity contributed by any iodine isotope quite variable with time (Figure 5-4). For about the first three hours after fission, ^{134}I ($t_{1/2}$ = 52.6 min) is the predominant radioiodine activity, then, for the next few hours, ^{135}I ($t_{1/2}$ = 6.6 h), then ^{133}I ($t_{1/2}$ = 55.4 min) at about 1 day, and finally, after about ten days, ^{131}I ($t_{1/2}$ = 8.05 days) (Kathren 1964). Thus, from a global fallout or long term environmental standpoint, only the 8.05 day half-life ^{131}I is significant (Eisenbud and Wrenn 1963). The dose delivered to the thyroid from ^{131}I—6.3 rad/ μCi (1.7 μGy/Bq) in the organ of an adult—is more than threefold greater than that from ^{133}I (1.8 rad/μCi or 0.5 μGy/Bq) and more than an order of magnitude greater than that of the other radioiodines (Kathren 1964).

The yield of ^{131}I, even though less than half that of the higher radio-iodines, is nonetheless great, being 64 KCi (2.4PBq)/KT. Thus, an estimated 2×10^{10} Ci (700 EBq) of ^{131}I have been introduced into the atmosphere from nuclear weapons tests. In the United States, where above ground testing of nuclear weapons was conducted in Nevada in the 1950's, ^{131}I was found in bovine thyroids and in cow's milk in concentrations in the nanocurie per liter (10 KBq/m^3) range; one experimenter reported a mean thyroid dose of 0.2 to 0.4 rad (2–4 mGy) to children in the United States during the period 1953–1958 in hearings before the Joint Committee on Atomic Energy in 1959 (Lewis 1959). Doses to the thyroids of children located downwind of the Nevada Test Site during the 1950's have exceeded 100 rad (1 Gy) according to one group of investigators (Tamplin and Fisher 1966), and others have found similar values. The 1980 report of the National Research Council Committee on the Biological Effects of Ionizing Radiations estimated the radiation doses to the thyroids of one group of 2,691 children studied for thyroid abnormalities at 30 to 240 rads (0.3–2.4 Gy), with a mean of 120 rad (1.2 Gy) (NRC 1980).

Even higher doses were observed in a group of 267 persons in the Marshall Islands located downwind from a thermonuclear weapons test in 1954. These individuals were exposed to fresh fallout, and incurred thyroid doses estimated at 220–450 rads (2.2–4.5 Gy) for adults and 700–1400 rads (7–14 Gy) for a 4 year old child (NRC 1980). As the fallout was fresh, a significant fraction of the thyroid doses were delivered by radioiodines with mass numbers 132–135. About a sixth of those exposed developed thyroid nodules, which were carcinomatous in seven instances.

Environmentally, ^{131}I deposits on foliage and is ingested by animals, which concentrate and excrete the iodine in their milk. Thus, fresh milk is the dominant source of intake and dose to man from ^{131}I, and an excellent correlation has been found between ^{131}I in milk and in human thyroids (Eisenbud 1962). The nuclide can also be taken up by inhalation, ingestion through water and leafy vegetables, and possibly even absorbed through the skin.

Of the radiocesiums present in fallout, three isotopes are environmentally important: those with mass numbers 134, 136, 137. Cesium is an alkaline metal whose chemical behavior in the body is somewhat similar to that of its Group 1A homologue potassium. Thus, cesium is relatively well dispersed throughout the soft tissues of animals and people, where it is found primarily inside the cells. Of the three cesium isotopes, 136 has the shortest half-life by far, at 13 days, and thus is not of particular concern from the standpoint of long term global fallout. The dose contribution from this nuclide, integrated with respect to time from the start of weapons testing until 1983 is only a fraction of a millirad ($<$10 μGy). Of the remaining two radiocesiums, ^{134}Cs is shorter lived with a half life of 2.06 years and the rather high fission yield of 7.19%; for ^{137}Cs, the corresponding values are 30 years and 6.15% respectively. Because of the long half-life of the latter, it is particularly significant as a long term global fallout source. An estimated 34 MCi (1.26 EBq) of ^{137}Cs have been injected into the atmosphere as a result of weapons tests to 1971 (Joseph et al. 1971), with perhaps another 10–20% added by subsequent atmospheric tests. The United Nations Scientific Committee estimates 960 PBq (UNSCEAR 1982). Fallout of cesium over land is more significant than fallout over the oceans, and in the temperate latitudes is greatest during the spring, and thus assumes somewhat greater significance as this is the time of rapid plant growth. Much of the global fallout comes from ^{137}Cs that has been introduced into the middle stratosphere by weapons testing, where it has a residence time of 6 to 12 months (Joseph et al. 1971). Deposition thus was greatest during the early 1960's, or within a year or so after an atmospheric weapons test.

Because of its 30 year half-life, ^{137}Cs is environmentally persistent. Concentrations in ocean surface waters have been determined to be on the order of 0.2 pCi/l (7.4 Bq/m^3), being lower in deep waters by a factor of 30 (Hodge, Folsom and Young 1973). On land, most is contained in the top few centimeters of the soil, with relatively little leaching occurring. Cesium-137 is widespread in foods, with concentrations ranging from 1 or 2 pCi/kg (0.037 − 0.074 Bq/kg) in fruits and vegetables to perhaps 10 times this concentration in meats, dairy products, and grain products. As the biological half-life is relatively short, equilibration with the dietary and environmental levels will be quickly reached, presuming these are constant.

Internal doses to people from ^{137}Cs in fallout have been rather variable, dependent to a large extent on the recent history of atmospheric testing. For the period 1953, when measurements were begun, through 1970, the total dose to the average American adult was 16 mrad (160 μGy) from internally deposited Cs-137. During this 18 year period, the annual dose rate varied from 0.04 to 2.8 mrad (4–28 μGy) (NCRP 1977). During the period of extensive atmospheric testing in the middle 1960's, concentrations in the body ranged from 17 pCi/g (0.63 Bq/g) of potassium to perhaps 10 times this amount in the U.S.A. In other parts of the world, concentrations were similar except among subarctic populations whose diets consisted largely of caribou or reindeer; in these populations, concentrations of several thousand pCi/g K (approximately 100 Bq/g K) were typical, and doses were correspondingly greater (UNSCEAR 1977).

As a 662 keV gamma ray is associated with the decay of ^{137}Cs-Ba, fallout of ^{137}Cs has produced a long-lived external gamma field which delivers a dose of 1.7×10^{-3} μrad/h per mCi/mi^2 (1.2 pGy/Bq-km^2). Using average values for ^{137}Cs in soil in the United States, this would deliver a mean whole body dose of about 1 mrad (10 μGy). A peak dose of about 6 mrad (60 μGy) was delivered during the high fallout year 1963 (Klement 1982).

Other fallout radionuclides with long term implications include ^{106}Ru, with a one year half-life, and the 285 day ^{144}Ce with its 1.7 min daughter Pm-144. These nuclides are primarily of interest environmentally from the standpoint of lung dose from inhalation of particulates. Doses to individuals have been generally on the order of a fraction of a millirad ($<$ 10 μGy) per year. Barium-140 (12.8 days) is commonly found in eqilibrium with its shorter lived daughter 40 hour ^{140}La. These nuclides are of concern environmentally primarily from the standpoint of close-in or tropospheric fallout, largely because of their short half-lives. This nuclide pair has been identified in milk a few days to weeks after weapons tests, producing a lifetime dose commitment to the bone of $<$0.1 mrad ($<$10 μGy) (UNSCEAR 1977).

Tritium, C-14 and Other Activation Products

Various radionuclides are produced by neutron activation of the materials used for the casing and other materials of the explosive, the atmosphere, and the ground, if the explosion is sufficently close to the surface. Two of the activation products are produced in relatively large amounts and are also both important cosmogenic radionuclides; these are tritium and carbon-14. Nuclear detonations are an important source of environmental tritium. Tritium is produced at the rate of about 10^{-4} atoms per fission by ternary fission, but a far more important and productive source has been the residue from thermonuclear reactions. The total environmental tritium inventory from atmospheric testing since 1946 has been estimated at 8×10^9 Ci, with about 90% of the total from thermonuclear reactions (NCRP 1979). Most of this tritium—approximately 75–80%—has been introduced into the northern hemispere.

Tritium, an isotope of hydrogen, is a pure beta emitter with a mean energy of 5.69 keV (E_{max} = 18.6 keV) and a half life of 12.3 years (Lederer et al. 1977). It is readily oxidized or exchanges with other hydrogen isotopes forming HTO or tritium oxide (T$_2$O) which can remain in the atmosphere for a time or precipitate to earth and into the hydrosphere. Tritium is thus ubiquitous, and found in all living things. Comparison of the amount of tritium in the environment from nuclear detonations with that from natural sources reveals the former to be far more significant; it should also be noted that only atmospheric detonations are significant as tritium from underground testing has not added appreciably to the amount in the atmosphere and surface waters (NCRP 1979).

An estimated 8 $\times$ 10^9 Ci (3 $\times$ 10^{20} Bq) had been added to the atmosphere by testing through 1962 (Miskel 1973), perhaps increasing by about 10% by subsequent tests. As most of the testing has been in the northern hemisphere and there is little mixing between hemispheres, about 80% of the weapons produced tritium is assumed to have been in the northern hemisphere (UNSCEAR 1982). More than 99% of the tritium produced by atmospheric weapons tests is attributable to fusion which produces 20 MCi (740 PBq) of tritium per megaton. By contrast, only about 700 Ci (26 TBq) are produced per megaton of fission yield. As the yield of fusion devices is also greater than fission devices, most— about 75%—of the tritium produced has been injected into the stratosphere.

Concentrations of tritium in surface waters in the world have been greatly increased by weapons testing. In 1963, mean concentrations of H–3 in surface streams in the United States were on the order of 4000 pCi/l (67 KBq/m^3) about 400 times the contribution from naturally occurring tritium (NCRP 1979) and an order of magnitude greater than the concentration typically found in environmental waters in 1983. During the early 1960's when atmospheric nuclear testing was at its peak, the mean dose commitment from tritium to persons in the northern hemisphere has been estimated as 2–3 mrad (20-30 μGy), with the corresponding value for the southern hemisphere an order of magnitude less (UNSCEAR 1977, NCRP 1979). The dose rate to man reached a peak of 0.2 (2 μGy) mrad in 1963, and by 1979 had dropped to 0.01 (0-1 μ Gy) mrad annually (NCRP 1979).

Of the activation products, ^{14}C is environmentally the most significant. Carbon-14, a pure beta emitter with an average energy of 49.5 keV (E$_{max}$ = 156 keV) and a 5730 year half life, is produced primarily by the (n,p) reaction with ^{14}N, which constitutes nearly 80% of the atmosphere. As a result of nuclear detonations in the atmosphere, the worldwide ^{14}C inventory increased dramatically, with the concentration in the troposphere peaking at 70% greater than that from natural sources in 1965, and then tapering off (Figure 5–5). Approximately 6 MCi (220 PBq) have been introduced into the atmosphere by nuclear testing, which can be com-

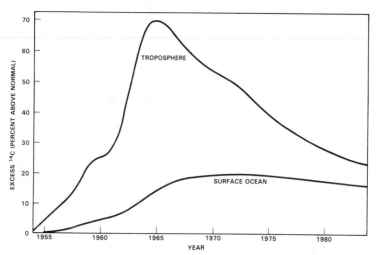

FIGURE 5–5   Excess Atmospheric C-14 Produced by Weapons Tests. (Adapted from UNSCEAR 1977, p. 119).

pared with the natural rate of production of about 27 KCi (1 PBq) per year (UNSCEAR 1982). This distortion of the natural level of ^{14}C has had a significant effect on radiocarbon dating techniques (Chapter 17).

Since carbon is so widely dispersed in the biosphere, the additional burden of ^{14}C introduced by nuclear testing is of considerable biological significance. The problem of assessing the impact is complicated by the fact that large quantities of CO_2 depleted in ^{14}C have also been introduced into the atmosphere by burning of fossil fuels. Thus, the ratio of $^{14}C/^{12}C$ has been increased on the one hand by nuclear testing, and decreased on the other by fossil fuel burning. It is this ratio that determines the dose from ^{14}C. Studies of this ratio in man indicate that equilibrium is reached between tissue and atmospheric CO_2 after about 1.4 year (Nydal, Lovseth, and Syrstad 1971). The internal dose commitment from ^{14}C from nuclear testing has been estimated to be 269 mrad (2.7 mGy) to the whole body, and 455 (4.6 mGy) mrad to the red bone marrow (UNSCEAR 1977).

Many other radionuclides, usually with relatively short half-lives are produced by neutron activation of the materials (e.g. the casing) of the device and of soil and rock below the point of detonation. With the exception of ^{55}Fe, discussed below, these have not contributed significantly to human exposure, and are of little environmental significance. During the 1960's, ^{54}Mn (310 days) was a significant component of fallout activity, and ^{45}Ca, ^{56}Mn, ^{58}Co, and ^{60}Co are among the many activation products that have been identified.

Both ^{55}Fe and ^{59}Fe are commonly found in fallout, being produced by neutron capture by stable isotopes of iron probably used in the structural components of the explosive. ^{59}Fe is produced in relatively small quantities, largely because the natural abundance of ^{58}Fe is only 0.31%, and this 44.6 day half-life nuclide is of little environmental significance. However, ^{55}Fe, produced from the 5.8% abundant ^{54}Fe, has a relatively long half-life (2.7 years) and appreciable activation cross-section. During the 1961–62 testing, an estimated 50 MCi (1.85 EBq) of this nuclide was produced, with concentrations falling to near zero by 1970 (UNSCEAR 1977). However, during the middle 1960's, appreciable body burdens of Fe-55 were accumulated in the norther temperate zones, exceeding 1 μCi (37 KBq) in persons in Alaska and Japan whose diet was rich in fish (Palmer and Beasley 1967). More typically, body burdens were on the order of 20–30 nCi (740–1110 Bq) in the mid-1960's, dropping to a few nCi by the end of the decade (Langford and Jenkins 1971). This nuclide was estimated to deliver a dose of 1 mrad (10 μGy) to the red blood cells (the critical tissue) of residents of New York City during the middle 1960's (Wrenn and Cohen 1967).

Plutonium and Other Transuranics

Several isotopes of plutonium as well as other transuranic elements are produced by nuclear explosions (Table 5–5). Of these, ^{239}Pu is the most important because of its long half-life and hence environmental persistance, and also because it has been produced in such great quantities. By 1970, an estimated 300 kCi (11 PBq) of ^{239}Pu had been produced by

TABLE 5–5   Radiological Properties of Transuranic Nuclides in Fallout.

Nuclide	Half-life	Decay Mode	Comments
Np-239	2.35 d	Beta	Parent of Pu-239; produced by neutron capture by U-238
Pu-239	24,360 y	Alpha	Daughter of Np-239
Pu-240	6540 y	Alpha	Produced by neutron capture by Pu-239
Pu-241	15 y	Beta	Produced by neutron capture by Pu-240; parent of Am-241
Am-241	433 y	Alpha	Daughter of Pu-241

nuclear testing, largely injected into the stratosphere from thermonuclear devices (Harley 1971). Given the specific activity of ^{239}Pu-239 as 0.06 Ci/g (2.2 GBq/g), on a mass basis this is equivalent to about 5,000 kg. By the end of 1973, the global deposition of ^{239,240}Pu was estimated at 320 KCi (12 TBq), of which approximately three quarters was in the northern hemisphere (Hardy 1974). Stratospheric levels were less than one kCi (370 GBq) by 1975 (UNSCEAR 1977). As there have been few atmospheric tests conducted since that time, relatively little additional plutonium has been added to the atmosphere.

Measurable concentrations of plutonium have been observed in numerous environmental media, including air, food, soils, and human and animal tissues. Mean air concentrations of 1.5×10^{-13} μCi/cm^3 (0.056 Bq/ m^3) were observed in Massachusetts during the middle 1960's, when atmospheric testing was near its peak (Magno, Kauffman, and Schleien 1967). Similar levels were observed in New York City, but by the early 1970's these had declined by about a factor of six (UNSCEAR 1977). A daily dietary intake of 7×10^{-3} pCi ($2.6 \times ^{-4}$ Bq) was also estimated, and a diet sampling program in New York City during 1972–4 showed a mean annual dietary intake of 1.6 pCi (0.06 Bq) (UNSCEAR 1977). The greatest concentrations were found in grains and fruits and vegetables. Human tissue analyses revealed the greatest concentrations of ^{239}Pu in the liver, which was found to contain an average of 0.78 pCi/kg (29 MBq/g), while lung tissue was found to contain 0.45 pCi/kg (17 MBq/ g). Plutonium-239, if taken into the body in soluble form, will ultimately end up in the bone where it irradiates the blood forming tissues, and in the liver. Insoluble plutonium bearing particulates tend to remain in the lung after inhalation, with translocation to the pulmonary lymph nodes. Residence time in bone and the pulmonary lymph nodes are long, and, when combined with the long half-life of ^{239}Pu, results in a continuous lifelong exposure to these tissues from the alpha particles associated with the decay of the plutonium.

Most of the plutonium injected into the atmosphere has fallen out onto the earth's surface. The cumulative deposition on soil in the New York area has been reported as 2.68 mCi/km^2 (Bennett 1976). Deposition would, of course, be less in the southern hemisphere, about one-fourth the value for New York. Plutonium is present in soil as an oxide. In sandy soils, about 70% is found in the top 4 cm of the soil with essentially all in the top 30 cm (UNSCEAR 1977). In other soils, somewhat smaller but still appreciable fractions are found in the top layers. Generally speaking, terrestrial plutonium is immobile, with relatively little uptake by plants and very small absorption through the gut of animals. In the marine environment, ^{239}Pu tends to move to greater depths than the fission products, with somewhat more settling out. During the peak years of testing, concentrations in the Pacific Ocean were on the order of 10^{-4} pCi/l (3.7×10^{-3} Bq/m^3) (Pillai, Smith,and Folsom 1964). Concentration by marine organisms at the low end of the food chain such as algae and plankton is in the range of 1000–3000, making this nuclide of additional significance from the standpoint of dose to man (Noshkin 1972).

Plutonium-239 intake and doses to man have been extensively studied. Estimated body burdens peaked at about 10pCi (0.37 Bq) during 1964, stabilizing at about 2 pCi (0.074 Bq) by 1980. Roughly half of the plutonium is found in the bone, with another 40% or so in the liver, and the remainder in the lymph nodes (Bennett 1976). Cumulative lifetime doses through the year 2000 have been estimated at 1.6 mrad to the lungs, 1.7 mrad (17 μGy) to the liver, and 1.5 mrad (15 μGy) to the cells lining the bone. The contribution from inhaled plutonium is expected to be minimal by 2000, assuming no significant atmospheric testing, and hence these values are considered to be reasonable approximations of dose commitments (UNSCEAR 1977). Note that these are lifetime doses and not dose equivalents which would be 10–20 fold greater depending on the quality factor selected.

About 26 kCi (960 GBq) of ^{238}Pu have been introduced into the environment, 9 from weapons testing (Hardy 1974) and another 17 kCi (630 GBq) from the reentry of a satellite in 1964, with most of the latter being deposited in the southern hemisphere (UNSECAR 1977). Additional ^{238}Pu may also have been introduced into the biosphere by other satellites, but the contribution from these has been relatively small. The dose commtment from ^{238}Pu has been estimated to be less than one tenth that from ^{239}Pu, with the doses from the higher mass number plutoniums and their daughters such as Am-241 lower still (Klement et al. 1972).

Not all satellites use ^{238}Pu isotopic power sources; the Soviet satelite that landed in Canada in 1982 contained portions of a small reactor, as was the case with the Soviet Cosmos 1402 in 1983. Neither ^{238}Pu nor the higher mass number plutoniums have the environmental significance of ^{239}Pu, largely because of their smaller quantity and frequently considerably shorter half-lives.

Underground Nuclear Detonations

As a result of the various test ban treaties, nearly all of the nuclear detonations, at least by U.S., U.S.S.R., and United Kingdom, since 1963 have been underground (Carter and Moghissi 1977; Carter 1978).

Underground testing involves the placement of the nuclear explosive into a hole several hundred feet below the surface of the ground, and detonating it there. The emplacement depth depends upon several factors, including the explosive yield of the device, the type of soil, and the information to be obtained from the test.

Radioactivity produced by underground testing is largely confined to the cavity produced by the detonation; indeed to do otherwise would obviate the purpose of performing the test underground and violate the test ban treaty. There is thus no fallout, except in the rather rare instance that venting occurs. The detonation produces a cavity which is effectively sealed from leakage of the radionuclides produced by the detonation. The tremendous temperatures produced by the detonation produce melting and fusing of the rock inside the cavity produced by the explosion, forming a ceramic-like matrix that is impervious to the radioactive species, many of which are incorporated into the cavity walls. Thus, the major environmental significance is the formation of the radioactive cavity, which, in addition to fission products, is rich in activation products produced from the elements in the surrounding soil and rock.

The large explosive yield of nuclear detonations led to their consideration for peaceful applications such as major excavation and similar engineering projects, including the building of harbors, highway and railroad cuts, and a new Panama Canal. Largely through the efforts and leadership of Harold Brown, Ernest Lawrence, and Edward Teller, the United States formally instituted the Plowshare Project for the peaceful applications of nuclear explosives in 1956; the first Plowshare Syposium was held a few months later to consider industrial uses of nuclear explosives.

The Plowshare Program was greatly concerned with cratering studies and related matters, and thus detonations were made with the nuclear explosive at a fairly shallow depth; some release of radioactivity was thus inevitable, even if venting did not occur. Such releases were generaly quite small and confined to the vicinity of the crater. However, from an environmental standpoint, not only does the immediate atmospheric injection need to be considered, but also the residual radioactivity left behind around the location of the detonation. As most detonations for cratering purposes were conducted under controlled conditions at the Nevada Test Site, these was little uncontrolled release of radioactivity into the general environment.

Although the Plowshare Program never moved out of the experimental phase, during its lifetime many imaginative engineering projects were proposed and studied, as were other novel ideas including the production of heavy elements by nuclear detonations, water resources management, geothermal energy production, mining and mineral exploitation, and oil and gas well stimulation. As an example of the latter, Project Gasbuggy was carried out in New Mexico during the late 1960's in an attempt to stimulate natural gas production; a 26 KT nuclear explosive was detonated some 4,000 feet (1.2 km) below the surface of the ground. Radioactivity released to the environment was undetectable (McBride and Hill 1969); however, the gas removed from the well subsequent to the explosion was found to contain both ^{85}Kr and tritium at concentrations of about 5×10^{-4} μCi/cm^3 (1.85×10^7 Bq/m^3) (Smith and Momyer 1968). Thus, use of natural gas produced by the Gasbuggy detonation or by the later Rio Blanco event in Colorado does have some environmental considerations, although in the main, these are relatively small, particularly when compared with the environmental aspects of nuclear detonations generally.

References

1. Bennett, , B. G. 1976. "Transuranic Element Pathways to Man", in *Transuranium Nuclides in the Environment*, International Atomic Energy Agency Publication STI/ PUB/410, pp. 367–381.
2. Carter, M. W. 1980. "Off-Site Health and Safety for Nuclear Weapons Tests", in *Health Physics: A Backward Glance*, R. L. Kathren and P. L. Ziemer, Eds. Pergamon Press, New York, pp. 197–215.
3. Carter, M. W., and A. A. Moghissi. 1977. "Three Decades of Nuclear Testing", *Health Phys.* 33:55.
4. Clark, H. M. 1954. "The Occurrence of an Unusually High Level Radioactive Rainout in the Area of Troy, N.Y.", *Science* 119:619.
5. Cline, J. F. 1981. "Aging Effects of the Availability of Strontium and Cesium to Plants", *Health Phys.* 41:293.
6. Eisenbud, M. 1973. *Environmental Radioactivity*, Academic Press, New York, p. 346.
7. Eisenbud, M., and M. E. Wrenn. 1963. "Biological Disposition of Radioiodine: A Review", *Health Phys.* 9:1133.
8. Farges, L., et al. 1974. "Activite du Krypton 85 dans l'air, Hemispheres Nord et Sud", *J. Radioanal. Chem.* 22:147.
9. Freiling, E. C. 1961. "Radionuclide Fractionation in Bomb Debris", *Science* 133:1991.
10. Gavini, M. B., J. N. Beck, and P. K. Kuroda. 1974. "Mean Residence Times of Long-lived Radon Daughters in the Atmosphere", *J. Geophys. Res.* 79:4447.
11. Glasstone, S. 1977. *The Effects of Nuclear Weapons*, U.S. Government Printing Office, Washington, D.C.

12. Glasstone, S. 1967. *Sourcebook on Atomic Energy*, Third Edition, Van Nostrand-Reinhold, New York.
13. Government of India. 1958. *Nuclear Explosions and Their Effects*, Revised and Enlarged Edition, The Publications Division, Delhi.
14. Hardy, E. P., Jr. 1974. "Worldwide Distribution of Plutonium", in *Plutonium and Other Transuranium Elements*, U.S. Atomic Energy Commission Report WASH–1539, Washington, D. C.
15. Harley, J. H. 1971. "Worldwide Plutonium Fallout from Weapons Tests", in Proceedings of Environmental Plutonium Symposium, Los Alamos Scientific Laboratory Report LA–4756.
16. Hodge, V. F., T. R. Folsom, and D. R. Young. 1973. "Retention of Fallout Constituents in Upper Layers of the Pacific Ocean as Estimated from Studies of a Tuna Population", in *Radioactive Contamination of the Marine Environment*, International Atomic Energy Agency Publication STI/PUB/313, pp. 263–276.
17. Joseph, A. B. et al. 1971. "Sources of Radioactivity and their Characteristics", in *Radioactivity in the Marine Environment*, National Acadmeny of Sciences, Washington, D.C., pp. 6–41.
18. Karol, I. L. "Numerical Model of the Global Transport of Radioactive Tracers for the Instantaneous Sources in the Lower Atmosphere", *J. Geophys. Res.* 73:3589.
19. Kathren, R. L. 1964. "Activity and Thyroid Dose from Radioiodines", *Nucleonics* 22(11):60.
20. Kellogg, W. W., et al. 1957. *Journal of Meteorology* 14:1.
21. Klement, A. W., Jr., et al. 1972. "Estimates of Ionizing Radiation Doses in the United States 1960–2000", U.S. Environmental Protection Agency ZPublication 0RP/CSD 72–1, Washington, D. C.
22. Klement, A. W., Jr. 1982. "Man-Made Sources of Environmental Radiation", in *Handbook of Environmental Radiation*, A. W. Klement, Jr, Ed., CRC Press, Boca Raton, pp. 23–27.
23. Langford, J. C., and C. E. Jenkins. 1971. "The Latitudinal Variations of Fe–55 in Man and Cattle", *Health Phys.* 21:71.
24. Lederer, C. M., et al. 1977. *Table of Isotopes*, Seventh Edition, Wiley, New York.
25. Lewis, E. B. 1959. Statement in *Fallout from Nuclear Weapons Tests*, Hearings Before the U.S. Joint Committee on Atomic Energy, pp. 1552–1554.
26. Machta, L. and R. J. List. 1960. "The Global Pattern of Fallout", in *Fallout*, J. M. Fowler, Ed., Basic Books, New York, pp. 26–36.
27. Magno, , P. J., P. E. Kauffman, and B. Schleien. 1967. "Plutonium in Environmental and Biological Media", *Health Phys.* 13:1325.
28. McBride, J. R., and D. Hill. 1969. "Offsite Radiological Surveillance for Project Gasbuggy", *Radiol. Health Data and Rep.* 10:543.
29. Miskel, J. A. 1973. "Production of Tritium by Nuclear Weapons", in *Tritium*, A. A. Moghissi and M. W. Carter, Eds., Messenger Graphics, Phoenix, pp. 79–85.
30. National Council on Radiation Protection and Measurements (NCRP). 1977. "Cesium-137 from the Environment to Man: Metabolism and Dose", NCRP Report No. 52, Washington, D.C.
31. National Council on Radiation Protection and Measurements (NCRP). 1979. "Tritium in the Environment", NCRP Report No. 62, Washington, D. C.
32. National Research Council. 1980. *The Effects on Populations of Exposure to Low Levels of Ionizing Radiation: 1980*, National Academy Press, Washington, D. C.

33. Noshkin, V. E. 1972. "Ecological Aspects of Plutonium Dissemination in Aquatic Environments", *Health Phys.* 22:537.

33. Nydal, R., K. Lovseth, and O. Syrstad. 1971. "Bomb C-14 in the Human Population", *Nature* 232:418.

34. Palmer, H. E. and T. M. Beasley. 1967. "Iron-55 in Man and the Biosphere", *Health Phys.* 13:889.

35. Pillai, K. C., R. C. Smith, and T. R. Folsom. 1964. "Plutonium in the Marine Environment," *Nature* 203:568.

36. Shurcliff, W. A. 1947. *Bombs at Bikini*, Wise and Company, New York.

37. Sisefsky, J. and G. Persson. 1970. "Fractionation Properties of Nuclear Debris from the Chinese Test", *Health Phys.* 18:347.

38. Smith, C. F., and F. Momyer. 1968. "Gas Quality Investigation Program Status Report for Project Gas Buggy", in *Proceedings of AIME 43rd Annual Fall Meeting*, CONF–680926–1.

39. Stewart, N. G. et al. 1957. "World-wide Deposition of Long-lived Fission Products from Nuclear Test Explosions", U.K. Atomic Energy Authority Report MP/R-2354

40. Tamplin, A. R. and H. L. Fisher. 1966. "Estimation of Dosage to Thyroids of Children in the U.S. from Nuclear Tests Conducted in Nevada during 1952 through 1955", Lawrence Radiation Laboratory Report UCRL-14707.

41. United Nations Scientific Committee on the Effects of Atomic Radiation (UNSCEAR). 1977. *Sources and Effects of Ionizing Radiation*, United Nations Publication E.77.IX.1, United Nations, New York (1977).

42. United Nations Scientific Committee on the Effects of Atomic Radiation (UNSCEAR). 1982. *Ionizing Radiation: Sources and Biological Effects*, United Nations Publication E.82.IX.8, United Nations, New York.

43. Volchok, H. L., et al. 1971. "Oceanic Distributions of Radionuclides from Nuclear Explosions", in *Radioactivity in the Marine Environment*, National Academy of Sciences, Washington, D. C., pp. 42–89.

44. Way, K, and E. P.Wigner. 1948. "The Rate of Decay of Fission Products", *Phys. Rev., 2nd Series,* 73:131844.

45. Wrenn, M. E., and N. Cohen. 1967. "Iron-55 from Nuclear Fallout in the Blood of Adults: Dosimetric Implications and Development of a Model to Predict Levels in Blood," *Health Phys.* 13:1075.

CHAPTER 6

The Front End of the Fuel Cycle

The reactor itself is only one portion of a complex cycle of activities that begins with the exploration and mining of ore from which the uranium that will fuel the reactor, continuing on through the purification and use of the metal as fuel in the reactor to the reclamation of the unused portion of the uranium and its plutonium byproduct from the spent fuel, and concluding with the treatment and disposal of the radioactive wastes generated by the reactor. There are many different types of nuclear reactors, with many different applications. Those that produce the largest amounts of radioactivity and have the greatest environmental significance are used for the production of electrical power, or for the manufacture of plutonium—the so-called production reactors—for use as reactor fuel or in nuclear explosives.

In the United States, most reactors used for production of electrical power are of the light water type in which ordinary water is used for both cooling and for neutron moderation. The light water reactor (LWR) fuel cycle is shown in Figure 6–1. Fuel cycles for other types of reactors are similar. However, the light water reactor fuel cycle commands the greatest interest, largely because so many of the reactors in the world are of this type. Examination of the world list of power reactors for 1983 shows a total of 521 power reactors operational, under construction, or on order. Of this number, 277 are operational, with 215 or 78% of these being either cooled, moderated, or having both functions performed by ordinary water (ANS 1983). On the basis of power output, and hence radioactivity generation, the lightwater reactors accounted for more than 80% of the electrical generating capability of all power reactors. As of 1983 there were 83 licensed power reactors in the United States with a total of 65.7 gigawatts of electrical generating capability; all but one were of the light water type (AIF 1983).

Uranium Mining

The first step in the fuel cycle is, logically, the exploration for deposits of uranium ore with commercial value. This is frequently done by aerial

THE NUCLEAR FUEL CYCLE

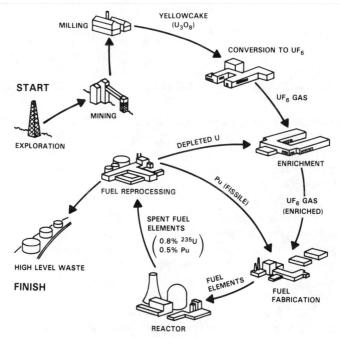

FIGURE 6-1   The Uranium Fuel Cycle.

surveying, followed by ground surveys and sampling of potential viable ore bodies. Sampling may involve core drilling, which may result in a slight release of radon gas in the vicinity of drill rigs. However, quantities are insignificant from an environmental standpoint, and well below the levels of concern occupationally (Miller and Scott 1981).

Uranium ores are generally categorized as high, medium, and low grade. High grade ores are those that contain a few per cent of uranium and are chiefly found in Zaire (Katanga) and Canada. In these rich ores, which typically contain 1–4% uranium, the uranium is present in the form of mineral oxides, chiefly uraninite, which is largely UO_2 and various higher oxides, and pitchblende.

Medium grade ores contain from 0.1 to 1% uranium and are found in many parts of the world. Major deposits exist in the Colorado Plateau or Rocky Mountain regions of the United States, particularly in Colorado, New Mexico, Utah, and Wyoming; deposits with commercial value have also been found in southern California, Nevada, Texas, and Washington.

Elsewhere in the world, significant deposits of medium grade ores have been found in the Northwest Territory and Province of Saskatchewan of Canada, and in Australia and Czechoslovakia. Medium grade ores are mostly oxides of uranium, usually carnotite ($K_2(UO_2)_2(VO_4)_2.3H_2O$), thorianite ($UO_2$, ThO_2), and some phosphates and carbonates. The medium grade ores are in some cases economically advantageous in that they may be easily concentrated by chemical means to yield at least 70–80% U_3O_8.

Low grade ores, generally containing less than 0.1% uranium, are found in many parts of the world. In recent years, these ores have attained commercial feasibility, and in fact most ore mined averages between 0.04 and 0.42% (ERDA 1977). In addition to these ores, certain byproducts of other mining operations such as phosphates in Florida and Idaho and gold in South Africa (especially the Witwatersrand mines) also contain recoverable quantities of uranium. Even though these residues may contain only a few parts per million or even less or uranium, recovey is nonetheless practical because a large portion of the recovery cost is allocated to the primary purpose of phosphate or gold extraction.

Although there have been some open pit mines, notably the large Shinkolobwe pit in Zaire and in the western United States, most uranium mining worldwide is so-called hardrock mining done by shaft mining techniques. A small fraction of uranium mining has been carried out by *in situ* leaching techniques, and this method appears to be gaining popularity (Brown and Smith 1981). In the United States, there is a greater fraction of open pit mining, with about an equal mix of open pit and deep mines (USERDA 1977). Surface mining or open pit methods are carried out where the ore deposits lie within a few hundred feet of the surface, under relatively easily removed overburden. Open pit mining results in increased releases of radon and daughters to the atmosphere. If the pit extends below the water table, groundwater must be removed to permit mining operations to continue, and this is usually accomplished by pumping with discharge into the ground or nearby bodies of water. Mine water is generally enriched in radioactivity, and its introduction into surface water bodies may produce measurable increases in radioactivity levels.

Shaft mining or *in situ* leaching methods are used for deeper lying ore bodies, or where the ore lies under a hard rock stratum. In addition to the normal hazards associated with hardrock mining, uranium mining presents the additional occupational hazard of exposure to airborne radioactivity from radon and daughters and to small external ambient radiation levels from the uranium and daughters in the ore. To remove the buildup of radioactive dusts and radon, uranium mines are well ventilated, with the exhausted air discharged directly to the atmosphere. The volume of air discharged may be quite large, and concentrations of radon and daughters may run to several hundred picocuries per liter in the mine air and in the discharge. Typically, uranium mines exhaust 10^3–10^6 cubic feet of air per minute, containing 0.5–20 μCi/1000 ft^3 (650–2600 Bq/m^3). In a mine exhausting a million cubic feet of air per minute (472 m^3/s) containing only 1×10^{-8} μCi/cm^3 (370 Bq/m^3) of radon, the discharge would be

$$1 \times 10^{-8}\mu\text{Ci/cm}^3 \times 2.8 \times 10^4 \text{ cm}^3/\text{ft}^3 \times 10^6 \text{ ft}^3/\text{min} = 280 \ \mu\text{Ci/min}$$

which corresponds to 1.7×10^5 Bq/m^3 or to a daily release of 0.4 Ci (3.7 GBq). For only 200 days per year of operation, this would result in a release of 80 Ci (3 TBq) of radon into the atmosphere annually.

The annual fuel requirement (AFR) for a light water reactor is defined as the quantity of fuel necessary to produce 1100 MWe, based on a reactor capacity factor 0f 80%. Shaft mining operations to meet one AFR for a light water reactor, will result in the discharge of 4000 Ci (150 TBq) of radon annually (USEPA 1973).

In addition to gaseous discharges, uranium mining may produce liquid releases as well. Waste waters from open pit mines are typically one to two orders of magnitude greater in volume and in radioactivity concentration than those from shaft or underground mines (AEC 1974). Such waters are slightly alkaline, with pH below 9 and generally around 8; alpha radioactivity concentrations are in the range of 10^{-7} μCi/ml (3700 Bq/m^3) for open pit mines and about two orders of magnitude greater for surface mines. A typical open pit mine may discharge a million gallons of water daily, or on the order of 500 μCi (18.5 MBq) per day or about 200 mCi (7.4 GBq) per year.

Mining Hazards and the Working Level Concept

The primary radiological concern of uranium mining relates to the occupational exposure of the miners. Miners are exposed to external radiation fields as well as radon and radioactivity bearing dusts. Exposure to the latter is exacerbated by combination with silica and the damp mine atmospheres, and in particular by cigarette smoking. External radiation exposure is largely from gamma rays emitted by the radionuclides of the uranium and thorium series. As the mean energy of these photons is about 1 MeV, they are quite penetrating and a significant fraction escapes from the mine walls, floor, and ceiling. The contribution from the activity in the rock around the mine shaft more than offsets the reduction in cosmic ray flux that occurs deep underground. In shafts and tunnels through barren rock in uranium mines, exposure rates are generally less than 0.1 mR/h. For ores containing 0.2% U_3O_8, levels are in the range of 0.5–1.5 mR/h. Generally, the exposure rate is proportional to the percentage of uranium in the rock, with levels to about 100 mR/h having been observed in particularly rich mines having ore lenses containing 20–30% U_3O_8. Dose rates from beta rays are significantly lower, being about 1.3 mrad/h (13 $\mu Gy/h$) near the mine surfaces (ICRP 1977).

The relationship between hardrock mining and lung cancer has been known for at least 100 years, having been noted as early as 1878 in silver miners in the Schneeberg area of southern Germany and the adjacent St. Joachinsthal area of Czechoslovakia (Teleky 1937). These mines have been in continuous operation since 1410, largely mined for silver, later becoming the first sources of uranium. It is from these mines that Madame Curie obtained the pitchblende from which she separated radium. About 30% of the miners from the mines in this area were observed to have cancer of the lung at autopsy (Teleky 1937). Average concentrations of radon in these mines has been estimated at 1×10^{-6} $\mu Ci/cm^3$ (1000 pCi/l or 37 KBq/m^3) and on this basis a maximum permissible concentration of 1×10^{-8} uCi/cm^3 (10 pCi/l or 370 Bq/m^3) was suggested (Evans and Goodman 1940). In addition to the radon and daughters, arsenic, another known tumorigen, was also present in the mine atmosphere as were high concentrations of free silica (Lorenz 1944).

Radon-222 is an alpha emitting noble gas with a half-life of 3.825 days that readily diffuses through rock and emanates continuously from the walls, floor and ceilings of the mine shaft. The radon decay chain includes 3 short lived alpha emitters (Table 6–1) which attach to dust particles in

TABLE 6–1   Short Lived Radon Daughters.

Nuclide	Historical Name	Half-life	Type of Decay	Energy
^{218}Po	Ra-A	3.05 min	Alpha	5.998 Mev
^{218}At*		1.3 s	Alpha	6.63 MeV
^{214}Pb	Ra-B	26.8 min	Beta	0.69, 0.74, 1.04 MeV
^{214}Bi	Ra-C	19.7 min	Beta	3.28 Mev
^{214}Po	Ra-C′	164 us	Alpha	7.68 MeV
^{210}Pb	Ra-D	22 y	Beta	0.17, 0.61 MeV
MeV				

*Branching ratio 0.02%

the respirable range and produce most of the exposure to the lung. Indeed, at equilibrium, the dose delivered by the radon daughters is about 500 times greater than that from radon (Shapiro 1956). In 1953, the National Committee on Radiation Protection recognized the contribution of the daughter products and established a maximum permissible concentration of 1×10^{-8} μCi/cm^3 (10 pCi/1 or 370 Bq/m^3) of radon in equilibrium with daughter products. Exposure to this level forty hours per week was thought to deliver a dose equivalent of 300 mrem/week (3 mSv) to the lungs.

The same year, representatives from the United States, Britain, and Canada, meeting at the Tripartite Conference, recommended an increase in the maximum permissible concentration for radon in equilibrium with daughters to 100 pCi/1 (3.7 KBq/m^3). Based on the Tripartite Conference recommendation, the Seven States Mining Conference adopted this level at its meeting in Salt Lake City in February 1956, referring to it as the *working level* (WL). Although conceptually good, the working level presented great practical difficulties as the degree of equilibrium between radon and its short lived daughters was rarely known, and seldom was complete. Thus, the following year, the working level was redefined by the U.S. Public Health Service, expressing it in terms of alpha energy released by the decay of the radon daughters, this being related to the dose delivered to the lungs.

The modern definition of the working level is any combination of short-lived radon daughters in air that will result in the ultimate emission of 1.3×10^5 MeV of alpha particle energy. This is exactly equal to the alpha energy from 100 pCi/1 (3.7 KBq/m^3) of short-lived radon radon daughters in equilibrium, but the working level, as defined, is wholly independent of the degree of equilibrium or of the concentration of any one nuclide in the radon decay chain. As the concentration of radon in equilibrium with daughters, expressed in units of pCi/1 is 100 WL, an equilibrium factor, F, can be calculated by.

$$F = 100(WL)/C_{Rn}$$

in which WL is the number of working levels and C_{Rn} the pCi/1 of radon in eqilibrium with its short lived daughter products. The equilibrium factor is of some use in calculating a value known as equilibrium equivalent radon (EER), which permits working levels to be compared with the ICRP standard of 30 pCi/1 (1.1 KBq/m^3) for radon in equilibrium with daughters (ICRP 1977).

The working level is commonly applied in terms of the *working level month* (WLM), or the product of the number of months worked and the working level. One WLM is equal to exposure to one WL for 170 (some authorities cite 173) hours. Total exposure is sometimes expressed as cumulative working level months (CWLM). Current limits, established by the U.S. Environmental Protection Agency, are 4 WLM in any 12 month period for occupational exposure; this is based on the observation that lung cancer has rarely been observed in miners with a lifetime exposure of less than 120 WLM (USEPA 1973).

The relationship between the dose or dose equivalent delivered to the lung and the working level is uncertain. The dose delivered to different portions of the bronchial epithelium is quite variable, ranging from less than 0.1 rad/WLM (1 mGy/WLM) to 20 rad/WLM (200 mGy/WLM) (Haque and Collinson 1967). Many factors, including the state of the equilibrium, the fraction of the daughters available as free ions, clearance rate, particle size distribution, and thickness of the mucous layer will affect the dose delivered. The National Academy of Sciences/National Research Council Committee on the Biological Effects of Ionizing Radiation—the BEIR Committee—suggested a dose of 0.5 rad/WLM (5 mGy/WLM) or a dose equivalent of 5 rem/WLM (50 mSv/WLM) in its inital report (BEIR 1972). In its 1980 report, the BEIR Committee noted that the dose delivered to the lungs was a function of the fraction of free ions compared to those attached to dust particles (usually very low in mines), the breathing pattern, and whether or not the individual was mouth or nose breathing. On the basis of these uncertainties, a range of 0.4–0.8 rad/WLM (4–8 mGy/WLM) was given, along with a single value for dose equivalent of 6 rem/WLM (60 mSv/WLM) (BEIR 1980).

The working level has been criticized on both conceptual and practical grounds. The major criticism, which has some validity, relates to the dependency of its measurement on the state of equilibrium. Other criticisms relate to the difficulties in measurement, and to its relationship with dose, as well as the difficulty of relating the WL to the standard activity per volume nomenclature used for other radionuclides. However, the concept has been successfully applied to estimates of bronchial cancer risk in uranium and other hardrock miners as well as other (fluorospar, iron, zinc, silver, vanadium) miners who may be exposed to elevated levels of radon and daughters (BEIR 1980).

Exposure to elevated levels of radon has resulted in excessive bronchiogenic cancer in hardrock miners of all types, with a 40 fold increase in lung cancer seen in miners who smoked as compared with nonminers who were nonsmokers, a tenfold increase over nonminers who smoke, and a six fold increase over nonsmoking miners. Smoking also appears to shorten the latent period for cancer development. The risk, based on evaluation of the most reliable studies of miners, appears to be about 18 cases per million person-years per WLM (Archer, Wagoner and Lundin 1973; BEIR 1980).

Uranium Milling

Uranium milling is the process of benefication and purification of the ore, and is generally carried out close to the mine site to minimize transportation costs. It is the most significant step in the front end of the fuel cycle from a radiological point of view. The basic procedure calls for crushing and grinding the ore, and mixing it with water to form a slurry. The slurry is then leached with acid or alkali, depending on the ore, a process which extracts most of the uranium along with vanadium, usually found in association with carnotite, molybdenum, and iron. The leach liquid is then separated from the residue, perhaps with the aid of a flocculent. The residue thus removed is the sand-like mill tailings and contains virtually all the radium isotopes, lead, and much of the thorium; it is washed with acid or alkali and pumped as a slurry to the tailings pond for drying.

Common practice is to remove the uranium from the leach liquid by organic solvent extraction, leaving behind most of the impurities in an aqueous residue known as raffinate. The uranium is precipitated by ammonia, and the precipitate separated and calcined (heated) to yield yellowcake, the powdery yellow sodium or ammonium diuranate which is assayed as U_3O_8. Typically, yellowcake from the milling operation is 70–90% U_3O_8, but 95% concentration is not uncommon.

If processing ore containing $>$ 0.05 weight per cent of uranium, mills in the United States are required to have a license from the Nuclear Regulatory Commission in accordance with the provisions of Title 10, Code of Federal Regulations, Part 40. Since passage of the National Environmental Policy Act (NEPA) in 1969, an environmental impact statement is also required. Guidance for the preparation of the environmental impact statement, which must include information on the pre-existing radiological conditions in the vicinity of the mill, as well as the radiological aspects of operation and waste disposal, is provided by NRC Regulatory Guide 3.5, "Guide to the Contents of Applications for Uranium Milling Licenses". In accordance with other provisions of the Atomic Energy of 1954, as amended, the NRC is able to transfer the authority for regulation of uranium mills to various agreement states

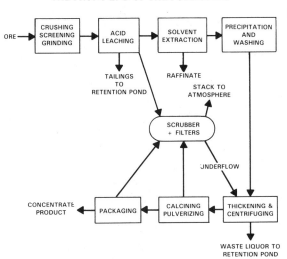

FIGURE 6-2   Uranium Milling Process.

which have demonstrated appropriate capabilty; four states—Colorado, New Mexico, Texas, and Washington—have entered into agreements to date, and have assumed the responsibility for licensing about half of the operating mills.

The occupational health hazards associated with uranium milling are largely from radioactivity bearing dusts. These generally occur during the early stages of the milling, and contain not only uranium, but the accumulated daughters as well. The primary nuclides of concern from the standpoint of internal exposure are thus ^{226}Ra and ^{230}Th, which is the major source of exposure from inhaled natural uranium which has not been separated from its daughters (Harley and Pasternack 1979).

The environmental radiological aspects of uranium milling are far more significant than those of mining. Radioactive wastes are produced at several steps in the milling process. The initial crushing and grinding process, even if done wet, generates radioactivity bearing dusts, as does the pulverizing and packaging of the finished, dried yellowcake. The acid or alkali and solvent extraction procedures produce solid tailings, and liquid radioactivity bearing wastes result from the precipitation step.

Mills are commonly located in sparsely populated areas, usually near the mine, and processes about 600,000 tons of ore annually. Average mill lifetimes have been about 15 years. The mill will occupy about about 300 acres, with about 225 acres devoted to tailings piles and other waste disposal. As one ton of ore typically contains about 0.2% uranium, only about 4 pounds of uranium per ton of ore are recovered, and the tailings produced are in quantity about as much as the material removed from the mine.

The radioactivity content of the tailings is highly variable, dependent largely on the uranium contant of the ore and efficacy of the milling process. Based on an assumed processing of 30,000 tons per day of 0.2% ore, an estimated 15 Ci (555 GBq) each of ^{226}Ra and ^{230}Th enter the tailings (Whicker and Schultz 1982); this is a concentration of about 500 pCi/g (18.5 Bq/g) fcr each. This is consistent with other estimates (Eisenbud, 1973; USEPA 1973).

The solid tailings contain essentially all the ^{230}Th and ^{226}Ra originally present in the ore; hence they produce as much radon as the ore did originally, before the separation process. The rate of radon generation is controlled by the 80,000 year half-life of the ^{230}Th which will exist in secular equilibrium with its 1620 year half-lived daughter ^{226}Ra, assuming that both are removed in approximately equal percentages by the milling. The 3.825 day half life ^{222}Rn will very quickly come to secular equilibrium with its long lived parent ^{226}Ra. Thus, for all practical purposes, the emanation of radon decreases with a half-life of about 80,000 years, and is for all practical purposes constant on a human time scale.

Control of tailings piles is achieved by stabilization and by keeping them well above groundwater levels. However, the best control is probably returning them to the mine shaft for disposal. In past years there have been problems of blowing dust from unstabilized tailings piles. To control this problem, stabilization consisting of covering the tailings piles with a foot of clean earth has been tried. Although this remedied the dust blowing problem, it had little effect on the evolution of radon, reducing the emanation rate by about 15% (USEPA 1973). Each three foot layer of soil reduces the radon concentrations over the tailings pile by about a factor of two; an overburden of 20 feet deep is needed to reduce emission to about twice background (Shiager 1974) and plastics, mastics, concrete, and asphalt covers in various compositions and thicknesses have been tried with varying success.

Overall, the aggregate release of radon from mill tailings and associated mill wastes is relatively small when looked at on a national scale. By the year 2000, an estimated 10^9 tons of tailings will have been accumulated in the United States, producing an annual radon release to the atmosphere of 400,000 curies (14.8 PBq), or about 4% of that released naturally. Milling for one AFR results in the release of an estimated 170 Ci (6.3 TBq) of radon per year from the tailings along with 100 mCi (3.7 GBq) each of ^{226}Ra and ^{230}Th (Sears et al. 1975). These latter radionuclides are subject to wind resuspension and hence may find their way into the atmosphere. Most of the tailings piles are located within a small area of the Rocky Mountain States, and only localized elevated external radiation fields as well as radon concentrations are associated with the them. Dose rates on the order of a few tenths of a mrad/hr (few μGy/h) to perhaps several mrad/h (several tens of μGy/h) may be obtained at waist height (3 feet) above the piles, even when covered with overburden (Snelling 1969,1970). Concentrations of radon and daughters, measured above or in the vicinity of the piles may be several times normal levels in the area; a typical tailings pile containing 560 pCi/g (21 Bq/g) of ^{226}Ra will release about 500 pCi (18.5 Bq)/m^2-s of^{222}Rn (USEPA 1973). The release rate is not only dictated by the stabilization covering, but also by the various factors that affect radon emanation from ordinary soil (Chapter 3). About 90 per cent of the radon emanation is from the first few meters depth of the tailings, with the release rate changed little or not at all by depths greater than about 10 feet (USEPA 1973).

Liquid releases from the mills have been implicated in contamination of surface waters and nearby wells and groundwater. The tailings liquids generally contain relatively little uranium but significant amounts of radium (Table 6–2). However, the tailings liquids are not the only source, nor may they even be the most significant source, of radioactivity released to surface or ground waters. Leaching of radium from the tailings pile by precipitation may be far more significant. Increased concentrations of radium attributed to mill tailings have been observed downstream from uranium mill operations in several locations. A classic case is the Animas River in southwestern Colorado, which showed a tenfold increase in radium concentrations downstream of Durango, where a mill was located, resulting in a daily intake of radium in towns using this river for their water supply in excess of that recommended by the International Commission on Radiological Protection. In addition, the water was also used for irrigation, resulting in additional exposure from food crops as well as a small increase in ambient levels. A two order of magnitude elevation in radium levels in the Colorado River has been observed downstream of Grand Junction, a uranium mining and milling center (Eisenbud 1973).

Possibly the most interesting dose pathway associated with uranium

TABLE 6–2   Radioactivity Content of Tailings Liquid.

U_3O_8	7 ppm
^{Nat}Th	2 ppm
^{226}Ra	10^{-8}–2×10^{-6} μCi/ml
^{230}Th	10^{-10} μCi/ml

mining and milling occurred about 1950, uranium tailings from a mill in Grand Junction, Colorado, were distributed to contractors and the general public for use as fill and other construction purposes. An estimated quarter million tons of tailings were given away and used for the construction of roads, sidewalks, and other paved areas, with another 50,000 tons used as fill for residential and other construction. The practice was halted by the Colorado Department of Health in 1966, but not until an estimated 3300 buildings, about 3000 of which are homes, were built on such fill in the Grand Junction area, with perhaps 5000–7000 buildings *in toto* built on fill in the entire state. Radioactivity concentrations are greatest in basements and in the lower levels, and in general do not exceed 0.1 WL or 0.2 mrad/h (2 μGy/h) although levels have approached 1 WL and 1 mrad/h (10 μGy/h) in the extreme. A survey in 1968 of about 100 building locations having used more than 20 cubic yards of the tailings revealed levels of > 0.01 WL in more than two-fifths of the residences and three-fifths of the businesses, with a peak value of 1.88 WL. Although even critics point out that these levels are generally not very great, they nonetheless could pose a significant health risk to long term residents (Hollocher and MacKenzie 1975). Exposure to an average of 0.1 WL only 12 hours per day would produce an annual exposure of about 2.5 WLM; clearly in some cases the occupational exposure limit has been exceeded among residents in the general population. Remedial action, including removal of the tailings and the use of sealants has been undertaken, funded by the federal government (75%) and the State of Colorado (25%).

The widespread use of tailings both for construction fill and for mixing with concrete used in roads and sidewalks, coupled with the large tailings piles nearby, produced generally elevated levels of radon and daughters in the city of Grand Junction. Air concentrations were about an order of magnitude higher than in the surrounding countryside, averaging about 8×10^{-10} μCi/cm^3 or 0.8 pCi/l (30 Bq/m^3) (USPHS 1969).

Production and Enrichment of Uranium Hexafluoride

Yellowcake is converted to uranium hexafluoride by one of two basic processes, wet and dry. The dry process begins with the reduction of U_3O_8 to UO_2 by heat in the presence of ammonia followed by hydrofluorination (addition of HF) to the UO_2 to UF_4. The tetrafluoride thus produced is then treated with fluorine to produce UF_6 which is purified by distillation.

Gaseous wastes are produced in all the steps along with solid wastes from the distillation operation. The radioactivity content of these wastes is relatively small, however, as most of the high specific activity nuclides have been removed in the milling operation.

The wet conversion process begins with the digestion of the yellowcake in nitric acid, followed by organic solvent extraction. Raffinate is produced at this step. The uranium is then reextracted as pure uranyl nitrate with water, and heated to convert the nitrate to UO_3. Reduction with hydrogen gas produces the dioxide which is then treated with hydrofluoric acid as in the dry process to produce UF_4. The final two steps, fluorination and distillation, are similar to those in the dry process. Gaseous and solid wastes are produced as in the dry process.

From an environmental standpoint, control of gaseous wastes is the primary concern in the conversion step. However, the activity in these waste gases is small and is largely natural uranium; only 150 μCi (5.6 MBq) of uranium are released annually from the conversion process for a typical 1000 MWe light water reactor (USAEC 1974). Raffinate production is relatively small, and is acceptably handled by placing it in a lined pond with a sealed bottom in which it is chemically neutralized with lime, which precipitates out the radionuclides—largely ^{232}Th, ^{238}U, and their daughters—as sludge which is relatively easily removed and disposed of. For one AFR, the UF_6 conversion process produces solid wastes containing less than a curie of the various activities. Liquid wastes resulting from conversion of uranium for one AFR contain 44 mCi (1.6 GBq) of natural uranium, and less than a tenth that amount of ^{226}Ra. The ^{230}Th content is even smaller, on the order of 1.5 mCi (555 MBq) (USAEC 1974).

The UF_6 produced in the conversion process contains the natural abundance of 0.714 per cent ^{235}U and is enriched by gaseous diffusion to 2–4% ^{235}U for use in light water reactors. Gaseous diffusion involves the diffusion of the gaseous UF_6 through a series of porous barriers. The lighter $^{235}UF_6$ passes more readily through the barriers, each one producing an enrichment of 1.0043; thus about 1700 barrier stages are required to produce an enrichment of 4%. In the United States, the enrichment process is carried out at one of three gaseous diffusion plants, all located in approximately the same area of the country, one each in Oak Ridge, Tennessee, Portsmouth, Ohio, and Paducah, Kentucky. Radiological releases to the environment from the gaseous diffusion process are minimal. As most of the radium and other daughters of the uranium series have been removed in the milling process, and relatively little time has elapsed to permit ingrowth, the major relase is uranium. An estimated 20 mCi (740 MBq) are produced in the form of solid wastes per AFR, plus about an additional one-tenth this amount in gaseous wastes (USAEC 1974).

Enrichment is a highly energy consumptive process, requiring approximately 98% of the energy used in the entire LWR fuel cycle. One AFR requires about 310,000 MWh of electricity which is largely produced by coal fired generating plants. Hence, there is a secondary radiological aspect, viz. that relating to the radiological releases from coal combustion.

Fuel Fabrication

The final step in the front end of the fuel cycle is the fabrication of the fuel. The enriched UF_6 is hydrolyzed to form uranyl oxyfluoride as shown:

$$UF_6 + H_2O \rightarrow UO_2F_2$$

This is followed by treatment with aqueous ammonia solution, producing a precipitate of ammonium diuranate which is centrifuged or filtered and dried. The dried solid is calcined and reduced in the presence of hydrogen to yield ammonia and uranium dioxide, in accordance with the following:

$$(NH_4)_2.U_2O_7 + H_2 \rightarrow NH_3 + UO_2$$

The uranium dioxide powder thus produced is pelletized, sintered, and encapsulated in zirconium-tin alloy (zircaloy) or stainless steel tubes for use in the reactor. A typical 1000 MWe LWR may contain more than 40,000 tubes, perhaps ten feet in length, containing many times that number of individual pellets and having a total weight of uranium in the neighborhood of 100 tons.

Fuel pellets must be manufactured to exacting tolerances, and hence there may be many rejections. Rejects are dissolved in nitric acid and extracted with organics, producing raffinate. The extract is then reextracted with water, and the UO_4 precipitated with ammonia and hydrogen peroxide and finally reduced to uranium dioxide. The environmental radioactivity implications of fuel fabrication are minimal; one annual fuel requirement for a light water reactor produces less than 250 mCi (9.3 GBq) of uranium in the solid wastes, and 20 mCi (740 MBq) of uranium plus smaller amounts of other naturally occurring members of the uranium series, largely ^{230}Th (10 mCi or 370 MBq), in liquid wastes. Gaseous releases for one AFR would contain a fraction of a millicurie of uranium.

References

1. Advisory Committee on the Biological Effects of Ionizing Radiations (BEIR). 1972. *The Effects on Populations of Exposures to Low Levels of Ionizing Radiation*, National Academy of Sciences/National Research Council, Washington, D.C. p. 148.

2. Advisory Committee on the Biological Effects of Ionizing Radiations (BEIR). 1980. *The Effects on Populations of Exposures to Low Levels of Ionizing Radiation: 1980*, National Academy of Sciences, Washington, D.C., p. 327.

3. American National Standards Institute (ANSI). 1973. "Radiation Protection in Uranium Mines, ANSI N13.8-1973, American National Standards Institute, New York.

4. American Nuclear Society (ANS). 1983. "World List of Nuclear Power Plants", *Nuclear News* 26(2):72.

5. Archer, V. E., J. K. Wagoner, and F. E. Lundin. 1973. "Lung Cancer Among Uranium Miners in the United States", *Health Phys.* 25:351.

6. Atomic Industrial Forum (AIF). 1983. "Nuclear Reactor Status Report", *INFO*, No 172, p. 8.

7. Brown, S. H., and R. C. Smith. 1981. "A Model for Determining the Overall Radon Release Rate and Annual Source Term for a Commercial In-Situ Leach Uranium Facility", in *Radiation Hazards in Mining"*, M. Gomez, Ed., Society of Mining Engineers, NewYork, pp 794–800.

8. Eisenbud, M. 1973. *Environmental Radioactivity*, Academic Press, New York, pp. 210–1.

9. Evans, R. D., and C. Goodman. 1940. "Determination of Thoron in Air and Its Bearing on Lung Cancer Hazards in Industry", *J. Ind. Hyg. Toxicol.* 22:89.

10. Harley, N. H., and B. S. Pasternack. 1979. "Potential Carcinogenic Effects of Actinides in the Atmosphere", *Health Phys.* 37:291.

11. Haque, A. K. M., and A. J. L. Collinson. 1967. "Radiation Dose to the Respiratory System Due to Radon and Its Daughter Products", *Health Phys.* 13:431.

12. Hollocher, T. C., and J. J. MacKenzie. 1975. "Radiation Hazards Associated with Uranium Mill Operations", Ch. 3 in *The Nuclear Fuel Cycle*, Revised Edition, MIT Press, Cambridge, pp. 41–69.

13. International Commission on Radiological Protection (ICRP). 1977. "Radiation Protection in Uranium and other Mines", *Annals of the ICRP* 1:1.

14. Lorenz, E. 1944. "Radioactivity and Lung Cancer: A Critical Review of Lung Cancer in the Miners of Schneeberg and Joachimsthal", *J. Nat. Cancer Inst* 5:1.

15. Miller, H. T. and L. M. Scott. 1981. "Radiation Exposures Associated with Explo-

ration, Mining, Milling, and Shipping Uranium", in *Radiation Hazards in Mining*, M. Gomez, Ed., Society of Mining Engineers, New York, pp. 753–9.

16. Sears, M. B., et al. 1975. "Correlation of Radioactive Waste Treatment Costs and the Environmental Impact of Waste Effluents in the Nuclear Fuel Cycly for Use in Establishing 'As Low as Practical' Guides—Milling of Uranium Ores", Oak Ridge National Laboratory Report ORNL-TM–4903/1.

17. Shapiro, J. 1956. "Radiation Dosage from Breathing Radon and Its Daughter Products", *AMA Arch. Industrial Health* 14:169.

18. Shiager, K. J. 1975. "Analysis of Radiation Exposures on or Near Uranium Mill Tailings Piles", *Radiol. Health Data Rep.* 11:411.

19. Snelling, R. N.. 1969. "Environmental Survey of Uranium Mill Tailings Pile, Tuba City, Arizona", *Radiol. Health Data Repts.* 10:475.

20. Snelling, R. N. 1971. "Environmental Survey of Uranium Mill Tailings Pile, Mexican Hat, Utah", *Radiol. Health Data Repts.* 12:17.

21. Teleky, L. 1937. "Occupational Cancer of the Lung", *J. Ind. Hyg. Toxicol.* 19:73.

22. U.S. Atomic Energy Commission (AEC). 1974. "Environmental Survey of the Uranium Fuel Cycle", USAEC Report WASH–1248.

23. U.S. Energy Research and Development Agency (ERDA). 1977. "Statistical Data of the Uranium Industry, January 1, 1977", USERDA Report GJO–100(77).

24. U.S. Environmental Protection Agency (EPA). 1973. "Environmental Analysis of the Uranium Fuel Cycle", USEPA Report EPA–520/9–73–003B.

25. U. S. Public Health Service (USPHS). 1969. "Evaluation of Radon-222 Near Uranium Tailings Piles", USPHS Report DER 69–1.

26. Whicker, F. W., and V. Schultz. 1982. *Radioecology: Nuclear Energy and the Environment*, Vol. 1, CRC Press, Boca Raton, p. 90.

CHAPTER 7

Radioactivity from Nuclear Reactors

The reactor is the dominant feature of the nuclear fuel cycle and represents a large man-made reservoir of radioactivity. There are many kinds of nuclear reactors, of varying size, configuration, and application. The largest, at least from the standpoint of production of radioactivity, are power reactors, used for the production of electricity. As of July 1983, twenty-five countries had one or more operating power reactors, with units under construction or planned in another eighteen (Table 7-1). The largest number was in the United States, which had 83 operational reactors, with another 58 under construction plus four more planned for a total of 145, or slightly more than one-fifth of the world total of 738 (USDOE 1983). World-wide, the U.S. total of 83 represents more than a fourth of the total number of operating plants and well over a third of the power generating capacity. Numerically, the Soviet Union is in second place with 40 operating power reactors and another 37 under construction, followed closely by the United Kingdom with 35 operating plants and France with 34. France had an additional 32 units under construction and the UK only seven (USDOE 1983).

In addition to the power reactors used for central station production of electricity, there are a number of other reactors used for a variety of other purposes, including materials testing, education, isotope production, research, and development. These are generally very much lower in power and hence radioactivity production. There are more than 150 reactors of this type in the United States alone (USCG 1977). There are also 174 military reactors in operation or under construction, mostly used to power nuclear submarines and other naval vessels, plus a few production reactors with relatively high power level used for the production of plutonium largely for military applications.

The production of radioactivity is a function of the energy produced in the reactor, which is simply the product of the reactor power output and operating time. Reactor power is usually expressed in megawatts or kilowatts of heat, correctly and precisely symbolized by MW(th) or KW(th),

TABLE 7-1   World Inventory of Power Reactors.

Country	Operating Reactors	MWe	Under Construction, Planned or On Order	MWe
Algeria			1	600
Argentina	2	935	4	2498
Austria			1	692
Bangladesh			1	300
Belgium	6	3470	4	4612
Brazil	1	626	8	9960
Bulgaria	4	1644	8	8000
Canada	13	7031	12	8597
China (People's Republic)			8	5700
Cuba			4	1680
Czechoslovakia	2	800	18	11406
Egypt			8	8600
Finland	4	2160	1	1000
France	34	25086	59	60190
Germany (Democratic Rep.)	5	1694	14	6896
Germany (Federal Rep.)	15	9832	22	25712
Hungary	1	408	7	5224
India	4	808	15	3776
Iraq			1	900
Israel			1	950
Italy	3	1232	11	9996
Japan	25	16615	28	24404
Korea	3	1798	10	9146
Libya			2	816
Luxemburg			1	1000
Mexico			2	1308
Morocco			4	2400
Netherlands	2	499		
Pakistan	1	125	3	2737
Philippines			1	621
Poland			6	5760
Portugal			1	1000
Rumania			6	5400
South Africa			4	3844

TABLE 7-1 (*Continued*)

Country	Operating Reactors	MWe	Under Construction, Planned or On Order	MWe
Spain	5	2855	14	13166
Sweden	10	7327	2	2120
Switzerland	4	1940	3	3007
Taiwan	4	3110	8	7714
Turkey			2	1800
United Kingdom	35	8829	9	6651
United States	83*	66831*	62	67875
USSR	40	17886	58	61670
Yugoslavia	1	632	4	4000
TOTALS	307	184173	431	303988

*Includes inoperative Three Mile Island 2 plant (906 MWe)
Sources: AIF 1983; ANS 1983; USDOE 1983.

respectively. However, in the case of power reactors, the power level is commonly given as the electrical output, or MWe, which in modern reactors is about one-third the total power produced. Because in general non-power reactors as a class operate at low power levels and are usually are operated infrequently or periodically, their potential contribution as a source of environmental radioactivity is relatively small.

A typical non-power reactor will produce a few tens to hundreds of kilowatts, with a relatively small number exceeding a few megawatts. By contrast, a modern large power reactor is operated continuously for a period of months at full power, and may produce on the order of 4,000 MW(th); even older and smaller power reactors produce several hundred MW(th). Power reactors are shut down periodically for refueling and maintenance, and may occasionally operate at reduced power levels but historical data show that they operate at about 55–60% of total capacity, which is defined as the product of the rated electrical output of the reactor and the time available for operation. Thus, power reactors are a very much larger and important source of radioactivity, and the study of the environmental radiological aspects of reactors must quite properly focus on them, although the same principles generally apply to non-power reactors, but generally on a considerably reduced scale. A typical operating 1000 MWe light water reactor (LWR) will contain about 1.5×10^{10} Ci

(5.55 $\times$ 10^{20} Bq) of activity, mostly short-lived fission gases, during operation; within a few seconds after shutdown, this will have decreased by a factor of 3.

Basic Thermal Reactor Design

Common to all nuclear reactors, whatever their application are the following features or systems:

1. Fuel, which consists of a fissile nuclide such as ^{233,235}U,^{239}Pu.

2. A coolant such as water to remove the heat produced in the fission process.

3. A moderator, which slows down the neutrons produced in fission, making them more efficient at producing more fissions.

4. A control system, usually of neutron absorbing material by which the rate of the reaction can be regulated.

5. Safety system, by which the reactor can be shut off rapidly or otherwise controlled in the event of a failure of the control system or coolant system.

6. Auxiliary systems, which include demineralizers and other methods of removing impurities from the coolant and moderator.

In addition, power reactors have a gas or steam turbine electrical generating system.

Specific reactor designs are many and frequently unique. Most of the smaller reactors used for research and educational purposes are pool or tank type reactors in which the core containing the fuel and control rods sits within a pool or ordinary water which is used as both the coolant and moderator. Depending on the size (i.e. power output) of the reactor, these are fairly simple, and produce relatively small amounts of radioactivity.

Power reactors are usually classified by the type of coolant or moderator that they use. As a class, light water reactors are those that use ordinary water for both cooling and moderation, although a few hybrid types are occasionally included in this category. There are two basic types of light water reactors. In the pressurized water reactor (PWR) (Figure 7-1) in which the water in the primary coolant system is maintained under high pressure and not permitted to boil, although it may reach temperatures well in excess of the boiling point of water at normal atmospheric pressure. The water in the primary system is passed through a type of heat exchanger known as a steam generator, which is essentially a series of coaxial tubes in which the heat from the primary system is

PRESSURIZED WATER REACTOR (PWR)

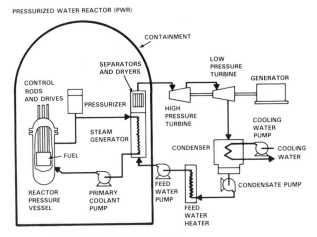

- PRESSURE OF PRIMARY SYSTEM 2,250 psi
- REACTOR OUTLET TEMPERATURE 605°F

FIGURE 7-1   Pressurized Water Reactor.

transferred to the water in the secondary system, which is permitted to boil, with no mixing of the water in the two systems. The steam thus produced is used to drive a turbine, which in turn is cooled by a third separate cooling system is used to cool the turbine.

In the boiling water reactor (BWR) (Figure 7-2), the primary coolant

BOILING WATER REACTOR (BWR)

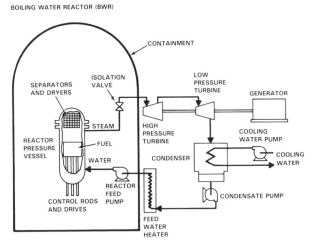

- PRESSURE OF PRIMARY SYSTEM 1000 psi
- OUTLET TEMPERATURE 550°F

FIGURE 7-2   Boiling Water Reactor.

is circulated through the reactor core and is permitted to boil within the reactor vessel above the core under pressure. The resultant high temperature steam is used to drive the turbine directly. The steam is cooled in a condenser, and recirculated through the reactor core. A separate secondary coolant system is provided for the turbine condenser.

With two exceptions, all nuclear power reactors in use or under construction in the United States are either PWRs or BWRs. One exception is the N Reactor, located in southeastern Washington state, which is unique both in design and in that it is used both for generation of electricity and for production of plutonium. The N reactor is graphite moderated and light water cooled. The other exception is the high temperature gas cooled power reactor (HTGCR), the Fort St. Vrain plant, located near Denver, Colorado.

Although there is only one gas cooled power reactor in United States, this type of power reactor is popular in the United Kingdom, which uses them exclusively, and in France, which has eight such plants operational. Single units are operational in Italy, Japan, and Spain, bringing the world total to 44. The basic design of a gas cooled reactor is similar to that of a LWR (Figure 7–3). In the gas cooled reactor, carbon dioxide or helium is used as the coolant and graphite as the moderator. As graphite absorbs fewer neutrons than light water, these reactors can use natural uranium as fuel instead of the slightly enriched uranium required by LWRs, although as a practical matter, commercial gas cooled reac-

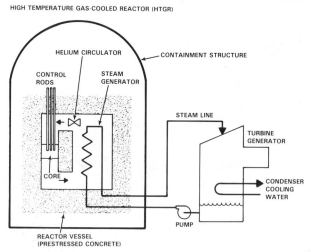

FIGURE 7–3   Gas Cooled Reactor.

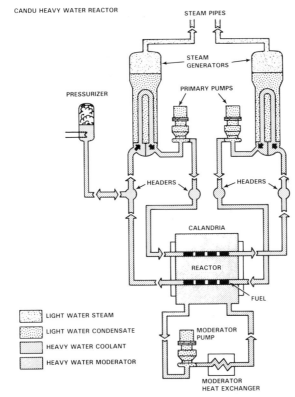

CANDU HEAVY WATER REACTOR

STEAM PIPES

STEAM GENERATORS

PRESSURIZER

PRIMARY PUMPS

HEADERS   HEADERS

CALANDRIA

REACTOR

FUEL

LIGHT WATER STEAM
LIGHT WATER CONDENSATE
HEAVY WATER COOLANT
HEAVY WATER MODERATOR

MODERATOR PUMP

MODERATOR HEAT EXCHANGER

FIGURE 7-4   CANDU Reactor.

tors such as the high temperature gas cooled reactor (HTGCR) and advanced gas reactor (AGR) use enriched fuel. Also, the use of gas as coolant permits operation at higher temperature and thus provides a higher theoretical efficiency for conversion of heat to electricity. Despite these apparent advantages and a long and successful operating history, no gas cooled reactors are under construction outside of the United Kingdom, and even Britain is considering adopting the light water reactor for its nuclear power program.

Heavy water moderated reactors (Figure 7-4) were pioneered in Canada, and are also used in Argentina, India, Pakistan, and Rumania (under construction). The Canadian design is known as the CANDU, for "Canadian Deuterium-Uranium" reactor, but this made-up word could also be considered a pun (i.e."can do") suggestive of the ability of this relatively sparsely populated nation in the nuclear reactor business. The heavy water moderated and cooled reactor differs from the PWR not only in that heavy water is used instead of ordinary water as the neutron moderator, but also in physical design as well. In the CANDU type reactors, the fuel is arranged horizontally rather than vertically as in a LWR permitting the fuel to be replaced while the reactor is operating. Like gas cooled reactors, heavy water moderated reactors can be fueled with natural uranium because the heavy water absorbs fewer neutrons than ordinary light water. Any of several cooling liquids may be used; the Canadian reactors with one exception use the D_2O for cooling as well.

In the Soviet Union, the dominant type of reactor is a light water cooled, graphite moderated reactor (LGR); 18 of this type are known to be operational with several others under construction (ANS 1983). Other power reactors in operation include the liquid metal fast breeder reactor (LMFBR) which uses liquid sodium as the coolant; two such plants are producing electricity in the Soviet Union, and another in France. A large scale test LMFBR, the Fast Flux Text Facility (400 MW_{th}), began operation at the Hanford site in southeastern Washington state in 1982, and an actual 330 MWe demonstration plant—the Clinch River Breeder Reactor—is under construction near Oak Ridge, Tennessee, although there is some question regarding the auhorization of federal funds for completion. The 1200 MWe Superphenix reactor in France, is a breeder reactor being built jointly with Italy and West Germany, and is scheduled to come on line in 1984. In addition, there are few one-of-a-kind power reactors, but more than 95% of the world nuclear electrical generating capability is produced by the basic types described, with the light water reactor, especially the PWR, becoming even more dominant. France now has the world's largest PWR construction program and is expected to have a nuclear generating capacity of 55 to 60 GWe by 1990 (Carle 1983).

Production of Fission Products

There are three basic sources of radioactivity in a nuclear reactor. The first is the fuel itself, which is naturally radioactive, although relatively insignificant when compared with the other two sources, fission products and activation product. Most reactors are fueled with uranium of varying enrichments of ^{235}U. Uranium, having been separated from its daughters during the milling and enrichment procedure is a relatively weak source in itself because of its low specific activity. A large LWR fueled with 100 tons of 3.5% enriched uranium would only contain 6.8 Ci (252 GBq) of ^{235}U and 29.3 Ci (1.1 TBq) of ^{238}U.

Other reactor fuels are ^{239}Pu and ^{233}U. The latter is not used as a reactor fuel except in a few small experimental reactors. Reactors fueled with ^{239}Pu are also relatively rare; despite the economic advantages of recycling the plutonium produced in a uranium fueled reactor, ^{239}Pu currently cannot be used as fuel in operating power reactors in the United States, although it has been successfully used in smaller experimental and testing reactors.

The other two sources of radioactivity result from operation of the reactor and are fission products and activation products. The fission reaction in a reactor is no different than that of a nuclear explosive except that it proceeds at a slower rate. Thus, an operating reactor will produce more than 300 fission products, mostly radioactive, during operation. As there are only about 40 ways for binary fission to occur, there are only about 80 primary fission products; the remainder grow in from chains, and, because of the long operating time of the reactor, appreciable quantities of long lived fission products may build up with time. The net rate of accumulation of any individual fission product, dA/dt, is simply equal to the rate of formation less the rate of removal by radioactive decay and by neutron capture or other neutron interations, as shown:

$$\frac{dA}{dt} = Y\Sigma_f\phi - \lambda_A A - \sigma_A\phi A \qquad (7-1)$$

in which Y is the fission yield,

Σ_f the macroscopic fission cross-section (cm^{-1}),

ϕ the neutron fluence rate $(n/cm^2\text{-s})$,

λ_A the decay constant of the nuclide (s^{-1}),

σ_A microscopic absorption cross-section (cm^2),

and A is the number of atoms per cm^3 of the nuclide at any time.

The right hand side of Equation (7-1) can be broken into three terms. The first, made up of three variables, describes the rate of formation; the middle, containing two variables, the loss through radioactive decay; and the last term, also containing three variables, the loss by neutron capture. Equation (7-1) can be rearranged to solve for the rate of formation as shown:

$$\frac{dA}{dt} + (\lambda_A + \sigma_A\phi)A = Y\Sigma_f\phi \qquad (7-2)$$

As each nuclide has a different rate of decay, the rate of accumulation will vary, but at some time an equilibrium situation will be reached, in which the rate of production is equal to the rate of removal by radioactive decay and neutron capture. Given a constant power level, and hence rate of formation, this will, for all practical purposes, occur after about seven half-lives have passed. Thus, for a relatively short-lived nuclide like ^{131}I, equilibrium will be reached about two months after the start of the reactor. However, in the case of long-lived fission products such as ^{137}Cs (30 y), equilibrium will not be reached until after more than 200 years of continuous operation. Thus the mixture of radionuclides in the reactor, as well as the total inventory of fission products or source term, will be a function of not only the power level, but of the operating time and operating history of the reactor as well as the type of fuel and components (Table 7–2).

The activity and mix will also change with time after shutdown of the reactor (Table 7–3, Figure 7–5). Although the activity theoretically can be calculated for any combination of operating power levels (i.e. fission rate) and operating times, such computations are exceedingly complex. However, the rate of beta emission for a given time t after fission can be approximated to within about a factor of two by the following empirical expression (Way and Wigner 1948).

$$\text{Betas/s-fission} = 3.8 \times 10^{-6}t^{-1.2} \qquad (7\text{–}3)$$

or

$$\text{Bq-fission} = 3.8 \times 10^{-6} \qquad (7\text{–}3a)$$

and the rate of gammas by $t^{-1.2}$

$$\text{Photons/s-fission} = 1.9 \times 10^{-6}t^{-1.2} \qquad (7\text{–}4)$$

In both Equation (7–3) and (7–4) the time after shutdown, t, is in units of days. Since one curie is 3.7×10^{10} disintegrations, Equation (7–3) can be expressed in units of curies, A, if the number of fissions, F, is known, by.

$$A = 1.03 \times 10^{-16}t^{-1.2}F \qquad (7\text{–}5)$$

The above equations hold for times of a few seconds to a few months after fission has occurred.

TABLE 7–2 Buildup of Important Fission Product Activities in an Operating Light Water Reactor.

Nuclide	Thermal Fission Yield (%)	Half Life	Inventory in Curies at Indicated Time Assuming Continuous Operation at 1000 MW(th)		
			100 days	1 year	5 years
Kr-85	1.33	10.73y	53	191	818
Sr-89	4.79	50.5d	28,200	38,200	38,500
Sr-90	5.9	29y	402	1,430	6,700
Y-90	0.22	64h	402	1,430	1,430
Y-91	5.4	58.6d	34,800	48,900	49,500
Zr-95	6.2	65.5d	32,900	49,200	50,300
Nb-95	6.2	35.1d	20,900	48,200	50,500
Nb-95m	6.2	3.61d	446	687	704
Ru-103	3.0	39.6d	25,100	30,900	31,000
Ru-106	0.38	396d	753	2,180	4,220
Rh-103	3.0	56m	25,100	30,900	31,000
Rh-106	0.38	2.18h	753	2,180	4,220
Ag-111	0.024	7.47d	151	151	151
Sn-125	0.013	9.65d	100	101	101
Sb-127	0.13	3.8d	787	787	787
Te-127	0.035	109d	146	260	277
Te-127m	0.035	9.4h	808	922	939
Te-129	0.35	33.4d	1,410	1,590	1,590
Te-129m	0.35	79m	1,410	1,590	1,590
Te-132	4.7	78h	36,900	36,900	36,900
I-131	3.1	8.04d	25,200	25,200	25,200
I-132	4.7	2.29h	36,900	36,900	36,900
I-133	6.9	20.8h	54,200	54,200	54,200
I-134	7.8	52.6m	61,270	60,270	60,270
I-135	6.1	6.59h	49,600	49,600	49,600
Xe-131	2.93	12d	23,800	23,800	23,800
Xe-133	6.62	5.29d	53,800	53,800	53,800
Xe-135	6.3	9.17h	51,200	51,200	51,200
Cs-136	6.41	13d	52	52	52
Cs-137	6.15	30.1y	300	1,080	5,170
Ba-137	6.15	2.55m	285	1,030	4,910
Ba-140	6.36	12.79d	51,500	51,700	51,700
La-140	6.36	40.23h	51,300	51,700	51,700
Ce-141	6.0	32.53d	43,000	47,800	47,800
Ce-144	6.0	284.4d	9,860	26,700	44,000
Pr-143	5.7	13.58d	45,000	45,300	45,300
Pr-144	6.0	17.28m	9,860	26,700	44,000
Nd-147	2.7	10.99d	21,800	21,800	21,800
Pm-147	2.7	2.62y	1,290	4,900	16,000
Total	200		564,000	694,000	768,000

TABLE 7-3 Important Radionuclide Activity as a Function of Time After Shutdown for 1000 MWe (3100 MW(th)) Reactor.*

Days After Shutdown	Iodines	Noble Gases	All Fission Products	Actinides $(Z > 92)$	Activation Products	Total
0	1,435	1,240	13,800	3,450	10.6	17,250
1	265	221	2,890	1,330	9.2	4,230
5	101	105	1,870	432	8.4	2,310
15	28	29	1,280	40	7.4	1,330
30	6.7	4.8	947	9.4	6.4	963
60	0.5	0.78	656	6.3	4.8	666
120	0.003	0.65	401	5.9	2.8	410
365	—	0.63	146	5.6	1.4	152
1,095	—	0.55	47	4.5	0.3	52
3,652	—	0.35	18	3.3	0.1	22

Activity in Megacuries

*Fueled with 82 tonnes 3.3% enriched uranium; total specific power 33,000 MWD/tonne.

148

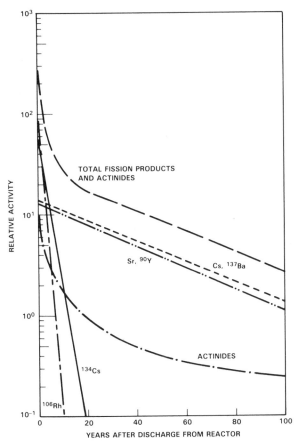

FIGURE 7-5   Fission Product Activity After Shutdown.

Since the time interval t in Equations (7–3), (7–4), and (7–5) refers to the time after fission has occurred, these equations can only be used where the fissions have occurred over a short period of time, or essentially instantaneously, as in a nuclear explosion. However, a reactor operates over long periods of time, and this must be taken into account when calculating the activity at shutdown or any time t thereafter. Such computations are also highly complex, but can be approximated from Equation (7–5) by considering a reactor operating at a constant power level for T days. Since the decay time, or time after shutdown is given as t days, the time from the commencement of operation to t days after shutdown is simply T + t, which can be represented by τ. Given an arbitrary time increment dT during operation, Equation (7–5) can then be written as.

$$A = (1.03 \times 10^{-16})(2.68 \times 10^{15})P(\tau\text{-}T)^{-1.2} \, dT \qquad (7\text{-}6)$$

in which A is the activity in curies remaining t days after shutdown for a reactor operating dT days at a constant power level of P watts, and 2.68×10^{15} is the number of fissions per day per watt. Integrating Equation (7–6) over T, the time of operation in days, yields.

$$A = 27.6 \, P \int_0^T (\tau - T)^{-1.2} \, dt$$

$$A = 1.4P\{(\tau - T)^{-0.2} - \tau^{-0.2}\}$$

which provides the total activity A in curies (or if the numerical constant of 3.8×10^{-11} is used in place of 1.4, in becquerels) at τ days after startup for a reactor operating at a constant power level of P watts for T days. The important fission products are shown in Table 7–4; other than tritium, all are produced directly by binary fission.

TABLE 7–4   Important Fission Products.

Nuclide	Half-life
Gases	
H-3	12.3y
Kr-85	10.8y
Xe-133m	2.3d
Xe-133	5.3d
Xe-135	9.1h
Solids	
Sr-89	53.0d
Sr-90	28.0y
Y-90	2.7d
Y-91	59.0d
I-131	8.1d
I-133	20.8h
I-135	6.7h
Cs-134	2.0d
Cs-136	14.0d
Cs-137	30.0y
Ba-140	13.0d
La-140	1.7d
Ce-144	290.0d

Activation Products

The final source of radioactivity in the reactor is from neutron activation of the materials from which the reactor is constructed, including the fuel, moderator, coolant, and other reactor components. The activity from neutron activation can be calculated from

$$A = N\sigma\phi(1 - e^{-\lambda T})e^{-\lambda t} \qquad (7\text{-}7)$$

in which A is the activity at any time t after shutdown,
   N is the number of atoms of the target atoms,
   σ is the activation cross-section, in barns/atom,
   ϕ is the neutron fluence rate, in $n/cm^2\text{-}s$,
   λ is the decay constant of the radionuclide,
and τ is the operating time of the reactor.

Activation products (Table 7–5) are produced by interactions of neutrons with hydrogen and oxygen in the coolant water, oxygen, nitrogen, and argon in air dissolved in the coolant, and any materials used in construction. The most significant activation products are ^{58,60}Co, ^{65}Zn, and ^{59}Fe (Table 7–5). Activation products are produced in locations where materials are subjected to high neutron fluence rates, such as the primary coolant system. In LWRs, the primary coolant system is subjected to high temperatures and pressures, which may result in selective dissolution of various elements or temperature gradient transfer. Also, tiny plaques of corrosion or scale will occur and may flake off the inner wall of the piping or valves and circulate within the primary system. Corrosion

TABLE 7–5   Important Activation Products.

Nuclide	Half-life
N-16	7s
Ar-41	1.8h
Cr-51	28d
Mn-54	300d
Mn-56	2.6h
Co-58	72d
Co-60	5.4y
Fe-59	45d

TABLE 7-6   Important Transuranic Elements Produced in Reactors.

Nuclide	Half-life	MCi in 1000 MWe Reactor
U-239	6.75d	1,708
Pu-238	86.4y	0.138
Pu-239	24,360y	0.032
Pu-240	6,580y	0.050
Pu-241	13.2y	12.4
Pu-243	5y	22.4
Am + Cm isotopes		1.14
All Actinides		3,614

Source: Nero 1979, p. 37.

may be is a significant source of activation products, which are produced when the ablated material passes through the high neutron fluence rates in the reactor.

Activation products are also produced by neutron capture in the fuel, and result in the production of neptunium, plutonium, and the higher actinides. Plutonium-239 is produced in every reactor containing ^{238}U by the following reaction:

$$^{238}\text{U} + {}_0\text{n}^1 \rightarrow {}^{239}\text{U} \rightarrow {}^{239}\text{Np} \rightarrow {}^{239}\text{Pu}$$

A 1000 MWe uranium fueled LWR with produce on the order of 30,000 Ci (1.1×10^{15} Bq) of ^{239}Pu or several hundred pounds during a year of operation. This fissile nuclide can be separated from the spent fuel and itself used as a reactor fuel. Production of actinides in a 1000 MWe light water reactor is given in Table 7-6.

Tritium and ^{14}C

Tritium is a very important radioactive species produced in nuclear reactors, both by ternary fission and by various nuclear reactions, including activation of deuterium. About one in 10,000 uranium fissions and one in 5,000 plutonium fissions are of the ternary type, producing tritium. As the tritium nucleus is very small, it has the ability to diffuse through the cladding and into the primary coolant of the reactor; as much as 1% of the tritium in the fuel will pass through zircalloy cladding, while about 80% will pass through stainless steel cladding (Kouts and Long 1973).

About 20 Ci/y (740 GBq/y) are produced from ternary fission in a 1000 MWe LWR (Erdman and Reynolds 1975).

Tritium is also an important activation product, particularly in PWRs and in heavy water reactors. Some Tritium is produced in LWRs by activation of the deuterium in ordinary water. The natural abundance of deuterium is 0.015%, or about 1 part in 7000 of naturally occurring hydrogen, and the activation cross-section is only 0.52 mb. For a 1000 MWe LWR, about 10 Ci (3.7×10^{11} Bq) are produced annually, with the actual amount a function the quantity of water used for cooling and the neutron fluence rates in the core. The effect is very much magnified in D_2O moderated reactors, and tritium production may run to several hundred curies (GBq) per year from activation of the deuterium.

In pressurized water reactors, boron, which has a high capture cross-section for neutrons, is used as a chemical shim to regulate or control the reactivity of the reactor by introducing it directly into the primary coolant in the form of boric acid. Tritium is thus produced by the $(n,2\alpha)$ on ^{10}B, which has a natural abundance of 19.8%. Several hundred curies of tritium may be produced annually in a 1000 MWe PWR from this source.

Tritium is also produced in small amounts in BWRs, which use BC_4 control rods; this source of tritium may be several curies annually, although it appears to be well contained in the BC_4 matrix. Production of up to a few hundred millicuries per year may also result in graphite moderated reactors such as those of the gas cooled type from activation of lithium impurities in the graphite.

Carbon-14 is produced in both LWRs and heavy water moderated reactors by the (n,) reaction on ^{17}O, which is present in the coolant and moderator, and also by the (n,p) reaction with nitrogen impurities in the fuel and elsewhere. Production is fairly small, typically on the order of a few tens of millicuries (MBq) per year in large LWRs and CANDU type reactors (Davis 1977).

Radioactive Gas Partitioning

Radioactivity is released to some extent from nuclear reactors under both abnormal (i.e accident) and normal operating conditions, through any of a number of pathways. The chief source of radioactivity is the fuel, which is in intimate contact with the primary coolant, separated from actual direct contact only by the fuel cladding. Thus the primary coolant is of primary significance in the transport of radioactivity from the fuel to the environment outside the reactor.

In addition to fission product activity which may enter the primary coolant through cladding leaks or failed fuel, radioactivity is produced in the primary coolant of a LWR by activation, either of the coolant itself, corrosion products within the coolant, or of additives to the coolant such as boron in the boric acid added as a chemical shim. Fission products also appear in the coolant under normal operating conditions from the fissioning of tramp uranium, which is the tiny amount of uranium than remains on the external surfaces of the cladding despite the best efforts at cleaning prior to use of the fuel. During operation of the reactor, the tramp uranium fissions or burns up very quickly, producing fission products directly in the primary coolant. This is in addition to the fission products from cladding failure or failed fuel, which is 0.2% or less under normal operation, although a value of 1% failed fuel is usually assumed in safety analysis work and for prediction of radioactivity releases to the environment by operating power reactors under normal conditions.

Dissolved solids are removed from the LWR primary coolant by continuously flow through ion exchange resins or demineralizers or by evaporation and recondensation of the coolant. However, neither of these methods remove tritium, and demineralization resins are slow to remove isotopes of cesium, yttrium, and molybdenum. Gaseous activity is removed by stripping the gases and allowing them to decay, usually in pressurized holdup tanks (PWRs) or charcoal beds (BWRs) before release to the atmosphere. Filters may be used for particulates and radioiodines.

As most of the fission product activity is in the form of radioactive gases such as the radiokryptons, radioxenons, and radioiodines, the escape of these gases from coolant removed from the system for chemical analysis or that escapes by leakage is important from an environmental standpoint. In reactor hazards evaluation, the partition factor concept is used to evaluate such releases. The partition factor, F, is the fraction of a volatile material such as radioiodines, noble gases, or ruthenium that is transferred across a liquid:gas, liquid:solid, or gas:solid interface. Partition factors are commonly used for radioiodine or noble gas transfer across a liquid:gas interface and are calculated by

$$F = \frac{(G/L) \div P}{1 + (G/L) P}$$

in which G/L is the gas to liquid ratio, either in terms of volume or flow, and P is the partition coefficient. The partition coefficient, P, the equilibrium ratio of the concentration of the radionuclide in the liquid phase to that in the gaseous phase and is a function of both temperature and concentration. For noble gases, P is approximately 0.1 at room temperature, decreasing with increasing temperature to a minimum of about 0.03 at 100 °C, and then increasing again with temperature. As noble gases are nonreactive, the partition coefficient is not affected by pH. A typical value for the partition factor for noble gases is 0.1 in ordinary circumstances, although it will approach unity during blowdown.

The partition factor for the radioiodines is dependent on chemical species, and is very much greater than that for the noble gases. As the radioiodines undergo chemical reactions with the coolant, the chemical species will change, and along with it the partion factor. pH, temperature, and concentration of iodine all affect the partition coefficient, which may become quite large in highly alkaline environments, exceeding 10,000. However, for most situations, a value of 100 is reasonable.

Management of PWR Wastes

In pressurized water reactors, the radioactive liquids from the primary coolant are handled in the chemical and volume control system (CVCS) which consists basically of two subsystems, one for reactor primary coolant cleanup and the other for boron recovery. In the reactor coolant water cleanup subsystem, a portion of the primary coolant is continually fed through a bypass system through a cleanup system, usually demineralizers which remove most of the radioactivity, except for tritium and the noble gases. In the boron recovery subsystem, some of the purified water from the demineralizers is removed, evaporated, and condensed, leaving boric acid behind in the evaporators. The boric acid can then be reinjected into the primary coolant, if desired, or it can be treated as solid radioactive waste as it is likely to be contaminated with radioactive impurities.

A distinction is made between clean (high purity) and dirty (low purity) radioactive wastes. Clean wastes are usually from the primary system, originating from samples taken for chemical and radiological analysis, excess from the CVCS, pump seal and valve leakage, and similar sources free of grease and dirt. These wastes are usually held for a month or so to allow short-lived nuclides to decay, and then filtered, decontaminated, and reused as primary coolant, or diluted and discharged to the environment. A typical treatment scheme (Figure 7–6), and shows the waste first collected in tanks for holdup, followed by treat-

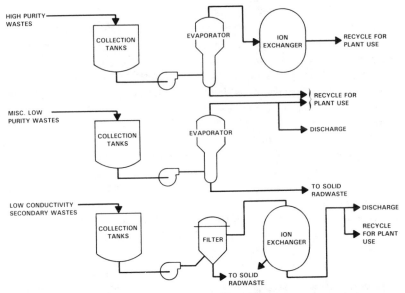

FIGURE 7-6   PWR Waste Teatment.

ment in an evaporator and then by ion exchange prior to recycling for plant use.

Dirty liquid radioactive wastes include liquids collected from floor drains, sumps, laboratory drains, and cleanup and decontamination activities. These ordinarily have low levels of activity, and are usually held for a time before discharge to the environment. Dilution may be necessary in some instances. A typical treatment scheme is to collect the dirty or low purity wastes in tanks, treatment in an evaporator, followed by recycling for plant use or discharge to the environment, with dilution being accomplished as needed to meet discharge limits. Low conductivity wastes may simply be filtered prior to discharge or recycle.

Steam generator blowdown is a major source of radioactive liquid discharge to the environment. Blowdown is essentially a cleanout or rapid withdrawal of water from the secondary system, and is performed to prevent the accumulation of salts and other materials, including algae, in the steam generator tubes which would reduce the effectiveness of heat transfer. Blowdown is usually discharged directly to the receiving body such as a river, and a relatively low rate, perhaps on the order of 10 gallons per minute. Unless blowdown is used to dilute other liquid wastes, or there has been steam generator leakage, it will normally be free of radioactivity.

The treatment of contaminated liquids produces solid radioactive wastes. Evaporator bottoms, or the solid residue left behind from the evaporation treatment, filters containing radioactivity, and ion exchange resins fall into this category and contain the bulk of the radioactivity removed from the liquid effluents. In addition, in some cases the liquids may be solidified directly be incorporation in an appropriate hygroscopic medium or by treatment with cement.

Steam generator leakage is a significant problem in pressurized water reactors. To facilitate heat transfer, the tube walls must be very thin, thus increasing the probability that small holes will develop allowing the radioactivity bearing primary coolant to mix with the normally nonradioactive secondary system. Although the tubes are ordinarily plugged when leakage becomes significant, in normal PWR operation, leaks of up to 20 gallons per day can be tolerated. Even this small amount of leakage can serve as a significant source of radioactive escape from the primary system of a pressurized water reactor to the environment.

Gases dissolved in the primary coolant of a PWR are vented from the CVCS and liquid wastes holdup tanks and directed to pressurized storage tanks where they will be held from two to four months prior to release to the atmosphere through a bank of high efficiency particle filters. Gaseous hydrogen and oxygen will also be present from the radiolysis of the primary coolant, and are recombined to minimize the hazard from explosion. The primary nuclides released after the holding period are ^{85}Kr (10.8 y) with some ^{133}Xe (5.3 d) and ^{131}I (8.05 d). In some plants, the release is passed through charcoal beds or filters to remove the radioiodine.

Secondary system gases are those released during steam generator blowdown, air ejector gases removed from the low pressure side of the turbine, steam being used to seal the turbine gland to prevent entry of air. These gases are usually not radioactive unless steam generator leakage has been large. Discharge to the atmosphere is thus ordinarily direct, through the general building ventilation air system without treatment. Alternatively, these gaseous wastes may be filtered through high efficiency air filters and charcoal and discharged through a separate roof vent. Sometimes only the gases from steam generator blowdown are treated in this fashion.

Under normal conditions, general building ventilation is not radioactive, and thus may be discharged directly to the environment, along with the gland seal effluent. The containment is purged a few times annually, and these discharges are passed through a filter system prior to discharge to the atmosphere.

Solid radioactive wastes include the spent ion exchange resins, filters of various types, including those from the ventilation systems, evaporator residues, and papers, glassware, rags, tools and other decontamination materials, and similar items. The ion exchange resins are in slurry form when removed from the demineralizers, and are ordinarily solidified with vermiculite or cement. The solid wastes are packed into 55 gallon drums or similar containers approved by the U.S Department of Transportation or various international bodies such as the International Atomic Energy Agency and shipped offsite for disposal in a licensed low level disposal facility.

Tritium bearing liquid wastes are of special concern in PWRs. Several hundred curies (TBq) per year of tritium may be disposed of from the primary system. Tritium bearing liquid wastes are ordinarily bled off, diluted, and discharged directly to a receiving body.

BWR Waste Management

There are significant differences in the design and operation of BWRs as compared with PWRs that affect their radiological aspects. Unlike PWRs, boiling water reactors do not use boron for reactivity control purposes, and hence have much less tritium in the primary coolant. Since the primary coolant is allowed to boil, fission gases and radioiodines are present in gaseous form. Hydrogen cannot be used as a cover gas in BWRs to prevent radiolytic breakdown of the coolant, and hence there is more gaseous hydrogen and oxygen present.

As is the case with PWRs, some of the primary coolant is bypassed through a system of demineralizers, which remove the ionic forms of the various radionuclides. Two types of wastes are produced: the high purity wastes, which have low solid content (<100 ppm) and are more or less analagous to the clean radwaste of the PWR, and the low purity or high solid content wastes, which are comparable to the PWR dirty radwastes. In addition, there are liquid laundry wastes, which may be dimineralized, stored for a time to allow the short-lived activity to decay, diluted if necessary, and discharged to the environment; and various chemical wastes, which may require chemical neutralization as well as decontamination before release to the environment.

Gaseous radioactive wastes from a BWR include air ejector gases, which are gases removed by a steam jet from the exhaust side of the turbine. About 99% of all the BWR gaseous radioactivity is contained in these. Air ejector wastes are treated by catalytic recombination of the radiolytic hydrogen and oxygen, which significantly reduces their volume, as much as 80%. After a holdup or delay of about a half hour, accomplished by passage through a convoluted pipe, the remaining gases are passed through high efficiency particulate air (HEPA) filters, iodine removal systems (usually charcoal filters or beds), and discharged through a stack.

Older BWRs were distinguishable from PWRs by their characteristic tall discharge stacks. In general, the older BWRs discharged significantly greater quantities of gaseous radioactivity than PWRs. However, the newer BWRs are equipped with large charcoal beds that adsorb noble gases and radioiodines, gradually releasing at least the former and thus serving as an effective holdup system. Other gaseous releases from BWRs include leakage from the turbine gland seal, which will ordinarily contain greater amounts of radioactivity than in a PWR, although still a relatively small amount. This is steam that is condensed and discharged directly to the atmosphere. Large volumes of building air may also be discharged, but the radioactivity content is small and the dilution great. Solid radwaste in BWRs is similar to that in PWRs.

Radiological Aspects of Normal Reactor Operations

Under normal operating conditions, every reactor produces radioactivity and releases a fraction of its production to the environment. The radiological consequences of reactor operation depend to a large extent on the size or power level of the reactor and its age, along with its specific design and operating characteristics. Even among various classes or categories of reactors there is a diversity of operating experience with regard to radiological releases to the environment. However, in general, releases from non-power reactors, with the exception of production reactors, are relatively small, largely because of the low power levels and frequently intermitant operation. Thus, power reactors are the most important from an environmental point of view.

The radioactive effluents from the various types of power reactors can be categorized as follows:

Releases to the Atmosphere

1. Noble gas fission products

2. Activation gases, including tritium,

3. Radioiodines

4. Carbon-14

5. Particulates

Liquid Effluents

1. Tritium

2. Mixed fission products.

Radioactive noble gases are the major source of activity released from operating nuclear power reactors. Both the quantity and constituents of the release is highly variable from plant to plant; in general, the greatest releases are from boiling water reactors, and in particular the older units. In a BWR, the noble gases are continuously removed from the primary coolant and discharged through a tall stack via the main condenser air ejector system. Noble gas releases from a typical BWR are variable, even when normalized to unit power production ranging from less than 100 to more than 10,000 Ci (3.7->370 TBq)/MWe (UNSCEAR 1977, 1982). Using 1977 as a typical year, the noble gas activity released ranges from a few thousand to perhaps a hundred thousand curies annually (about 10 TBq to perhaps 3.7 PBq) (USNRC 1979). The typical value declined from 1500 Ci/MWe-y (55.5 TBq/MWe-y) in 1970 to about 1000 Ci (37 TBq)/MWe-y in 1974, and still further to less than 800 Ci(30 TBq)/MWe-y, largely because of increased holdup time in the effluent gas treatment systems in new plants. The activity released is a mixture of several radiokryptons and radioxenons, with the short lived activities predominating.

Noble gas releases from PWRs are in general considerably lower than for BWRs, and consist largely of the longer lived fractions, particularly the 5.3 day ^{135}Xe and 10.76 year ^{85}Kr. For a typical large PWR, the total noble gas activity released may be one to two orders of magnitude lower than a BWR of comparble power. Thus, the normalized releases run on the order of 10 to 20 Ci(370–740 GBq)/MWe-y (UNSCEAR 1977, 1982). Noble gas releases from gas cooled reactors are small, generally less than 10 Ci(370 GBq)/MWe-y, while those from heavy water moderated reactors are quite variable, and when normalized to Ci (or Bq)/MWe-y may approach those of BWRs in some instances, although are usually very much less.

Most of the activity from noble gas nuclides released to the atmosphere is from short-lived nuclides, with half-lives ranging from several minutes to several days. The dose contribution from these is relatively small as they tend to decay away rather quickly. However, the 10.76 year ^{85}Kr is also emitted, and this nuclide with its long half-life tends to build up in the atmosphere. In 1970, the concentration of this nuclide in the atmosphere in the northern hemisphere was on the order of $10^{-11} \mu Ci/cm^3$ (0.37 Bq/m^3) with a 100-fold increase projected by the year 2000 (Klement et al. 1972). The skin dose was estimated at 0.02 mrad/y (0.2 $\mu Gy/y$) in 1970, rising to 1.6 mrad (16 μGy)/y in 2000. The scenario on which these estimates were based has proven to be optimistic, as cutbacks in nuclear plant construction coupled with improved effluent controls have significantly reduced the buildup of this nuclide.

Activation gas releases are associated with gas cooled reactors almost exclusively, and the only significant nuclide is ^{41}Ar, which is produced by the (n, γ) reaction with ^{40}Ar in the reactor coolant and shield cooling air. Normalized releases are on the order of 100–300 Ci (3.7–11.1 GBq)/ MWe-y (UNSCEAR 1977, 1982).

The radioiodines are an important group of gaseous radionuclides released from operating nuclear power plants. Ordinarily, iodine is considered to be present in elemental or organic form (i.e. methyl iodide) although many other chemical forms may also be present, including hypoiodous acid (HIO), hydriodic acid (HI), iodic acid (HIO$_3$), and metal iodides. In gaseous reactor effluents, about three-fourths is organic, a fifth hypoiodous acid, and the remaining 5% elemental (Pelletier, Cline, and Keller 1974). The major iodine isotope of concern is mass number 131 (8.05 d), and to a lesser extent 133 (20.8 h). Some concern has been expressed about ^{129}I, but the specific activity and yield of this nuclide is so low that it is of little consequence despite its enormous half-life of 16 million years. Typically, BWRs emit a few mCi (about 100 KBq)/ MWe-y in gaseous effluents, and other power reactors one to two orders of magnitude less (UNSCEAR 1977). For earlier BWRs, the normalized output of ^{131}I has been estimated at about 5 mCi (185 KBq)/MWe-y based on actual operation UNSCEAR 1982), although modern effluent treatment systems have reduced this by about two orders of magnitude to about the same level as that from other types of power reactors.

Carbon-14 is produced in power reactors by the (n, α) reaction with ^{17}O, the (n,p) reaction with ^{14}N, and, in graphite moderated reactors, by neutron capture by ^{13}C, which comprises 1.1% of naturally occurring carbon. Production is relatively low, and releases lower still. Available data suggest that the release of ^{14}C is on the order of 20 mCi (740 KBq)/ MWe-y for both PWRs and BWRs (Davis 1977). The resultant dose is very small relative to that from naturally occurring ^{14}C (UNSCEAR 1977,1982; USNRC 1979). Similarly, radioactivity bearing particulates, although identified in reactor gaseous effluents, are usually a small source of activity released to the environment, and largely contain mixed fission products.

The major radioactive constituent of liquid releases to the environment from reactors is tritium. PWRs, largely because of their use of boron in the primary coolant for reactivity control, and HWRs because of deuterium activation, are the major sources of reactor produced tritium. A typical modern large PWR will discharge several hundred to perhaps a thousand or so curies (about 10 TBq) of tritium annually; discharges from large heavy water moderated reactors may be ten-fold larger. Gas cooled reactors typically discharge a few tens of curies (hundreds of GBq) of tritium annually. Normalized to power production, HWRs discharge 20 Ci (740 TBq)/MWe-y as compared with 1 Ci(37 TBq)/MWe-y for PWRs and 0.2 (7.4) and 0.3 (11.1) for BWRs and GCRs, respectively (UNSCEAR 1977).

Liquid releases of mixed fission and activation products have been relatively small, generally averaging less than a curie (37 GBq) per year for most large American power reactors, although a few reactors have discharged several curies annually (USNRC 1979). Typical experience elsewhere in the world is higher discharge quantities, perhaps exceeding a hundred curies (several TBq) per year. However, when examined on a normalized basis, the values are less than 100 mCi (3.7 GBq)/MWe-y for BWRs and PWRs, and about 150 mCi(5.6 GBq)/MWe-y for gas cooled reactors (UNSCEAR 1977). In LWRs, the corrosion products may predominate. There is, however, great variability both in quantity and radionuclide content of liquid discharges from power reactors, even among a given class of reactor, and thus generalizations are of little validity.

In the United States, regulatory requirements limit reactor discharges of radioactive effluents to "as low as reasonably achievable" (ALARA). This has been defined for design purposes as limiting atmospheric releases to producing annual doses to persons in unrestricted areas of less than 10 mrad (100 μGy) from gamma radiation in liquid effluents, or 20 mrad (200 μGy) from beta radiation (CFR 1983). Similarly, liquid releases are limited to 5 Ci (185 GTBq)annually, exclusive of tritium, or to that quantity that will deliver an annual dose equivalent of no more than 5 mrem (50 μSv).

Reactor Accidents

Although there have been several accidents involving reactors and critical assemblies, only three are of major significance from an environmental standpoint. The first of these occurred in October 1957 at Windscale No. 1, a graphite moderated production reactor located in northwest England. When bombarded with fast neutrons at temperatures below about 300 °C, graphite undergoes extensive radiation damage which results in increased size, decreased thermal and electrical conductivity, and the ability to store energy. If the graphite is then heated slowly, the stored energy will be released in the form of additional heat. Thus, heating neutron irradiated graphite results in the release of more heat. This energy storage phenomenon was first explained by the Hungarian-American physicist Eugene P. Wigner, and is known as Wigner energy, or sometimes, Wigner's disease.

The Wigner energy in graphite reactors is commonly released by annealing the graphite through a slow heating process which gradually releases the stored energy and reverses the radiation damage. This gradual release of energy will increase the temperature of the reactor and is called Szilard's fever in the vernacular, after another Hungarian-American physicist, Leo Szilard, who identified the effect.

At about one in the morning on Monday, the seventh of October, the Windscale reactor was shut down. Some eighteen hours later, the reactor was started up again and run at low power to produce the desired low temperature annealing, and again shut down several hours later, in the early morning hours of October 8. The operators, however, suspected that the desired Wigner energy release had not occurred, and late Tuesday morning again started the reactor, which was run for only fifteen minutes before again being shut down. By that time, however, the stage had been set for a rapid and massive release of Wigner energy.

Graphite temperatures rose throughout the remainder of the day, and most of the next day. On Thursday, October 10, the dampers were opened in an attempt to provide cooling for the reactor, and high levels of radioactivity were observed on the stack monitors. Later that night, an inspection was made inside the reactor, and red hot uranium cartridges were observed. Attempts to discharge these hot cartridges failed as they were distorted from the excessive heat, and the reactor was finally cooled with fire hoses the following day. The reactor was a total loss, and the integity of at least one uranium cartridge was breached with the resultant release of fission products to the environment. An estimated 20,000 Ci (740 TBq) of ^{131}I were released along with 600 Ci (22 TBq) of ^{137}Cs, 80 Ci (3 TBq) of ^{89}Sr, and 9 Ci (3.3 GBq) of^{90}Sr (Loutit, Marley and Russell 1960).

The release of this quantity of radioiodine to the environment resulted in the contamination of milk as well as in the exposure of persons from inhalation. The 96 workers in the plant received a mean thyroid dose of 0.4 rad (4 mGy), largely from inhalation. In the general population in the plant environs, the mean thyroid dose in adults ranged from 0.3 to 1.8 rad (3–18 mGy) , with a maximum of 9.5 rad (95 mGy). In children, the mean dose ranged from 0.8 to 12.2 rad (80–122 mGy), with a maximum of 16 rad (160 mGy)(Loutit, Marley and Russell 1960).

The primary source of exposure to the general population was milk contaminated with radioiodine. Based on the risk of tumor production in the thyroid, the British Medical Research Council established a level of concern, or cutoff level of 0.1 μCi/l 3.7 $\times$ 10^6 Bq/m^3 for radioiodine in milk (UKAEA 1957). Milk from a 200 square mile area downwind of the reactor was confiscated for several days after the accident. Nearly a month after the accident, milk from a smaller area up to 12 miles from the reactor was still being confiscated; confiscation of milk ceased on November 23, some six weeks after the accident. The milk was not a total loss, as it could in some cases be diverted to cheese production, during which time the short-lived radioiodine would decay away.

Radioactivity from the reactor was detected in numerous samples collected in the environment. Air samples showed highly variable concentrations of activity, ranging up to 4.5 $\times$ 10^{-9} μCi/cm^3 (167 Bq/m^3) of ^{131}I. Radioiodine, radiocesium, and radiostrontiums were detected in nearby bodies of water, although in concentrations below those considered to be of concern. Analysis of foodstuffs revealed no increased levels of radioactivity attributable to the accident.

The second major reactor accident occured at the SL-1 Reactor at the National Reactor Testing Station near Idaho Falls, Idaho, in January 1961. The SL-1 was an experimental natural recirculation boiling water power reactor fueled with enriched uranium and designed to operate a continuous power level of 3 MW(th). The reactor had had a checkered operating history, including sticking control rods, and deterioration of boron-aluminum poison used for reactivity control within the reactor. Moreover, the reactor was designed such that it could achieve criticality with the partial removal of its central control rod.

After operation of nearly 1000 MWd, the reactor was shut down on December 23, 1960, and a 12 day maintenance program completed. On the night of January 3, a three man crew was assigned the task of reassembling the control rod drives on top of the reactor. One of the crew was apparently attempting to remove the central control rod, possibly for inspection. It is probable that this rod was sticking, and, in his attempts to free it, the crew member removed the rod too far and too fast, creating an explosive supercriticality (IDO 1962). An estimated 133 ± 10 MW-s was released by the nuclear transient, plus another 24 ± 10 MW-s from chemical (metal-water) reactions; the instantaneous peak power was 19 GW.

The explosive force was sufficent to kill the three man crew, and to destroy the reactor. One man was driven upward to the roof of the building by the explosion and remained suspended there. Most of the coolant water was ejected from the reactor vessel, carrying with it fission product activity from the core. The reactor vessel and internals, including the fuel, were severely damaged and stressed. An estimated 5–10% of the total fission product inventory escaped from the reactor vessel, which contained about a megacurie (37 PBq) of long-lived fission products. Only about 0.01% of the total inventory escaped from the reactor building, largely fission gases and radioiodines. However, less than 0.5% of the radioiodines and a neglibile fraction of the nonvolatile fission fission products were released to the environment (IDO 1959; Thompson 1964).

High direct radiation levels were detected in and around the reactor building following the accident; exposure rates in excess of 500 R/h were observed inside the reactor building about a half hour after the accident by rescue personnel, and levels in adjacent buildings ranged from several tens of mR/h to a few R/h. At the security fence about 150 feet from the reactor building, exposure rates were approximately 200 mR/h, dropping to 5 mR/r at 1000 feet and less than 2 mR/h about 2000 feet from the reactor (Horan and Gammill 1963). Initial emergency response efforts, including recovery of the bodies of the three men were carried out despite these high levels, and were completed with 23 persons receiving exposures in the range of 3 to 27 R (Thompson 1964).

Environmental monitoring was initiated immediately after the accident. Small increases in gross radiation levels were detected by aerial monitoring out to a distance of 3 to 4 miles from the reactor. Other than radioiodine and the noble gases, the fission product activity was virtually all contained in a three acre plot around the reactor. Within the first few hours of the accident, an estimated 10 Ci (370 GBq) of ^{131}I had been released, with a total of 20 Ci (740 GBq) released in the first 24 hours. About 70 Ci (2.6 TBq) were released during the four weeks following the accident. Contamination followed the prevailing wind patterns almost due south of the reactor, with radioiodine levels of about twice background observed in vegetation in an area approaching 100 square miles, and as far as 75 miles away. Levels of radioidine in vegetation $\geq$ ten times background were confined to a much smaller area, extending no more than about 30 miles from the southern boundary of the reactor site (Horan and Gammill 1963). Air concentrations of ^{131}I were generally less than 7.4 $\times$ 10^{-11} μCi/cm^3 (2.7 Bq/m^3) off-site.

The third major reactor accident occurred in Three Mile Island Unit 2, an 880 MWe pressurized water reactor near Harrisburg, Pennsylvania (Rogovin and Frampton 1980). In the early morning hours of March 28, 1979, while the reactor was operating at 97 to 98 per cent of full power, cooling to the steam generators was lost, and the reactor and turbine were automatically shut down. Three auxiliary feedwater pumps automatically started, but could not provide the needed cooling to the steam generators as the valves in the feed lines had been improperly left closed. The steam generators thus went dry, and were unable to cool the primary coolant, which resulted in increased temperature and pressure in the primary loop. The overpressure caused an automatic pressure relief valve to open, but the valve failed to close allowing some 32,000 gallons—more than one-third—of the primary coolant to escape from the system.

The reduction in pressure continued, and two minutes after the initiating event, the emergency core cooling system came on, injecting fresh cooling water into the primary system. The plant operators, however, mistakenly believed that the plant had gone "solid", i.e. the primary system was totally filled with water. Thus they shut off the high pressure emergency injection system a few minutes later, and about an hour and a half later compounded the problem even further by switching off all four main coolant to avoid damage to them, thus relying on natural circulation to achieve cooling of the shutdown reactor which was still generating 46 MW of heat an hour after the shutdown from the decay of fission products.

Some two hours and eight minutes after the start of the accident, the operators discovered the stuck open pressurizer valve and closed it, ending the loss of coolant phase of the accident. A sustained injection of high pressure cooling water ended the overheating of the core at nearly three and one-half hours after the accident began, but the addition of the cold water to the extremely hot core may have produced additional damage from thermal shock. At this point, many radiation monitors began to alarm, and shortly thereafter a general emergency was declared.

Although the core sustained serious damage, including a partial meltdown, relatively little radioactivity was released to the environment. Ambient radiation levels in the containment, however, were quite high, perhaps exceeding 10,000 R/h above the damaged reactor. Some of the activity released from the damaged fuel was transferred to the coolant, which was transported to the auxiliary building via the letdown line.

TABLE 7-7 Radionuclides Released to the Environment from the TMI-2 Accident.

Nuclide	Half-life	Curies Released	Fraction of Total
Kr-88	2.8h	3.8×10^4	0.15
Xe-133	5.2d	1.5×10^6	0.63
Xe-133m	2.2d	2.3×10^5	0.09
Xe-135	9.1h	3.0×10^5	0.12
Xe-135m	15.3m	2.5×10^4	0.01
I-131	8.1d	15	—

Source: Rogovin and Frampton, Vol. II, Part 2, p. 344.

Activity in the primary coolant, which was 0.4 μCi/cm^3 (1.5 $\times$ 10^{10} Bq/ m^3) before the accident, increased by about five orders of magnitude to more than 20,000 uCi/cm^3 (7.4 $\times$ 10^{14} Bq/m^3) after the accident. Gaseous activity was primarily released from the coolant, and was passed through a series of filters, including charcoal to remove radioiodines, before release to the environment.

The total quantity of activity released is estimated at 2.5 million curies (92.5 PBq), almost entirely short-lived noble gases (Table 7-7). About sixty per cent of the activity was ^{133}Xe, which has a half-life of 5.2 days. Only 15 Ci (55.5 GBq) of ^{131}I were released although both radioiodines with mass numbers 131 and 133 were identified in air samples collected the day of the accident (USNRC 1979). An extensive environmental monitoring program was carried out, with several governmental agenicies and other organizations participating. The maximum levels of ^{131}I observed in air were 3.2 $\times$ 10^{-11} μCi/cm^3 (1.2 Bq/m^3) during the first two weeks following the accident, with levels up to threefold greater being observed on April 16 as a result of filter changes at the plant. Typical air concentrations offsite were below 10^{-12} μCi/cm^3 (0.037 Bq/m^3) for ^{131}I; no particulate activity was detected, although several radioxenon isotopes were found in small concentrations (Rogovin and Frampton 1980).

Small concentrations of radioiodine—largely ^{131}I—were detected in milk samples offsite. The highest concentration observed was in a sample of goat milk taken on March 30 1.2 miles north of the site; the concentration was 41 pCi/l (1517 Bq/m^3). In cow's milk, the highest levels were 36 pCi/l (1332 Bq/m^3) of ^{131}I. Other fission products in milk were attributable to fallout from nuclear weapons tests (Rogovin and Frampton 1980). Levels of radioactivity in water were likewise low, being in all cases less than 1 pCi/l (37 Bq/m^3). Only ten of the many vegetation samples collected yielded positive results, with most of these occurring as a result of the filter changes on April 12. Radioactivity concentrations in all samples were significantly below regulatory limits and confirmed that the release of activity was relatively small.

Dose estimates to persons in the vicinity of the site were largely made from data obtained with thermoluminescent dosimeters. The collective population dose equivalent was estimated by several independent bodies, and ranged from 300 to 3500 person-rem (3–35 person-seivert), with the lower values estimated from computer models. The TMI Special Inquiry Group analysis indicated that the population dose equivalent was probably in the vicinity of 2000 person-rem (20 person-seivert). The dose equivalent to the maximum exposed individual was estimated at less than 100 mrem (1 mSv). Internal exposure was largely incurred from intake of ^{131}I, and was estimated at 6.9 mrem (69 μSv) to the thyroid of a newborn child. The maximum onsite thyroid dose equivalent to an adult was estimated at 53 mrem (530 μSv). The maximum internal whole body dose equivalent from noble gases was estimated at 0.3 mrem (3 μSv), and the lung dose equivalent as 3 mrem (30 μSv). The doses incurred were low, leading to the finding by the TMI Special Inquiry Group that "the effects on the population as a whole, if any, will certainly be nonmeasurable and nondetectable" (Rogovin and Frampton 1980).

In addition to the three major reactor accidents described, there have been several other small accidents in both reactors and critical assemblies, but these have had no or negligible environmental consequences. These include the partial meltdown at the Enrico Fermi plant, a 200 MW(th) sodium cooled breeder reactor in 1966, in which an estimated 10,000 Ci (370 TBq) were released to the containment. No radioactivity, however, escaped to the environment. A series of destructive reactor tests was carried out at the Nevada Test Site during the 1950's. These were the BORAX and SPERT tests, and resulted in release of fission product activity, largely noble gases, to the test site environment. Activity was almost wholly confined to the immediate area of the test.

Other accidents involving radioactivity have had more serious impacts, and include the accidental contamination of the environment from a fire involving plutonium at the Rocky Flats plant near Denver in 1969, which resulted in the release of about 6 Ci of plutonium, and a smaller accidental release of plutonium at the Oak Ridge National Laboratory in 1959. In both instances, the environmental contamination was confined to the immediate area of the site, although in the Rocky Flats case the area involved may have been several square miles. Plutonium contamination of the environment also resulted from two accidents involving nuclear weapons. The first of these occurred in Spain in January 1966, and the other near Thule, Greenland in two years later. In both these cases, contamination was limited to a small area and apparently removed by heroic decontamination measures. In general, plutonium contamination from other than weapons accidents does not become widely dispersed in the environment, and remains in the top several centimeters in the soil.

References

1. American Nuclear Society (ANS). 1983. "World List of Nuclear Power Plants", *Nuclear News* 26(2):71.
2. Atomic Industrial Forum (AIF). 1983. "Nuclear Reactor Status Report", *INFO*, No 172, p. 8.
3. Carle, Remy, "How France Went Nuclear". 1983. *New Scientist* 97(1340):84 (13 January 1983).
4. Code of Federal Regulations (CFR). 1983. Title 10, Energy, Part 50, Appendix I.
5. Davis, W. 1977. "Carbon-14 Production in Nuclear Reactors", ORNL/NUREG/TM–12.
6. Erdman, C. A., and A. B. Reynolds. 1975. "Radionuclide Behavior During Normal Operation of Liquid Metal Cooled Fast Breeder Reactors", *Nucl. Saf.* 16:43
7. Horan, J. R. amd W. P. Gammill. 1963. "The Health Physics Aspects of the SL-1 Accident", *Health Phys.* 9:177
8. Idaho Operations Office (IDO) of U.S. Atomic Energy Commission. 1962. "IDO Report on the Nuclear Incident at the SL-1 Reactor, January 3, 1959, at the National Reactor Testing Station", USAEC Report IDO–19302
9. Klement, A. W., Jr., et al. 1972. "Estimates of Ionizing Radiation Doses in the United States 1960–2000", U.S. Environmental Protection Agency Report ORP/CSD 72–1.
10. Kouts, H. and J. Long. 1973. "Tritium Production in Nuclear Reactors", in *Tritium*, A. A. Moghissi and M. W. Carter, Eds., Messenger Graphics, Phoenix
11. Loutit, J. F., W. G. Marley, and R. S. Russell. 1960. "The Nuclear Reactor Accident at Windscale, October 1957: Environmental Aspects", in *The Hazards to Man of Nuclear and Allied Radiation*, Her Majesty's Stationary Office Report No. Cmnd 1225, London.
12. Nero, A. V. 1979. *A Guidebook to Nuclear Reactors"*, University of California Press, Berkeley.
13. Pelletier, C. A., J. E. Cline, and J. H. Keller. 1974. "Measurement of Sources of Iodine–131 Release to the Atmosphere from Nuclear Power Plants", *IEEE Transactions on Nucl. Sci.* NS–21:478
14. Rogovin, M. I. and G. T. Frampton. 1980. *Three Mile Island. A Report to the Commissioners and to the Public*, Report of Special Inquiry Group, Nuclear Regulatory Commission, Washington

15. Thompson, T. J. 1964. "Accidents and Destructive Tests", Chapter 11 in *The Technology of Nuclear Reactor Safety. Volume 1. Reactor Physics and Control.*, T. J. Thompson and W. F. Beckerley, Eds., The M.I.T. Press, Cambridge

16. United States Department of Energy (USDOE). 1983. "UPDATE—Nuclear Power Program Information and Data, April-June 1983", Report Number DOE/NE-0048/3, August 1983.

17. United Kingdom Atomic Energy Authority (UKAEA). 1957. "Accident at Windscale No. 1 Pile on October 10,1957", Her Majesty's Stationary Office, London

18. United Nations Scientific Committee on the Effects of Atomic Radiation (UNSCEAR). 1977. "Sources and Effects of Ionizing Radiation", United Nations Publication E.77.IX.1, New York

19. United States Comptroller General (USCG). 1977. "Cleaning Up the Remains of Nuclear Facilities—A Multibillion Dollar Problem", Report to the Congress EMD-77-46, Washington.

20. United Stated Nuclear Regulatory Commission (USNRC). 1979. "Radioactive Materials Released from Nuclear Power Plants", Report NUREG-0521.

21. United States Nuclear Regulatory Commission (USNRC). 1979. "Investigation Into the March 28, 1979 Three Mile Island Accident by Office of Inspection and Enforcement", Report NUREG-0600.

22. Way, K. and E. P. Wigner. 1948. "The Rate of Decay of Fission Products", *Phys. Rev., 2nd Series* 73:1318.

Processing and Radioactive

Classification of Radioactive Wastes

In the broad sense, virtually anything containing radioactivity that is no longer needed or that cannot be economically reclaimed may be considered as radioactive waste, including liquid and gaseous effluents from nuclear power plants and other radiological operations. However, the term radioactive waste is often used in a much more limited sense; hence, radioactive wastes are usually considered to equipment and material from nuclear operations that are in themselves radioactive or are contaminated with radioactivity and for which there is no further use.

Unfortunately, there is no standard and commonly accepted scheme for categorizing radioactive wastes, although several have been proposed (USNRC 1981b). One of the most common and simple classifications is according to physical state: liquid, solid, or gaseous. Categorization is also commonly made in terms of activity content, as low, intermediate, and high level. These designations are usually qualitative, but attempts have been made to quantify them (Table 8–1). Low level wastes (LLW) constitute the bulk of the radioactive waste volume, but contain relatively little of the activity. They are wastes with low hazard potential which may be released to the environment without treatment or for which a dilution of perhaps 1000 may be required for uncontrolled release. The term low level waste is quite broad, and generally includes wastes from a variety of sources. It accounts for upwards of 90% and probably closer to 99% of the total volume of radioactive wastes generated, but contains only 1% or less of the radioactivity.

Low level wastes include a wide variety of materials contaminated with or suspected of contamination with various radionuclides, including protective clothing, tools, air filters, and other materials from various operations with radioactivity, as well as animal carcasses, scintillation vials

TABLE 8-1   A Classification Scheme for Radioactive Waste.

Category	Gaseous (μ Ci/ml)	Liquid (Ci/l)	Solid (Ci/ft^3)
Low Level			
Alpha	$<10^{-13}$	$<10^{-3}$	<10
Beta-Photon	$<10^{-10}$	$<10^{-3}$	<10
Intermediate Level			
Alpha	10^{-13} -10^{-7}	0.001–1	10–1000
Beta-Photon	10^{-10} -10^{-4}	0.001–1	10–1000
High Level			
Alpha	$>10^{-7}$	>1	>1000
Beta-Photon	$>10^{-4}$	>1	>1000

Adapted from Fitzgerald 1970, p. 751.

and similar laboratory wastes from research laboratories and hospitals. Fuel cycle wastes account for about 65% of the volume and nearly all of the activity of the LLW sent to commercial burial sites (Overcamp 1982).

The category of intermediate level waste is sometimes applied to wastes that can be safely released to the environment after treatment (Fitzgerald 1970), or to wastes that require shielding to protect personnel from external radiation exposure (Overcamp 1982). In general, the intermediate waste category is not usually used.

High level wastes (HLW) are relatively small in volume but contain large quantities of radioactivity; thus these are also known as high specific activity wastes. If liquid, the heat produced by the radioactive decay may cause boiling to occur. HLW is usually thought of as the highly radioactive concentrated residues left behind after the reprocessing of reactor fuels or from the separation of plutonium from production reactor feedstocks. These wastes contain upwards of 90%, and perhaps as much as 99% of the radioactivity but only a small fraction of the volume—typically less than 1% when solidified.

Transuranic wastes (TRU) are those containing greater than 10 nCi (370 Bq) of transuranics per gram of waste material. The most common nuclides in TRU are long lived and include ^{239}Pu, along with ^{241}Am and other isotopes of americium and curium. Mill tailings also have a rather specific definition, and usually categorized separately from other forms of radioactive wastes.

TABLE 8-2   High Level Waste Volumes at USDOE Sites.

Location	Liquid	Salt Cake	Other	Total
Hanford	1377	3354	1730	6461
Idaho NEL	330	73	—	403
Savannah River	2112	932	371	3415

All values in thousands of cubic feet; adapted from USDOE 1981.

Wastes are also classified by their mode of production, e.g. reactor wastes, medical wastes, or by any one of a number of descriptive adjectives or phrases suitable to the purpose. Thus, many qualitative terms appear in the literature, including such combinations as primary wastes, secondary wastes, commercial wastes, evaporator bottoms, spent resins, and remedial action waste. To be properly understood, such descriptions or categorizations should be used only in an appropriate context.

Great quantities of high level radioactive wastes —both in volume and activity—have been produced, largely from production of plutonium for military purposes (Table 8–2). In addition, almost 10 million cubic feet ($\sim$ 250,000 m^3) of TRU has been produced (Table 8–3).

Low–Level Wastes

Low level radioactive wastes are produced in almost all activities involving radioactive materials. Thus, they are a byproduct of all steps of the fuel cycle as well as the various medical and commercial applications of

TABLE 8-3 Transuranic Waste Volumes at USDOE Sites.

Location	Millions of Cubic Feet
Hanford	5.9
Idaho NEL	2.02
Los Alamos	0.41
Nevada Test Site	<0.01
Oak Ridge	0.22
Savannah River	0.22

Adapted from USDOE 1981.

radioactivity. About half of the low level wastes are from nuclear power plants, with about another third coming from various non-fuel cycle activities associated with the use of radioactivity in hospitals, research laboratories and universities, and industry. The remainder are from a variety of sources associated with portions of the fuel cycle other than the nuclear power plant. During the twenty year period 1980–2000, and estimated 3.6 million cubic meters (128 million cubic feet) of low level waste will be produced (Anderson et al 1978 USNRC 1981b).

Low level nuclear power plant wastes may be liquid, solid, or gaseous. Gaseous wastes are ordinarily held for a time to permit the short lived nuclides to decay and then discharged to the atmosphere; in a few cases, cryogenic distillation may be carried out to remove the major conbtributors to radioactivity content prior to release. The active gas fraction may be stored in pressurized containers to allow time for decay.

Liquid low level reactor wastes include materials from floor drains and sumps, laundry and decontamination wastes, and wash liquids of various types. If sufficiently low in activity, these wastes may be directly discharged to the environment with or without dilution or chemical neutralization. Otherwise, these wastes are filtered, passed through ion exchange resins or evaporators, chemically neutralized, and the liquid portion diluted as needed prior to discharge to the environment. In some cases, small volumes of high activity liquids may be solidified by the addition of concrete, plaster of Paris, or similar materials.

Solid low level reactor wastes include spent resins from the ion exchangers used to purify the coolant, evaporator bottom wastes, compactable trash, including paper, rags, and disposable protective gloves and clothing, and bulk wastes including air and liquid filters, and occasionally even construction materials such as concrete, steel, and wood. Spent resins and evaporator residues are usually in slurry form, and may be solidified in the same manner as liquids. Solid wastes are compacted when possible, and packaged in 55 gallon (7.35 m^3) drums or similar container for offsite burial.

Solidified low level reactor wastes contain a mixture of fission and activition products. The solid wastes generated each year by a 1000 MWe PWR will contain about 6000 Ci (222 TBq), with about 90% of the activity from radiocesiums, and most of the remainder from radiocobalt and ^{55}Fe (Table 8–4). A similar size BWR will produce only about 1600 Ci (59 TBq), about half of the activity being from ^{55}Fe and most of the remainder from the radiocobalts, radiocesiums, and radiostrontiums (Table 8–4). A typical 1000 MWe LWR will produce on the order of 2000–4000 cubic feet or 55–110 cubic meters of solid low level waste

TABLE 8-4   Radioactivity in Reactor Solid
Waste.

Nuclide	Ci/yr from 1000 MWe	
	PWR	BWR*
^{89}Sr	2	49
^{90}Sr	2.7	90
^{95}Zr	0.64	1.1
^{106}Ru	1.7	—
^{134}Cs	2800	130
^{137}Cs	2700	120
^{144}Ce	4.2	5.1
^{54}Mn	22	13
^{55}Fe	200	800
^{59}Fe	1.8	6.6
^{58}Co	120	130
^{60}Co	280	200
Total	6100	1600

*With charcoal absorber system.
Adapted from USNRC 1973b.

annually, or the equivalent of several hundred drums per year (USNRC 1973a); thus the specific activity is on the order of a curie per cubic foot, or a few tens of microcuries per cubic centimeter (about 1 MBq/m^3). Various estimates of the volume of low level reactor wastes have been made; a recent (1979) study by the Environmental Protection Agency indicates an annual volume of about 100,000 cubic meters through the 1980's (Holcomb 1980).

Low level non-fuel cycle (institutional) wastes include laboratory materials, animal carcasses and other biological wastes, papers, disposable gloves and other protective clothing, and miscellaneous materials. The largest fraction of these wastes result from liquid scintillation counting which is widely used in medicine and research. Liquid scintillation counting fluids typically contain toluene or xylene contaminated with the low energy beta emitters tritium or ^{14}C. Technetium-99m, used in medicine, is another common constituent of institutional wastes. Together, these three nuclides account for about two-thirds of the approximately 131,000 Ci (5 PBq) of non-fuel cycle low level wastes disposed of annually (Table 8-5), with a volume of about 40,000 cubic meters (1.4

TABLE 8-5   Activity in Low Level
Wastes Sent to Commercial Sites
Annually.

Nuclide	Activity, Ci
^3H	36,023
^{14}C	10,709
^{32}P	6,274
^{35}S	2,440
^{51}Cr	1,891
^{67}Ga	132
^{99m}Tc	37,573
^{125}I	12,737
^{131}I	7,263
Others	15,490
Total	131,000

Source: USNRC 1978.

million cubic feet). The volume is expected to drop slightly to about 30,000 cubic meters (1.1 million cubic feet) by the year 2000 (Holcomb 1980).

Solid low level radioactive wastes are typically packaged in 30 or 55 gallon (4 or 7.35 m³) metal drums or in special fibreboard or cardboard containers and shipped to an approved facility for disposal by shallow land burial, or, in the case of TRU, temporary storage or burial in retrievable containers. In the United States, there are 20 repositories or low level waste disposal sites (Figure 8–1). Two sites also exist for shallow land burial in the United Kingdom; mines are used in Spain, Ger-

FIGURE 8–1   Location of Commercial Waste Disposal Sites in the U.S.

TABLE 8-6   Waste Inventories at USDOE Burial Sites.

Location	Cu Ft Buried	Ci Remaining	Uranium (kg)	TRU (kg)
Hanford	199,000	810,000	600,000	365
Idaho NEL	140,000	3,579,000	288,000	361
Los Alamos	225,000	160,000	251,000	13
Oak Ridge	179,000	60,000	100	13
Savannah R.	279,000	4,410,000	600,000	7

Adapted from Overcamp 1982, p. 224.

many, and Austria. Fourteen LLW disposal sites are operated by the U.S. Department of Energy and are located at or near government nuclear research, testing, or production facilities. These sites accepted commercial low level wastes from the civilian sector until 1962. More than a million cubic meters of solid waste containing nearly ten million curies of activity (400 PBq) plus more than a thousand metric tons of uranium and about 750 kg of TRU have been buried at the five major DOE sites (Table 8-6).

Commercial low level waste disposal operations commenced in 1962, when two sites were licensed. Three additional sites were licensed during the 1960's, and another in 1971. At the six commercial low level waste disposal sites, more than a half million cubic meters of waste containing 5.5 MCi (200 PBq) of activity, 2200 metric tons of unenriched uranium and thorium, and nearly 1000 kg of special nuclear material had been buried through 1978 (Table 8-7). Three of the six commercial sites— the Maxey Flats, West Valley, and Sheffield locations—are temporarily closed, the first two for economic and environmental reasons and the latter because it is full. In 1978, an estimated 2500 Ci (about 100 GBq) was sent to LLW disposal sites. About half of this activity was short-

TABLE 8-7   Waste Inventories at Commercial Burial Sites.

Location	Volume (m³)	Activity (Ci)
Barnwell, SC	205,439	1,122,823
Beatty, NV	67,365	160,819
Maxey Flats, KY	135,287	2,406,288
Richland, WA	23,659	711,837
Sheffield, IL	86,701	60,206
West Valley, NY	66,521	577,778

Source: Holcomb 1980.

lived, with about three-fourths of the remainder attributable to tritium and ^{14}C (Eisenbud 1981).

The disposal of low level radioactive wastes has been fraught with political as well as technical difficulties. Waste management activities at some of the sites have been poorly carried out, and packages have leaked and radioactivity has been detected in the water table and surrounding environment. At the Oak Ridge National Laboratory, LLW has been disposed of in trenches dug in shale, and refilled with the original soil overburden. The higher permeability of the fill overburden permitted rainfall to flood the trenches, resulting in leaching of radionuclides from the waste. In one location, transport of ^{90}Sr to a creek was studied, and found to be 1–2 Ci (37–74 GBq) annually (Duguid 1974).

At both the Savannah River site and Los Alamos National Laboratory, tritium migration via groundwater has been observed (Overcamp 1982). Although soil and rainfall conditions at these two sites are quite different, neither has experienced significant migration, either in quantity or rate. Disposal at the arid Hanford site has been in the form of shallow land burial, or temporary disposal of liquids into trenches or cribs. Biota have been a major force in radionuclide transport from these disposal sites (Wegele 1980). At the Idaho National Engineering Laboratory, radionuclide transport has resulted from infiltration of groundwater, and from flooding by snow melt, in addition to biolgical transport and wind resuspension (Overcamp 1982). In general, the migration of radioactivity from DOE low level waste disposal sites has not been of particular significance from an environmental standpoint.

At the six commercial sites, similar radionuclide migrations have been observed. The three closed sites have had migration of tritium and possibly other radionuclides in both ground and surface waters (Overcamp 1982). Radiostrontiums and tritium were detected in surface waters near the West Valley site in 1969, and a method developed to remove radiostrontiums and cesium was developed. Tritium, however, could not be removed, and the site closed down voluntarily in 1975 (Holcomb 1980). Minor tritium migration has been observed at the Sheffield site, which was closed in 1978. At the Maxey Flats site, plutonium and other radionuclides were detected in soil cores taken at depths of 75 meters as well as in solids and sediments from wells and surface waters. Concentrations were well below acceptable levels defined by regulations. The Maxey Flats site was closed in 1977 (Overcamp 1982).

Low level waste disposal sites have been plagued with a series of political and public concern problems, sometimes based on technical grounds. In 1979, commercial burial ground at Hanford was temporarily closed because of the ingress of contaminated shipments into the state. The largest and newest site, at Barnwell, South Carolina, has been restricted by the state to burying no more than 68,000 m^3 per year to avoid being filled up too rapidly, and also refuses to accept liquid scintillation wastes because of their high chemical toxicity (Margarrell 1980). States with disposal sites have threatened to refuse to accept wastes from states without sites, and political pressures of various kinds have been applied by both the haves and have nots.

In December 1980, enactment of the Low-Level Radioactive Waste Policy Act (PL 96–573) gave each state the authority to regulate low-level radioactive wastes (non-defense) produced within its borders, and authorized and encouraged the establishment of interstate compacts and regional waste control. In addition, regulatory action was taken by the Nuclear Regulatory Commission to reduce the volume of institutional waste by permitting the discharge of 5 Ci of tritium and 1 Ci (37 GBq) of ^{14}C per year into sewer systems, and by permitting scintillation fluids and animal carcasses containing less than 0.05 $\mu Ci/g$ (1850 Bq/g) of these nuclides to be treated as non-radioactive waste (USNRC 1981a).

In addition to land burial, low level radioactive wastes have been managed in other ways. Animal carcasses containing small quantities of tritium or ^{14}C have been incinerated, but the total amount of activity released to the environment from this mechanism has been slight. For example, during the heyday of this activity in the mid-1950's, the UCLA Medical School, a major user of radionuclides for research purposes, disposed of an average of 95 μCi per day, largely ^{131}I, by incineration (Silverman and Dickey 1956).

Sea burial of packaged radioactive wastes was carried out in both the Atlantic and Pacific Oceans by several countries, including the United States, United Kingdom, Germany, Netherlands, Belgium, and France, beginning in the mid-1940's and continuing for more than 20 years. About 135,000 Ci (5 PBq) of activity were disposed of in the Atlantic and 15,000 Ci (5.6 TBq) in the Pacific (Joseph et al. 1971). Included in this total were an estimated 33,000 Ci (1.2 PBq) of induced activity in the reactor pressure vessel of the USS Seawolf, an early nuclear submarine. Wastes were ordinarily packaged in 55 gallon drums or in concrete containers. The question of ocean disposal has periodically resurfaced, with concern being expressed over leaking containers near the Farralon Islands off San Francisco, and with a plan of the Navy to scuttle old nuclear submarine hulls in deep waters.

There have also been direct discharges of low level liquid wastes to the ocean or to rivers. The Columbia River has earned a reputation as the world's most radioactive river because of discharges from the single pass cooling production reactors at the Hanford site; the nuclides discharged were largely activation products with a peak discharge of more than 2000 Ci/d (74 TBq/d), mostly ^{51}Cr with lesser amounts of ^{65}Zn and ^{32}P reported (NAS 1962). The production reactors at Hanford began to close down in 1964, with all except the dual purpose N reactor ceasing operations by 1970. The N reactor, however, differs from the earlier pure production reactors in that it is cooled by recirculated demineralized water instead of single pass river water and thus discharges only small quantities of radionuclides to the Columbia (Foster 1974). Discharges of radioactivity from the Hanford reactors have been extensively studied, and have been measured in sediments, waters, and biota along the more than 200 miles of the river downstream from the site, as well as in the Pacific Ocean.

The Windscale production reactor in Britain has discharged liquid effluents into the Irish Sea, with a daily permissible discharge limit of 1000 Ci (37 TBq) established. In addition, nuclear powered vessels have discharged small quantities of wastes, largely comprised of corrosion product activity, into the ocean as part of their normal operation.

Closing the Fuel Cycle

The limiting factor on fuel lifetime in most power reactors is not the consumption of the fissile material, but rather the production of fission products which act as poisons and reduce the efficiency of operation. Thus, in most power reactors, fuel is changed after about a three year residence time; hence for a typical large (1000 MWe) LWR, about a third or 30 tons of spent fuel are removed from the reactor annually.

Economically, it is desirable to recover the spent fuel by removing the fission product poisons, reenriching it with ^{235}U, and reusing it. In contrast to fresh fuel, which contains 3.3 weight % ^{235}U and 96.7% ^{238}U, spent fuel has the following approximate composition (Glasstone 1982):

^{235}U	0.8%
^{236}U	0.4%
^{238}U	95%
Fissile Plutoniums*	0.65%
Nonfissile Plutoniums	0.25%
Fission Products	2.9%

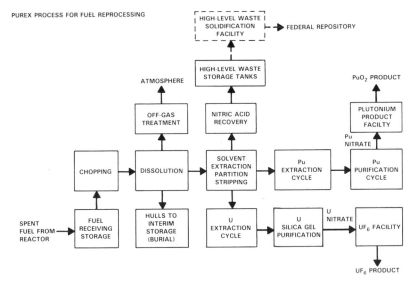

FIGURE 8-2   The Purex Process.

Actually, the spent fuel contains an appreciable amount of fissile material—about 1.45%—or about twice as much as natural uranium and thus recovery is economically feasible.

Fuel reprocessing is accomplished by allowing the fuel to cool for several months to allow the short-lived activity to decay, and then sending it to a reprocessing plant where the spent fuel elements are chopped into smaller, more manageable pieces, dissolved in nitric acid, and the uranium, plutonium, and fission products successively extracted by a modified Purex process (Figure 8-2). The method is similar whether done to recover spent fuel from power reactors or for plutonium recovery from production reactors. Reprocessing procedures must be done remotely because of the high levels of radioactivity from the fission product activity, and result in the production of concentrated high level radioactive wastes in solution. About 99.5% of the uranium and plutonium is removed by the reprocessing procedure, leaving behind the fission products, about 0.5% of the U and Pu, plus most of the other transuranic isotopes in the waste solution.

The activity and nuclidic composition of the high level wastes will vary according to the power produced from the fuel and the irradiation time in the reactor as well as the length of time between reprocessing and removal from the reactor. In general, the plan is for the fuel to be stored for several months at the reactor site after it is removed to allow the short-lived fission products to decay. Thus, the gross radioactivity inven-

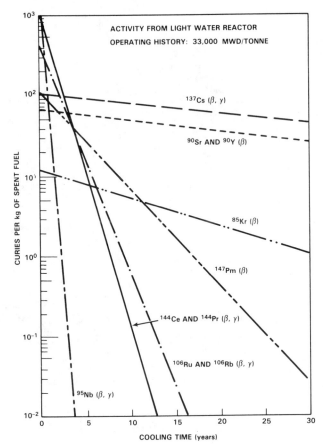

FIGURE 8-3   Activity-Time Relationships for Wastes from 1000 MWe LWR Reactor (Source: National Academy of Sciences 1975).

tory is greatly reduced, and the major activities in the high level waste initially are from ^{95}Nb, ^{106}Ru-Rh, ^{134}Cs, and ^{144}Ce-Pr (Figure 8–3). Within a few years, these decay away, and the major activity is from ^{90}Sr-Y and ^{137}Cs, giving the waste mixture the characteristic thirty year half-life of these dominant nuclides (Figure 8–4). After a few hundred years, the dominant activities and hazards are from actinides and ^{99}Tc (NAS 1975).

The hazard from high level radioactive wastes containing a mixture of radionuclides can be estimated with the Potential Hazard Index (PHI), which can be calculated for any number of radioactive constituents by (Gera and Jacobs 1972):

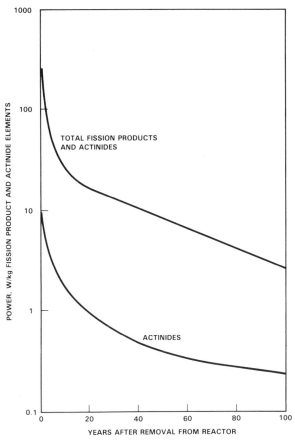

FIGURE 8-4   Heat Generation from High Level PWR Wastes (Source: National Academy of Sciences 1975).

$$PHI_i = \sum_0^i P_i \frac{Q_i}{ALI_i} \cdot \frac{T_i}{0.693}$$

in which P_i is the biological availability factor of the ith nuclide
   Q_i the activity of the ith nuclide in uCi (or Bq)
   ALI_i the annual level of intake in uCi (or Bq)
   (ICRP 1980) of the ith nuclide.
and T_i the physical half life of the ith nuclide in years.

The biological availability factor is essentially the probability of the nuclide reaching people once it has been placed in the disposal site, and is usually taken as unity for conservatism since no real data exist.

In addition to the high level wastes, fuel reprocessing plants produce low level wastes and release radioactivity to the environment. Gaseous wastes are produced during the fuel shearing and dissolution, and include noble gases, largely ^{85}Kr, radioiodines, and tritium. Processing fuel for a 1000 MWe LWR would result in an estimated release of a half million curies (18.5 PBq) of ^{85}Kr, 23,000 Ci (850 TBq) of tritium, and about 30 Ci (1.1 TBq) of ^{14}C (USNRC 1976). Releases of radioiodines would be small, these isotopes being removed by various gaseous treatment sytems such as silver zeolite or charcoal filtration or caustic scrubbing effective for iodine. Low level liquid wastes would be generated from laundry and decontamination operations, as well as other routine operations. The activity released to the environment from these wastes is small, estimated at less than a quarter of a curie (925 GBq) annually, and almost wholly from fission products (USNRC 1976).

Other than the Department of Energy plants for the plutonium pro-duction reactors at Hanford (Washington state) and Savannah River (South Carolina), plus a small non-commercial DOE facility at Idaho Falls for high enrichment fuels from naval and research reactors,, there is no fuel reprocessing in the United States. This is at least partially as a result of a 1977 federal moratorium on fuel reprocessing which requires power reactor licensees to store their spent fuel pending resolution of the reprocessing and waste disposal problems. Thus, there are no commercial reprocessing plants operational in the United States at present. A small plant operated at West Valley, New York, from 1966 to 1972, shutting down to improve radiogical controls and expand capacity; subsequently, it was decided to close the plant indefinately, and later to decommission it entirely. A second, larger facility, the Midwest Fuel Recovery Plant, was built in Illnois, but never became operational because of engineering difficulties. A third plant, near Barnwell, South Carolina, is near com-pletion but has run into funding problems as well as the uncertainty asso-ciated with the moratorium.

By regulation, high level wastes, are defined as " . . . aqueous wastes resulting from operation of the first cycle solvent extraction system, or equivalent, and concentrated wastes from subsequent extraction cycles in a facility for processing irradiated fuels". At commercial nuclear fuel reprocessing plants in the United States, such wastes are to be solidified within five years of processing, and shipped to a federal repository within ten years (CFR 1983). No such repository yet exists, although many plans have been put forth. The Department of Energy is responsible for developing high level waste storage and disposal options and sites that are environmentally safe, but largely due to political and social pressures has been unable to do so. Such a facilities must meet the standards for

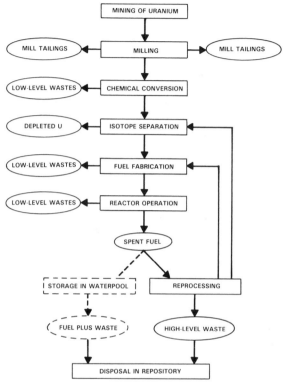

FIGURE 8-5 The Uranium Fuel Cycle showing the significant waste generation at each step.

radiological control put forth by the Environmental Protection Agency, and will be licensed by the Nuclear Regulatory Commission. Plans put forth under President Reagan call for completion of the first such facility in the early 1990's.

Nearly 100,000,000 gallons (13,370,000 cubic meters) of high level liquid radioactive wastes have been produced to date in the United States, mostly from the recovery of plutonium from production reactors. These highly acidic or alkaline wastes, containing radioactivity concentrations, mostly fission products, in the range of about 0.001–1 Ci/ml (0.37–37 PBq/m^3) have been stored in underground tanks near their point of generation. In addition, about 600,000 gallons (80,000 cubic meters) of waste have been produced from commercial reprocessing of fuel, these waste being almost entirely alkaline.

Aqueous high level wastes are mixtures of metallic nitrates in solution with some precipites. Most high level acidic wastes contain a precipitate of zirconium phosphomolybdate; the Zr and Mb are, of course, fission products while the phosphorous comes from the solvent used in reprocessing. This precipitate does not tend to cake. However, in alkaline solutions of waste, barium and strontium sulfates do precipitate out into a crystalline, cement-like cake, carrying with them sodium sulfates of the rare earths, depending on the molarity of the Ba and Sr. Neutralized wastes contain hydrated oxide precipitates of nearly all metallic fission products except the alkali metals. These precipitates are voluminous and flocculent.

High level wastes have been stored in about 200 large underground tanks pending final treatment and disposal. Most of the tank storage has been at the U.S. Department of Energy Hanford site in southeastern Washington state, with smaller amounts at the Savannah River Plant in South Carolina and at the Idaho National Engineering Laboratory. The commercial wastes are stored in tanks at the site of the now defunct West Valley, N. Y., fuel reprocessing plant. Acidic or alkaline wastes are stored in stainless steel tanks, while neutralized wastes can be stored in carbon steel tanks. The tanks are equipped with level indication devices, and may be enclosed in concrete vaults with a steel lined concrete collection 'saucer' underneath to collect any leakage from the tank. The storage tanks have been estimated to have usable lifetimes ranging from 15 to in excess of 50 years, necessitating periodic pumping or transfer of waste solutions as the design lifetime is reached or as leaks occur. Individual tank capacities range from about 0.3 to 1.3 million gallons (40,000–175,000 cubic meters).

The environment in the tanks is very harsh; pH may be low or high, and the waste liquors are thermally hot—in some cases boiling—due to their high concentrations of radioactivity. The combination of heat, radiation, acidity or alkalinity, high salt concentrations and corrosive solutions, plus mechanical stress, has led to a number of tank failures. The first occurred in 1958, and by 1970, 15 failures, all in carbon steel tanks, having been reported (USAEC 1972). By 1973, 16 tank failures had been reported at Hanford alone, including one which resulted in a release of about 350,000 Ci (13 PBq), mostly ^{137}Cs, to the ground below the tank. From an environmental dispersion standpoint, the significance of the release of this large a quantity of radioactivity is surprizingly small as the soil acted to sorb the radionuclides and little, if any, migration occurred a down to significant depths or laterally away from the tank site. Vertically, the radioactivity from small leaks has been confined to the 10–15 feet (3–5 m) immediately below the tank, with activity reaching a depth of perhaps 100 feet from the largest leak. Radionuclides from the tanks have remained well above the water table, thus minimizing the spread.

High–Level Waste Disposal Options

Numerous options for high level radioactive waste disposal and interim storage have been proposed and studied, some being quite imaginative. Storage, of course, implies retrievability, which might be desirable if a use is later found for the waste materials, or if it is necessary to recover damaged or leaking waste containers. In addition, arguments in favor of storage include the idea that at some future time, a superior method of high level waste disposal may be developed, thus making it desirable to retrieve old wastes and treat them with the improved method.

Perhaps the simplest and least innovative option is to simply continue to do as has been done, viz. underground storage of high level liquid wastes in tanks. This is at best a short term solution, and has the major disadvantages of potential tank failure along with the necessity for continual monitoring to assess tank condition in an attempt to predict and prevent tank failure. There is also the possiblity, admittedly rather remote, of tank failure following a catclysmic natural event such as major earthquake or volcanic eruption. Nonetheless, a scheme of double wall tanks, 50 feet (16 m) in diameter and 20 feet (6.5 m) high has been proposed (NAS 1975). Each tank would be equipped with a cooling system, and could handle 300,000 gallons (40,000 cubic meters) of non-boiling liquid HLW.

Most of the other proposed options involve solidification of the waste. Solidification would result in a considerable reduction in volume and ideally should be a simple, inexpensive process that produces a chemically inert (or at least highly resistant) and insoluble matrix with high resistance to internal heating and radiation damage from fission product decay, along with a fair degree of mechanical strength. Several waste solidification processes have been proposed and tried experimentally, and many appear promising. In-tank solidifcation of radioactive wastes is one possible method, the radionuclides in the waste solution being precipitated into solids containing most of the radioactivity right in the storage tank by the addition of appropriate chemicals. The solidified mass can then be left in the tank or removed, if desired, after pumping out the remaining supernatant, which can be treated as a low level liquid waste.

One of the earliest methods, begun in the 1960's, is calcination. Several calcination methods—spray, pot, and fluidized-bed calcination—have been developed, and result in the production of a powdery metalic oxide which can be further treated to form a glassified frit. Calcined wastes are up to 90% by weight fission product oxides. In general, calcined wastes have relatively low thermal conductivity and hence potentially poor resistance to internally generated heat. They are also relatively leachable, showing leach rates of 0.1–1 g/cm^2-day (USDOE 1979; USNRC 1978).

Glassification or vitrification of high level radioactive waste is a most promising option. Radionuclides can be relatively easily incorporated into a glass matrix which has good thermal conductivity and high chemical resistance. Leach rates are several orders of magnitude lower than those of calcined wastes; a typical glass will have a leach rate of 10^{-7} to 10^{-4} g/cm^2-day. Glass is limited, however, in how much radioactivity can be incorporated into it; about 25–50% by weight seems optimal. In Europe, calcined wastes are further treated to incorporate them into borosilcate glass, which has excellent properties of low leachability, high thermal conductivity, and high bulk density (USNRC 1978, USDOE 1980).

Planning in the United States calls for temporary management of the high level wastes by storage in retrievable surface-storage facilities (RSSF). Solidifed waste would be transported and stored in the RSSF in one or two foot diameter canisters 10 feet (3 m) long cooled by natural draft air flow (NAS 1975). Various other concepts, including forced cooling and greater radioactivity content of canisters, have also been put forth.

Ultimately, terminal storage will likely be in stable geologic formations with rock media with low intrusion of water. The site should be one with a history of low and minor seismic activity, with an absence of cracks or faults than might provide flow paths for ground water. Potential geologic media suitable for terminal storage include salt beds and domes, hard crytalline rock formations such as granites, limestone and dolomite beds, shale, and volcanic tuff. Investigation of salt beds as a potential disposal or terminal storage medium was spurred by a 1957 report of the National Academy of Sciences-National Research Council to the Atomic Energy Commission which recommended that disposal in salt was the most promising method of high level waste diisposal. Accordingly, the AEC commissioned the Oak Ridge National Laboratory to undertake a 10 year program of research and engineering studies known as Project Salt Vault to develop a viable procedure for waste disposal in salt.

Salt has many desirable features as a geologic disposal or repository medium. It is nearly impermeable as it is plastic—i. e. cracks and fissures are healed or sealed by pressure. Many large salt domes and beds, the remains of ancient seas, exist in the United States in areas of low seismic activity. Salt has has excellent physical properties; its thermal conductivity is good, and the presence of salt beds many hundreds of feet thick for tens of thousands of years is strongly suggestive of not only geologic stability but lack of contact with ground water. Radiation damage and thermal effects, while present, were not serious enough to cause concern.

The initial laboratory and field results were highly promising. A salt mine near Lyons, Kansas, was selected as the site of a demonstration waste repository, and several successful tests were conducted there. Regretably, there was opposition from the local population as well as some adverse technological aspects to the site, including the fact that the integrity of the salt bed was questionable because of nearby drilling for gas and oil. It was also learned that the salt had been removed for a time by solution mining, rasing doubts as to whether the mine was dry and indicating that there were openings for surface water to gain access to the salt. In addition, there were great communication difficulties between state and federal officials, and considerable mistrust developed. Thus, the project was dropped. Salt beds, however, are still a highly promising geologic medium for ultimate storage or disposal of high level wastes.

Various advanced or long range disposal concepts have also been put forth (USNRC 1979). These include deep land geologic disposal, which in one concept, the solidified waste, in canisters, would be placed into an array of holes drilled into a stable geologic formation several miles below the surface of the ground. Another concept calls for the emplacement of the waste in deep rock strata which would turn molten from the heat generated by the radioactivity and ultimately cool and fuse into an impermeable mass as the heat output drops from the radioactive decay. Yet another deep geologic disposal scheme calls for pumping liquid high level wastes directly into approporiate geologic strata which have been hydrofractured, and following this with an injection of cement to produce a solidified mass containing the high level waste in the interstices of the hydrofractured rock. A more imaginative scheme that proposes placement of high level radioactive wastes in shallow holes, with the decay heat melting the underlying rock and allowing the waste canister to gradually descend by gravity deep into the ground (Logan 1974).

Yet another proposed land disposal scheme would take advantage of a large underground cavity produced by an underground nuclear detonation (Cohen, Lewis, and Braun 1972). Liquid wastes could be injected directly into this already radioactive cavity of fused rock and allowed to boil dry and solidify. The decay heat of the wastes would melt the walls of the cavity which would then incorporate the radioanuclides into an insoluble impervious rock matrix. Even power generation from the steam produced in the cavity was suggested. Clearly, however, this innovative approach involves far too many unanswerable or unresolvable technical and social questions, including the detonation of a nuclear explosive near a waste reprocessing facility, although presumably this could be done in advance of any reprocessing activity.

Other concepts for HLW disposal include seabed geologic disposal, disposal in the polar ice sheet, and extraterrestrial disposal. While all are technologically feasible, there are a great many unanswered questions, as well as potential adverse environmental consequences. For example, extraterrestrial disposal raises the question of aborted rockets and potential dissenination of the wastes in the earth's atmosphere. Shooting wastes into the sun or even outer space could result in unknown consequences. Even orbiting radioactive wastes carry the potential for reentry with potentially serious environmental implications.

Polar ice sheet disposal would be accomplished in the thick permanent ice layers near either pole. Specific concepts include meltdown of the wastes into the polar ice sheet, where they ultimately would end up on solid bedrock. Another involves anchored emplacement of the waste bearing canisters, which could be recovered at some future time if desired. The simplest concept is the construction of a surface disposal facility which would take advantage of the cold temperatures and remote location of the site. The ice sheet disposal plan has many drawbacks, some technical, others economic and political. A major political limitation is an international treaty which requires the continent of Antarctica to be kept free of nuclear wastes. Technical limitations include inadequate geological knowlege of the icecaps themselves and the underlying geological strata.

Artificial transmutation of nuclear wastes has been suggested as a means of eliminating the high toxicity long lived actinide components. Transuranics would be partitioned from the wastes and subjected to neutron bombardment which would ultimately fission them or their activation products, producing shorter lived less toxic fission products. The concept, while interesting, would require expensive and difficult separation of the TRU material from the wastes.

For the present, land disposal or interim storage of high level nuclear wastes appears to offer the safest and technologically soundest available means. After much delay and political wrangling, the United States officially selected the land disposal option with the passage of a bill by the 97th Congress that was signed by the President in January 1983. The new law requires the Secretary of Energy to recommend three sites to the President by January 1, 1985, who would then select one of them by 1987. Any state chosen to house the site could exercise a veto over the selection, but this veto could be overriden by a vote of both houses of Congress. Five other sites would be nominated by July 1, 1989. To ensure that no waste repository is located near a heavily populated area, the law prohibits any burial facility from being built adjacent to a one square mile area having more than 1000 inhabitants.

References

1. Anderson, R. L. et al. 1978. *Institutional Radioactive Wastes*, USNRC Report NUREG/CR-0028.
2. Code of Federal Regulations (CFR). 1983. Title 10, *Energy*, Part 50, Appendix F.
3. Cohen, J. J., A. E. Lewis, and R. L. Braun. 1972. "*In situ* Incorporation of Nuclear Waste in Deep Molton Silicate Rock", *Nucl. Technol.* 14:76.
4. Duguid, J. O. 1974. "Groundwater Transport of Radionuclides from Buried Wastes: A Case Study at Oak Ridge National Laboratories", in *Proc. Second AEC Environ. Prot. Conf.*, USAEC Report WASH–1332(74), Vol. 1, pp. 511–529.
5. Eisenbud, M. 1981. "The Status of Radioactive Waste Management: Needs for Reassessment", *Health Phys.* 40:429.
6. Fitzgerald, J. J. 1970. *Applied Radiation Protection and Control*, Gordon and Breach, New York, Vol. 2, p. 751.
7. Foster, R. F. 1972. "The History of Hanford and Its Contribution of Radionuclides to the Columbia River", in *The Columbia River Estuary and Adjacent Ocean Waters*, A. T. Pruter and D. L. Alverson, Eds., University of Washington Press, Seattle, pp. 3–18.
8. Gera, F., and D. G. Jacobs. 1972. "Considerations in the Long-Term Management of High-Level Radioactive Wastes", USAEC Report ORNL-4762.
9. Glasstone, S. 1982. "Energy Deskbook", USDOE Report DOE/IR/05114–1.
10. Holcomb, W. F. 1980. "Inventory (1962–1978) and Projections (to 2000) of Shallow Land Burial of Radioactive Wastes at Commercial Sites", *Nucl. Safety* 21:380.
11. International Commission on Radiological Protection (ICRP). 1980. "Limits for Intakes of Radionuclides by Workers", ICRP Publication 30 *Annals of the ICRP* 2(3/4):1; Also supplements in *Annals of ICRP* 3(1–4):1 (1980); 4(3/4):1 (1981); 5(1–4):1 (1981).
12. Joseph, A. B., et al. 1971. "Sources of Radioactivity and Their Characteristics", in *Radioactivity in the Marine Environment"!5, National Academy of Sciences, Washington, D. C., pp. 6–41*.
13. Logan, S. E. 1974. "Deep Self-Burial of Wastes by Rock Melting Capsules", *Nucl. Technol.* 21:111.
14. Magarrell, J., "Universities Face New Curbs on Dumping Radioactive Wastes", *Chronicle in Higher Educ.* (May 29, 1979), p. 1.

15. National Academy of Sciences. 1962. "Disposal of Low-Level Waste into Pacific Coastal Waters", NAS-NRC Publication 985, Washington, D. C.
16. National Academy of Sciences. 1975. *Interim Storage of Solidified High Level Radioactive Wastes*, National Academy of Sciences, Washington, D. C.
17. Overcamp, T. J., "Low-Level Radioactive Waste Disposal by Shallow Land Burial" in *Handbook of Environmental Radiation*, A. W. Klement, Jr., Ed., CRC Press, Boca Raton (1982), pp. 207–267.
18. Silverman, L. B. and R. K. Dickey. 1956. "Reduction of Combustible, Low-Level Contaminated Wastes by Incineration", U.S. Atomic Energy Commission Report UCLA–368.
19. United States Department of Energy (USDOE). 1978. "Report of the Interagency Review Group on Nuclear Waste Management", Report No. TID–29442.
20. United States Department of Energy (USDOE). 1981. "Spent Fuel and Radioactive Waste Inventories and Projections as of December 31, 1980", Report No. DOE/NE–0017.
21. United States Nuclear Regulatory Commission (USAEC). 1973a. "The Safety of Nuclear Power Plants (Light Water Cooled) and Related Facilities", Report No. WASH–1250.
22. United States Nuclear Regulatory Commission (USNRC). 1973b. "Numerical Guides for Design Objectives and Limiting Conditions for Operation to Meet the Criterion 'As Low As Practicable'", Report No. WASH–1258.
23. United States Nuclear Regulatory Commission (USNRC). 1978. "Generic Environmental Impact Statement in Handling and Storage of Spent Fuel from Light-Water Power Reactors", ReportNo. NUREG–0404.
24. United States Nuclear Regulatory Commission (USNRC). 1981a. "1981 Annual Report", Washington, D. C.
25. United States Nuclear Regulatory Commission (USNRC). 1981b. "Draft Envionmental Impact Statement on 10 CFR Part 61 "Licensing Requirements for Land Disposal of Radioactive Waste", Report No. NUREG–0782.
26. Wegele, A. E. 1980. "Radioactive Waste Management at the Hanford Reservation", *Nucl. Safety* 20:434.

CHAPTER 9

Atmospheric Transport

Composition of the Atmosphere

By volume, air consists of 78.1% nitrogen and 20.9% oxygen, with 0.934% argon, 0.033% carbon dioxide, and lesser or trace amounts of numerous other gases. The total mass of the atmosphere is about 5.1×10^{18} kg distributed over the 5.1×10^{14} m^2 surface area of the earth. This produces an atmospheric pressure of 1000 g/cm^2 (10^4 kg/m^2) at sea level, which corresponds to 0.1 MPa or the more commonly used 760 mm Hg or 14.7 lb/in^2. Pressure decreases sharply with altitude, such that at 50 km above the surface of the earth it is only 0.001 that at the surface, further decreasing by a factor of 1000 at an altitude of 100 km (Figure 9–1). The density of air is approximately 0.001293 g/cm^3 at the surface of the earth.

Typically, the atmosphere contains 1.5×10^{17} kg of water vapor. Water vapor thus comprises about 3% by weight of the atmosphere. The amount is highly variable, and dependent on the temperature of the air, which determines its capacity to hold water. Water content is important from the standpoint of the specific thermodynamic properties of the air, as well as affecting radon diffusion from the ground and the removal of particulate radioactivity from the air by precipitation mechanisms.

The atmosphere has several relatively well defined regions (Figure 9–2), of which only the two lowest are of significance from the standpoint of transport of environmental radioactivity. The bottommost region is the troposphere, which extends to an altitude of about 11 km (7 miles) in the temperate zone, thinning to about 8 km (5 miles) over the poles and increasing to nearly 20 km over the equatorial regions. The height also varies with season to some extent. The troposphere contains about three-fourths of the atmosphere by weight and virtually all the moisture and pollution. Most of the weather occurs in this region. Typically, temperature in this region normally decreases with altitude at a rate of about

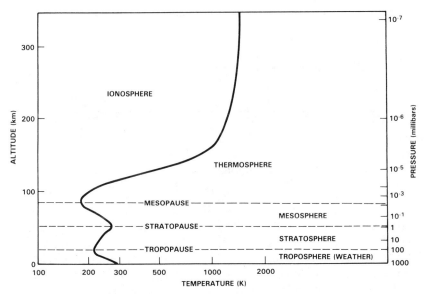

FIGURE 9–1   Atmospheric Temperature and Pressure.

−6.5 °C/km (−3.5 °F/1000 ft). This phenomenon is known as the temperature lapse rate, and is of great importance in atmospheric mixing and diffusion.

The troposphere is separated from the next higher atmospheric region, the stratosphere, by an imaginary boundary known as the tropopause. The stratosphere ranges from an altitude of about 11 to 32 km, and is an isothermal layer with a temperature of about −55°C. The stratosphere is dry and cloudless, with westerly winds in the winter in the northern hemisphere and easterly winds in the summer. It is in the stratosphere that meteors burn, and ozone is formed. The stratosphere is of great importance in that nuclear weapons testing has injected large quantities of long lived nuclides into it, and these settle back to earth, carried by stratospheric circulation patterns. Above the stratosphere lies the mesosphere or ozonosphere, which ranges in altitude from about 30 to 80 km. The temperature in this layer increases for a few km, levels off, and then decreases. At the mesopause it reverses again in the outermost region of the atmosphere, the ionosphere which is a strongly ionized region extending to perhaps 4000 km.

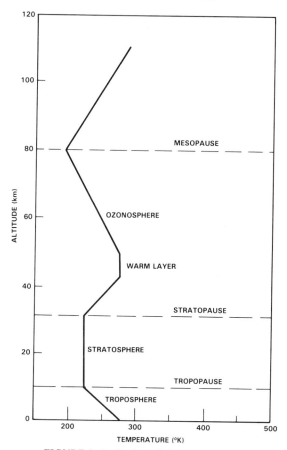

FIGURE 9-2   Regions of the Atmosphere.

General Global Circulation

Globally, air circulation is determined by the heat balance of the earth. Air near the equator is heated and rises, while air near the poles loses heat and sinks. This creates two convective cells, one each in the northern and southern hemispheres lying between the equator and the poles. Air movement in the two cells is modified by centrifugal and Coriolis forces. Thus, in the north temperate latitudes, surface winds generally blow from

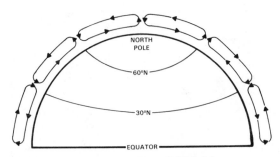

FIGURE 9-3   Three Cell Model.

southwest to northeast, with the direction being reversed in the southern temperate latitudes. This simple model of generalized global air circulation was first put forth in 1735 by George Hadley, and the convective cells are sometimes known as Hadley cells.

A more complete yet still highly simplified model of global air circulation reveals six bands of prevailing winds over the surface of the earth (Figure 9-4). The prevailing wind patterns are most noticeable over the oceans where heating and cooling are fairly uniform and the air movement is not affected by large landforms such as mountain ranges. The three prevailing wind bands in each hemisphere are explained in terms of a three cell model (Figure 9-3). As was the case with the original Hadley model, the warm air at the equator rises and the cold air at the pole sinks. In between, however, at approximately 30°, these cells have downward moving air, and at about 60° latitude, upward moving air. The midlatitude cells lying between 30 and 60° have warm air moving toward

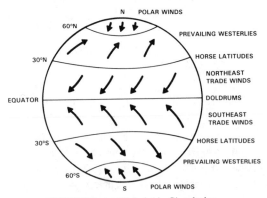

FIGURE 9-4   Global Air Circulation.

the poles at the surface of the earth rather than aloft as in the other two cells or in the simple Hadley model.

The three cell model, first proposed in 1941 by Rossby, provides a reasonably good explanation of the prevailing surface winds of the earth. It does not, however, account for other wind phenomena, including the jet streams and the generally west to east moving zones of low and high pressure that constitute weather systems. The jet streams are high altitude prevailing west to east winds that flow in well defined streams or rivers at speeds frequently exceeding 100 knots. There are two major streams, the subtropical jet stream which is found at an altitude of 30,000 to 40,000 feet (50 to 70 km) in the latitude 30 to 35°, and the circumpolar jet stream, found at an altitude of about 18,000 feet (30 km) at latitudes much further from the equator. Clearly these high speed winds have significant implications for dispersal of radioactive debris from nuclear weapons testing.

Locally, air movement near the surface is determined by ground contours, wind speeds imposed by global and local weather conditions, and local temperature gradients. Important to local wind movement are obstructions, such as hills or mountains, or even large solid objects such as buildings. An obstruction may also be a denser cooler air mass. Less dense warmer air, including any entrained pollutants, encountering such a mass may be lifted up over the cooler mass, which acts as a wedge. This effect of a colder dense air mass is particularly pronounced in a downhill flow situation, as shown in Figure 9–5. The warmer pollutant laden air

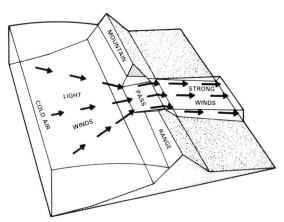

FIGURE 9–5   Downslope Circulation: Cool Air Mass Wedge Effect. (Redrawn from Slade 1968).

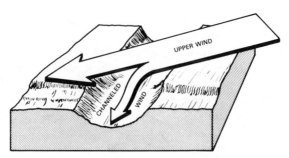

FIGURE 9-6   Channeling. (Redrawn from Slade 1968).

flowing down the slope is diverted or wedge upward by the colder and
more dense air mass lying in the valley. The effect of obstructions is
extremely important, especially for prediction of atmospheric dispersion.
For example, if a mass of air was moving in the downslope direction, and
a radioactivity concentration and wind velocity measurement were made
by an observer at point A, an erroneous prediction of what would occur
at points B and C might be made if the effect of the cold air mass lying
downslope was ignored.

Channeling may occur when the wind pattern is across a valley but not
at right angles to it (Figure 9-6). The far wall of the valley acts as an
obstruction to turn the surface wind down the valley, while the upper
winds continue in the same general direction. A moving air mass will be
subject to changes in shape or deformation as it encounters obstacles or
obstructions within its path. Thus, a mass of air moving between two hills
will necessarily narrow (Figure 9-7) and increase in velocity; increased

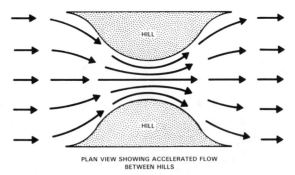

PLAN VIEW SHOWING ACCELERATED FLOW
BETWEEN HILLS

FIGURE 9-7   Deformation by Passage between Hills. (Redrawn from Slade 1968).

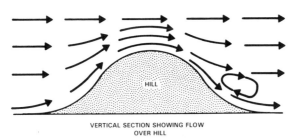

VERTICAL SECTION SHOWING FLOW
OVER HILL

FIGURE 9-8   Passage Over Hills. (Redrawn from Slade 1968).

velocity may provide greater carrying capacity and increased mixing of radioactivity and other pollutants. Passage of winds through mountain passes is an exaggerated case of this situation, and may produce strong downslope winds on the opposite side of the pass. Thus, light winds channeled through a mountain pass may become very strong winds on the other side of the pass, and is typical of the Santa Ana winds of Southern California or the Chinook (Columbia River gorge) winds in the Pacific Northwest. Similar deformation effects are noted with air masses moving across hills and valleys (Figures 9-8, 9-9). In the case of the hill, velocity is increased, with the converse being true for winds moving across a valley.

Wind roses are a graphical means of showing wind distribution over a long period of time. The frequency of winds of a given velocity are shown in the direction from which they occur, as shown in the example (Figure 9-10). The per cent calm is given in the center of the wind rose. Data can be presented on wind roses in a variety of ways; the total winds over a period of years may be summarized, or winds over a year or a season,

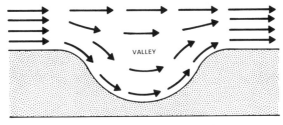

VERTICAL SECTION SHOWING DECELERATED
FLOW ACROSS A VALLEY

FIGURE 9-9   Passage Over Valleys.

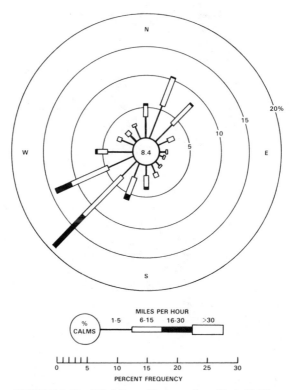

FIGURE 9-10   Wind Rose. (Redrawn from Slade 1968).

or a plot made of both day and nighttime winds can be shown on the same wind rose (Figure 9-11). Wind roses are excellent synoptic tools for wind prediction, and hence find application in evaluation of routine planned releases of radioactivity as well as theoretical prediction of accident consequences.

Two special local wind situations bear mention: the sea or lake breeze and valley slope circulation. In the sea or lake breeze situation, the stronger daytime heating of air over land leads to reduced density and rising of air, allowing the inflow of cooler air from over the water. In temperate latitudes, this occurs most frequently during spring and summer. The sea breeze ordinarily begins early in the day, with relatively low speed winds, and achieves its maximum in early afternoon. Speed and depth are reduced as the sea breeze moves inland; in general, the sea breeze only has a depth of a half mile or so from the shore. The land

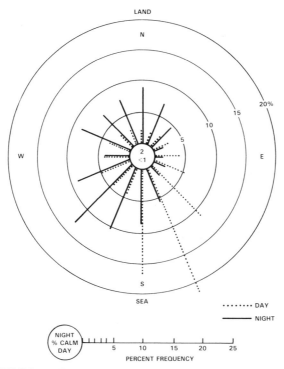

FIGURE 9–11   Day-Night Wind Rose. (Redrawn from Slade 1968).

breeze is the inverse or reverse effect from the sea breeze, beginning a few hours after sundown and resulting from the more rapid cooling of the land than the water body. The land breeze is usually not as pronounced as the sea breeze.

An obvious implication of the sea breeze relates to ambient air concentrations of radon and daughters. As radon emanation is much less over water, the sea breeze effectively dilutes or reduces concentrations over the affected land. Thus a location close to a large body of water may show larger diurnal variation and considerably lower daytime levels than a location a short distance inland, even though the two locations may have identical geology and soil emanation.

Valley-slope circulation is most frequent in mountainous areas, but can occur with small slopes also (Figure 9-12). During the day, air over the slope will be warmer than air at the same height over the valley, producing an upslope wind. The reverse is true at night. Valley mixing may

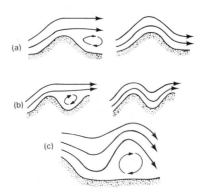

FIGURE 9-12   Valley Overturning. (Redrawn from Slade 1968).

occur in the situation where the oblique rays of the sun heat one side of the valley while the other side stays relatively cool, which may produce an overturning or generalized mixing of air in the valley.

Atmospheric Mixing and Turbulence

Turbulence, defined in the *Glossary of Meteorology* as ". . . a state of fluid flow in which the instantaneous velocities exhibit irregular and apparently random fluctuations so that in practice only statistical properties can be recognized and subjected to analysis" (Huschke 1955), is largely responsible for atmospheric mixing. As implied by the definition, turbulent velocity or air flow may be considered a randomly variable motion superimposed on the mean motion, and can thus be mathematically described in terms of a three dimensional coordinate system. Mathematical analysis of turbulent flow, however, is very difficult and complex, and for this reason classical hydrodynamic analyses have tended to ignore turbulence and to deal with smooth or laminar flow.

According to kinetic theory, the molecules of a gas are in constant random motion, with the energy of this motion determined by the temperature of the gas. Thus, if two different gases are brought together into a single container, mixing will occur as a result of the normal molecular motion; obviously, the higher the temperature, the more rapidly will the mixing occur. This natural mixing phenomenon is known as diffusion, a process that occurs in the atmosphere, although at a relatively slow rate in comparison to certain other mixing phenomena.

A major contributor to local mixing is eddy diffusivity. Eddies are individual motion systems that result from the transition of laminar flow to turbulent flow. Thus, a mass of laminar flowing air will produce eddies when encountering an obstacle such as a building, tree, or change in topography. Obstacles produce mechanical eddies by breaking the smooth flow of air. Hence, the greater the roughness of the ground surface, (produced by vegetation, uneveness of terrain, or the presence of buildings), the more eddies and therefore the greater mixing or dispersion of a pollutant such as radioactivity in the atmosphere.

Convection may also produce eddies and mixing, in addition to mechanical transport, and results from bubbles of heated air produced by uneven heating of the ground surface. These bubbles of less dense heated air rise, carrying with them entrapped pollutants and, because of their mechanical motion, create eddies and mixing as they move upward. Convection also occurs when a heated plume is released, resulting in a rise of the material and greater dispersion than would otherwise occur, or, it may bring the material to the ground in greater concentrations than otherwise expected. Turbulence and mixing may be produced by other phenomena, including the mixing or joining of masses of air of different temperatures. Gustiness, which is simply a random and usually rather abrupt variation in the force of the wind is physical evidence of eddy motion. The eddy velocity, and hence energy attributable to eddies, is simply the total wind velocity less the mean velocity.

The mean path of an eddy over which the eddy is maintained is the mixing length and is thus analogous to the mean free path of molecules in a gas. This permits a statistical treatment of turbulence, with the following six statistical properties (Neuberger, Panofsky and Sekera 1956):

1. Turbulent energy is greatest in unstable air, and least in stable air.

2. Horizontal turbulent energy is somewhat greater than vertical turbulent energy.

3. In a stable atmosphere, turbulent variations are of short duration, generally on the order of a few seconds or less. In an unstable atmosphere, longer durations, on the order of minutes, may occur as a result of convection.

4. Turbulence increases with altitude as a result of reduced ground damping; hence longer time periods for turbulent fluctuations are of more significance at higher elevation.

5. Turbulent energy increases with wind speed.

6. Turbulence is greater over rough ground than over smooth ground.

Transport and diffusion of material in the atmosphere depends on other factors than eddy diffusion. Wind velocity will affect dilution as well as determining the direction that the material will take. Thermal turbulence results from atmospheric stability, and is common on clear nights with light winds, when heat is radiated from the surface of the earth with resultant cooling of the ground and the air above it. Indeed, the extent of diffusion of material released from stacks or other point sources is primarily a function of atmospheric stability.

Atmospheric stability can be described by the change in temperature with altitude, known as the lapse rate or temperature profile. Atmospheric processes are ordinarily assumed to be adiabatic, that is, the temperature of a volume of air in motion rises or falls without an exchange of heat across its boundaries. Thus, the dry adiabatic lapse rate is that rate of temperature decrease with altitude that permits a parcel of air to move upwards or downwards such that it always has the same density as its environment; in other words, there is no exchange of heat with the surroundings. The value of the dry adiabatic lapse rate is -10 °C/km (-5.4 °F/1000 feet).

If the lapse rate exceeds adiabatic it is said to be superadiabatic, and exchanges heat with its surroundings. Thus, if rising, a parcel of air will have lower density than the surrounding air and thus continue to rise; conversely, if sinking, it will have greater density and continue to sink. Under superadiabatic conditions, vertical movements are accelerated and the atmosphere is unstable, perhaps with considerable windiness and unpredictability. If the lapse rate is less than adiabatic, the atmosphere is stable, and the parcel of air will tend to return to its original level. A subadiabatic lapse rate is known as an inversion, although this term is sometimes reserved for those situations in which the lapse rate is positive—i.e. the temperature increases with altitude. If the temperature does not change with altitude, the lapse rate is said to be isothermal. There is considerable diurnal variation in the lapse rate; at midday, it is ordinarily superadiabatic. At sunset, the ground cools rapidly, resulting in an intense inversion. At sunrise, the ground heats, causing the inversion to dissipate or burn off. Obviously, this generalized description is greatly simplified, but is nonetheless useful to the understanding of the factors that control the lapse rate and hence atmospheric stability.

The stability of the atmosphere is of primary import in determining the diffusion of material released from a stack. Six types of plume behaviour have been identified based on atmospheric stability (Church 1949,

Hewson 1960). These are (Figure 9-13)
1. looping
2. coning
3. fanning
4. lofting
5. fumigation
6. trapping

Looping occurs when the lapse rate is large and hence superadiabatic.

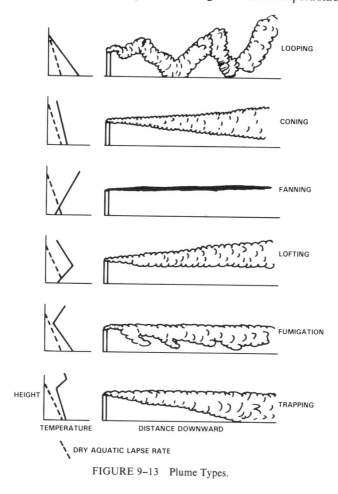

FIGURE 9–13   Plume Types.

This results in unstable air in which large thermal eddies develop resulting in the plume transitorily touching the ground at various points along its travel. The superadiabatic conditions necessary for looping occur only with strong solar heating of the ground and light wind. Cloud cover and high winds preclude the formation of such unstable conditions.

With a smaller but still superadiabatic lapse rate, coning will result. Coning occurs when the lapse rate lies between the dry adaibatic lapse rate and isothermal. This results in moderately or slightly unstable atmospheric conditions with both horizontal and vertical mixing, but of an intensity less than with looping. Coning is common under cloudy and windy conditions. Atmospheric diffusion models are most sucessfully applied to the coning type of plume or moderately unstable atmospheres.

When inversion conditions prevail, vertical turbulence and mixing are suppressed. Horizontal mixing, although somewhat reduced, still occurs giving rise to a fan shaped plume, which is essentially equivalent to a much flattened cone. Under inversion conditions, winds are usually light, and some horizontal meandering of the plume is not uncommon. As the plume is relatively small in volume, concentrations within it are relatively great, but unlike the looping plume that occurs with an unstable atmosphere, or even the coning type of plume, the fanning plume from an elevated source such as a stack does not touch the ground unless the inversion is broken as might occur from surface heating of the ground, or a hill or other sloping surface downwind is at the altitude of the fan. Fanning is most common encountered during clear nights with light winds.

If a low lying or surface inversion is overlain with a superadiabatic layer of air, the characteristic lofting plume will occur. In lofting, there is upward diffusion primarily. Downward diffusion is stifled because of the inability of the plume to penetrate the inversion. Lofting is most likely to be observed when the release is at or near the interface of the two lapse rates, and is usually observed near sunset under clear skies in open country. As the evening progresses, the inversion deepens, and lofting gives way to fanning; thus, lofting is ordinarily a transitory situation.

The inverse of lofting is fumigation, which occurs when a lapse is overlain by an inversion. This is a contributing cause of the classic smog of Los Angeles. Such a situation might occur during the day when the lower layers of the atmosphere are heated by the sun, producing thickening of the superadiabatic layer. When this layer reaches the elevation of the plume (which will be of the fanning type as previously an inversion existed), thermal turbulence occurs, bringing portions of the high concentration fanning plume to the surface. This situation is favored by clear daytime skies with good insolation and light winds; thus it is more likely to take place in summer.

Fumigation may also result in the early evening from the heat island effect of cities, which causes the air over them to rise and be replaced by the lower layers of stable air from the adjacent countryside where inversions are beginning to form. This, coupled with mechanical turbulence from surface roughness, produces a lapse in the lower layers of the air over the city which will persist until sufficient heat is lost from the city such that the lapse can no longer be maintained.

The final plume type is known as trapping, and is characteristic of an inversion overlying weak fumigation conditions. In this situation, a plume released below the inversion will be trapped below it, forcing it down to the ground surface. In many respects, trapping is similar to fumigation, and is sometimes considered a special case of the latter.

Winds, micrometeorological conditions, certain specific characteristics of the area such as the presence of nearby obstructions, and the temperature of the plume may cause deviation from the stylized description of plume shape and behavior described above. More importantly, there are diurnal variations attributable to changes in the lapse or inversion conditions as the day progresses. For example, on a clear summer or spring day, the ground level concentration at a point downwind from an elevated source continuously emitting a constant amount of radioactivity would begin to rise shortly after sunrise, peaking about mid-morning, largely from fumigation. The concentration would drop relatively rapidly, reaching a minimum about mid-afternoon due to the slight increase in atmospheric stability during the period of maximum heating. As the afternoon progresses, the lapse decreases giving rise to an evening inversion, accompanied by a rise in the concentration. Over a city, this rise will be more pronounced, and produces a secondary maximum in the early evening hours due to the heat island fumigation effect previously described.

Atmospheric Diffusion

To a great extent, diffusion of radioactivity or any pollutant or foreign material injected into the atmosphere near the earth's surface is determined by atmospheric stability. Several categorizations of stability have been proposed and are summarized in various works (Hanna, Briggs, and Hosker 1982; Gifford 1976; Slade 1968). Perhaps the simplest is to

TABLE 9-1   Atmospheric Stability Categories

Stability Category	Condition	Description	Wind Speed (m/s)	(deg)	Lapse Rate*
A	Highly unstable	Sunny summer weather	1	25	−1.9
B	Moderately unstable	Sunny and warm	2	20	−1.9 to −1.7
C	Slightly unstable	Average day	5	15	−1.7 to −1.5
D	Neutral	Overcast day or night	5	10	−1.5 to −0.5
E	Slightly stable	Average night	3	5	−0.5 to 1.5
F	Moderately stable	Clear night	2	2.5	1.5 to 4.0
G	Highly stable		—	1.7	>4.0

*°C/100 m.

describe the atmosphere as being either stable (inversion) or unstable (lapse), and for many situations this will suffice. A more widely used and practical scheme was empirically developed by Pasquill (Pasquill 1961, 1974) and later modified slightly (Turner 1967; Gifford 1976). In the Pasquill classification scheme (Table 9-1), seven atmospheric stability classes represented by the letters A-G and ranging from extremely stable to extremely unstable are put forth. These categories can be correlated with the standard deviation of the lateral wind direction distribution, σ_θ, which defines the spread of the plume in the horizontal direction and is more or less independent of downwind distance from the point of release. The categories can also be correlated with diffusion coefficients in both the horizontal and vertical directions as a function of distance downwind (Figures 9–14, 9–15).

The early development of atmospheric diffusion theory took place about 1920 and was largely the work of British mathematician Sir Geoffrey I. Taylor and Austrian meteorologist Wilhelm Schmidt who represented atmospheric motion as continuously variable, and who considered the movement of a particle with time (Taylor 1921, Schmidt 1925). This led to the independent derivation of the following theoretical differential equation by Schmidt and L. F. Richardson (Schmidt 1925; Richardson 1926)

$$\frac{d\chi}{dt} = \frac{\partial}{\partial x}\left(K_x \frac{\partial\chi}{\partial x}\right) + \frac{\partial}{\partial y}\left(K_y \frac{\partial\chi}{\partial y}\right) + \frac{\partial}{\partial z}\left(K_z \frac{\partial\chi}{\partial z}\right) \qquad (9-1)$$

in which χ is the concentration at any time t in the three dimensional coordinate system defined by x, y, and z.

K_x, K_y, and K_z are the coefficients of diffusion in the x, y, and z directions; these were termed 'austachkoeffizienten' or exchange coefficients by Schmidt. Values of these coefficients are not known and hence an exact solution of Equation 9–1 is not possible. It is known, however, that these coefficients may have a wide range of values, from 0.2 cm^2/sec for molecular diffusion to perhaps 10^{11} for very large storms (Richardson 1926).

Equation 9–1 is the general case of diffusion in three dimensions and was based on Fourier's law of heat conduction which had been applied by the German physiologist Adolph Fick to diffusion in 1855 (Fick

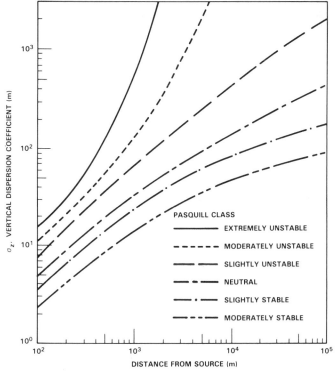

FIGURE 9–14   Vertical Diffusion Coefficients.

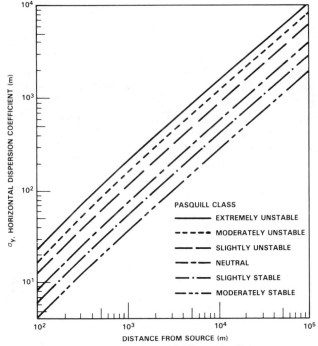

FIGURE 9–15   Lateral Diffusion Coefficients.

1855). Basically, what has come to be known as Fick's Law of Molecular Diffusion states that diffusion is in the direction of decreasing concentration and is proportional to the concentration gradient. Mathematically this is represented by Fick's Equation

$$\nabla^2\chi = \frac{1}{d}\frac{dx}{dt} \qquad (9-2)$$

in which d is the molecular diffusivity coefficient. Equation 9–1 is the general case of Fick's Equation, and if the K_x, K_y, and K_z are constants, the diffusion is termed Fickian. Several atmospheric diffusion theories have been developed, largely based on the fact that the Gaussian distribution function provides a fundamental solution to the Fickian diffusion equation.

Although much theoretical work has been done in developing mathematical models of atmospheric diffusion, most of the equations in use today are based on empirical observations. O. G. Sutton began his observations and analyses of smoke emissions from coal plants in the 1920's and continued them for more than three decades. From this work, Sutton evolved a model of averaged plume diffusion which in turn led to the well known Sutton Equations for calculation of atmospheric concentration under a variety of conditions. These equations are based on a model which assumes a Gaussian distribution in both the horizontal and vertical direction.

For a ground level release from a point source such as a ventilation stack, the generalized Gaussian diffusion equation has the form

$$\chi_{(x,y,z)} = \frac{Q}{\pi \sigma_y \sigma_z \bar{u}} \exp - \frac{1}{2} \left[\frac{y^2}{\sigma_y^2} + \frac{z^2}{\sigma_z^2} \right] \tag{9-3}$$

in which $\chi_{(x,y,z)}$ is the concentration at the point represented by x, y, and z, and σ_y and σ_z are the standard deviations of the plume width and height, respectively. The generalized Gaussian diffusion equation can be applied to a variety of situations, as shown in the following series of word equations, in which the following abbreviations are used:

MWS = Mean Wind Speed
HPS = Horizontal Plume Spread
VPS = Vertical Plume Spread
EEH = Effective Emission Height
DDR = Downwind Distance of Receptor

General Case

Downwing
Concentration = $\dfrac{\text{Source Emission Rate}}{\pi[\text{MWS}][\text{HPS}][\text{VPS}][\text{EEH}][\text{DDR}]}$
(point source)

Special Cases: Downwind Concentration

Ground Level Source
and Receptor on
Plume Center = $\dfrac{\text{Source Emission Rate}}{\pi[\text{MWS}][\text{HPS}][\text{VPS}]}$
Line

Elevated Source with
Ground Level
Receptor on =
Plume Center $\dfrac{\text{Source Emission Rate}}{\pi[\text{MWS}][\text{HPS}][\text{VPS}][\text{EEH}]}$
Line

Elevated Source with
Ground Level
Receptor not on =
Plume Center $\dfrac{\text{Source Emission Rate}}{\pi[\text{MWS}][\text{HPS}][\text{VPS}][\text{EEH}][\text{DDR}]}$
Line

The word equations are useful to gain a general understanding of the various meteorological factors affecting atmospheric diffusion. Inspection of the examples given shows that downwind concentration is in all cases directly proportional to the rate of emission, and inversely proportional to the mean wind speed and horizontal and vertical plume spread. For an elevated release, as from an exhaust stack, the effective emission height is important, and as indicated by the approporiate word equations, is inversely proportional to the downwind concentration at any point. This, perhaps, is intuitive in that the higher the release point, generally speaking, the lower the concentration at a given point on the ground downwind. However what is not so readily apparent is that changing the stack height also changes the downwind distance at which the maximum ground level concentration occurs, as can be seen from the figures in Appendix B. Indeed, it is possible to raise the stack and have the concentration increase at a given point downwind, although in general ground level concentrations will be lowered.

Perhaps somewhat surprisingly the downwind distance is only of importance when the receptor is not on the centerline of the plume. The assumption of a Gaussian plume inherently produces the classical ellipsoid shaped plumes or fallout patterns in which the shortest linear distance that a given concentration will be found from the point of release lies on the center line. Thus, the horizontal and vertical plume spread factors determine where this point will be. In other words, the general case equation calculates the maximum downwind concentration that will occur at some distance downwind from the point of release, and this point will always lie on the plume center line.

The word equations shown above apply only to point source releases over a period of time. However, the Gaussian type diffusion equations can be applied to many other situations, including instantaneous releases, and emission from other source geometries such as lines and areas. The equation for downwind concentration from an area source has the same form as the general case except that the horizontal plume spread factor must be appropriately adjusted to account for the areal source dimensions. For an infinite line source, the downwind concentration is inversely proportional to $\sqrt{2\pi}$ (instead of π) and to the mean wind speed and vertical plume spread.

The actual diffusion equations are somewhat more complex. Equation 9–4 can be used to calculate the ground level concentration from a continuous point source release:

$$\chi_{(x,y,z)} = \frac{Q}{\pi \bar{u} \sigma_y \sigma_z} \exp \left[\frac{h^2}{2\sigma_z^2} + \frac{y^2}{2\sigma_y^2} \right] \tag{9–4}$$

in which χ = the concentration, usually in units of Ci/m^3 or g/m^3 at a downwind point described by the coordinates x, y, and z,

Q = the source emission rate in Ci/s or g/s,

σ_y = the crosswind (i.e. horizontal) plume standard deviation, in meters,

σ_z = the vertical plume standard deviation, in meters,

$\bar{u}$ = the mean wind speed, in m/s, at the point of release

h = the effective stack height, in meters,

and x and y = the downwind and crosswind distances, respectively, in meters.

Equation 9–4 assumes total reflection of the plume at the earth's surface, which is valid if downwind diffusion is small as would be the case if the duration of the release were equal to or greater than the travel time to the point of interest.

The coefficients for both horizontal or crosswind plume spread (σ_y) and vertical plume spread (σ_2) are a function of the downwind distance x, and are thus in units of meters. These coefficients vary according to atmospheric stability and such factors as wind shear and ground roughness. The vertical plume spread also increases with distance, this being determined in part by vertical mixing. Values for these coefficients must be obtained from actual measurement, or can be estimated from Figures 9–14 and 9–15.

When y = 0, Equation 9–4 will calculate the center line concentration. If the release is at ground level, z = 0 also and Equation 9–4 reduces to

$$\chi_0 = \frac{Q}{\pi \bar{u} \sigma_y \sigma_z} = \frac{0.318Q}{\bar{u} \sigma_y \sigma_z} \qquad (9\text{–}5)$$

A plot of center line concentration as a function of distance can be readily obtained by manual calculation of a few points with the aid of Equation 9–5.

Calculation of the maximum radioactivity concentration at a point on the ground downwind from an elevated release point such as a stack is of importance in many environmental radioactivity studies, and can be accomplished by

$$\chi_{max} = \frac{2Q}{e\pi \bar{u} h} \cdot \frac{\sigma_z}{\sigma_y} \qquad (9\text{–}6)$$

in which the terms are defined as for Equation 9–4. As a rule of thumb, the downwind distance from the stack to the point of maximum concentration is usually in the range of 15–30 times the stack height. It is at this point that the plume first touches the ground.

An important consideration, mentioned in the definition of terms under Equation 9–4, is the effective stack height, h, which is simply the sum of the actual stack height, h_s and a term h to correct for plume buoyancy. Plume buoyancy results from the heat content and upward velocity component of the stack effluent. Several empirical or semi-empirical equations have been developed to calculate h (Slade 1968; Hanna, Briggs and Hosker 1982). Among the more commonly used are the Davidson-Bryant formula

$$\Delta h = d \left(\frac{W}{u}\right)^{1.4} \left(1 + \frac{T - T_s}{T_s}\right) \qquad (9\text{--}7)$$

in which d is the inside diameter of the stack, w the effluent velocity, and T and T_s the absolute temperature of the ambient air and stack effluent, respectively. Note that this model gives a negative value for effluents discharged at temperatures below that of ambient, thereby reducing the effective stack height.

Also in common use are empirical equations developed by Briggs and Holland. This Briggs equation is

$$\Delta h = \frac{1.6 F^{1/3} x^{2/3}}{\bar{u}} \qquad (9\text{--}8)$$

in which x is the downwind distance from the stack and F is the buoyancy coefficient which is calculated from.

$$F = 3.7 \times 10^{-5} QH \qquad (9\text{--}9)$$

in which Q_H is the heat emitted from the stack in units of calories per second. This equation is valid out to a distance of about 10 stack heights. Holland's equation

$$\Delta h = \frac{1.5wd + 4.5 \times 10^{-5} Q_H}{\bar{u}} \qquad (9\text{--}10)$$

takes into account additional factors such as the stack diameter, d, and the stack exit velocity, w. Units are in meters and seconds. Note that the additional height gained is inveresely proportional to the wind speed, which in a calm is zero, suggesting that the stack will have infinite or indeterminate height. In practice, however, this mathematical difficulty is resolved by assuming the wind speed is never less than 1 m/s.

More recently, Briggs proposed a series of plume rise equations to predict the final rise in the plume, taking into account various atmospheric conditions, as shown:

Neutral Atmosphere

$$\Delta h = \frac{400F}{\bar{u}^3} + 3r\frac{w}{\bar{u}} \qquad (9-11)$$

Stable Atmosphere with Wind

$$\Delta h = 2.6\left(\frac{F}{\bar{u}s}\right)^{1/3} \qquad (9-12)$$

Stable and Calm Atmosphere

$$\Delta h = 5.1F \quad S \qquad (9-13)$$

In these equations, the buoyancy coefficient, F, has units of m^4/s^3, and r is the inside radius of the stack. The stability parameter s is expressed by

$$s = \frac{g}{T}\left[\frac{dT}{d_z} + 9.8\right]$$

in which dT/dz is the lapse rate in units of °C/km, g the gravitational constant, and T the absolute ambient air temperature. The equation of choice is the one that gives the minimum value for Δh in a given situation.

The above equations for plume rise apply to a continuous source. For an instantaneous source, the following equation applies:

$$\Delta h = 2.66\left[\frac{Q_i}{C_p\rho\frac{d\theta}{dz}}\right]^{1/4} \qquad (9-14)$$

in which Q_i is the instantaneous heat release in calories, C_p the specific heat at constant pressure of the ambient air, σ the density of the air, and $\frac{d\theta}{d2}$ the potential temperature gradient in degrees Kelvin per meter.

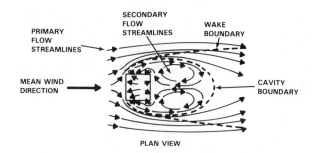

PLAN VIEW

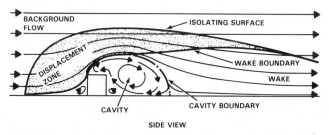

SIDE VIEW

FIGURE 9–16   Flow Around Buildings. (Adapted from Slade 1968).

Atmospheric diffusion is also affected locally by buildings and ground roughness factors. Streamlines of air flow are broken by buildings creating turbulence and hence greater mixing; flow is better around rounded rather than boxlike structures. Downwash of radioactive contaminants in the plume may result in the lee of buildings. In general, there are three distinct zones formed when a plume encounters an obstruction such as a building (Figure 9–16). The cavity zone may have secondary flow in the form of a vortex. Correction for wake effects can be made by

$$\chi = \frac{Q'}{Au} \qquad (9-15)$$

in which A is the projected area of the building in m^2, Q' the flow past the structure in Ci/s, u the wind speed in m/s, and x the concentration in the wake.

Particulates are subject to gravitational settling, with the velocity determined by Stokes Law

$$V_s = \frac{2gr(\rho_p - \rho_{air})}{9\eta} \qquad (9-16)$$

in which g is the gravitational constant, r the particle radius, ρ_p and ρ_{air} the densities of the particle and air, respectively, and η the viscosity of air. Stokes Law holds for particle diameters from 1–2 μm to 400 μm. For very small particles, typically those with diameters smaller than 1 μm, molecular motion offsets the gravitational force, thus leaving these particles suspended in the plume.

Particles are removed from the plume by settling or by impaction where the plume touches the ground. This mechanism is known as dry deposition or fallout. Precipitation scavenging is also an effective removal mechanism for particulate radioactive components from a plume or cloud of radioactivity, and follows the simple exponential equation

$$\chi = \chi_0 e^{-\Lambda t} \qquad (9-17)$$

in which χ is the concentration after rainout, χ_0 the original concentration, Λ the rainout or washout coefficient, and t the time of rainout or washout. The rainout coefficient typically has values on the order of 0.001/s and increases with particle size and rainfall rate.

Entrainment and resuspension mechanisms may add previously deposited material to the atmosphere (Sehmel 1980). Entrainment refers to the general pickup and suspension of particles from the surface by the wind, while resuspension refers to particles that have previously been airborne and deposited, and hence is a special case of entrainment. The term suspension is used to apply to particles, generally <50 μm in diameter, kept airborne for long distances by the force of the wind. In general, only resuspension is of significance, and is expressed in terms of a resuspension factor, RF, which has dimensions of recriprocal length and is simply the quotient of the air concentration and surface contamination level. Reported values of resuspension factors vary over at least 13 orders of magnitude, and are greatly affected by wind velocity and particle size. Typical values are on the order of 10^{-6} to 10^{-5} per meter. The resuspension factor is not a constant in that it appears to decrease with time (Anspaugh et al. 1975). This is thought to be an effect of weathering and is has been described in terms of an exponential decrease (Kathren 1968; USNRC 1980) and is also dependent on particle size (Smith, Whicker and Meyer 1982)

The factors affecting atmospheric diffusion and fallout are many and complex and therefore not accounted for by the simple Gaussian models used for estimating plume concentrations and transport. Among some of the more obvious sources of error are the assumptions implicit in the models that the plume travels in a straight line at constant wind speed, that atmospheric stability is constant over the distance travelled, the plume rise is affected only by heat and velocity, and that the terrain is perfectly flat. To remedy these defects, Gaussian trajectory models have been developed that predict the path of a continuous release of activity as wind velocity and atmospheric stability change with time. The most well known Gaussian trajectory models are the MESODIF computer programs (Powell, Wegley and Fox 1977); these and related models require real time data for wind velocity and atmospheric stability, generally updated on an hourly basis.

Also available are finite difference or grid models which involve numerical solutions of transport and diffusion equations by the method of finite differences. A grid is used to divide the area of interest into a network of constant volume cells, and the net flow of a pollutant is calculated across the cell boundaries taking into consideration turbulence, advection, and various removal mechanisms. Required input for these models is in the form of windfields which must include information re the effects of terrain. Computers are normally required for solution of finite difference models, which are usually applied to short term analysis of chemically reactive air pollutants (Strenge, Kennedy, and Corley 1982).

For non-reactive pollutants such as radioactivity, the particle-in-cell model is more useful, but also requires a computer for solution. Required input includes a definition of the velocity field, including turbulence, for the region of interest. These models need to track a large number of particles in order to obtain a reasonable degree of accuracy, and thus require large amounts of computer time and storage (Strenge, Kennedy, and Corley 1982).

Accurate prediction of atmospheric diffusion is at best hazardous, even with the aid of the most sophisticated and complex computer models currently available. Most models, including those provided in this chapter and in Appendix B, are probably accurate within a factor of 5 or 10; accuracy better than a factor of 2 to 3 is probably fortuitous. Indeed, Pasquill, based on tracer experiments, was able to conclude that the coefficients for both later and vertical plume spread were sufficiently related to wind, cloud cover, and solar radiation to permit use of the following very much simplified equation for calculation of concentration on the plume center line or axis:

$$\chi = \frac{0.0028Q}{\bar{u}x\theta h} \qquad (9\text{--}18)$$

in which χ is the concentration in Ci/m^3 (Bq/m^3),
u the mean wind speed in m/s,

Q the release in Ci/s (Bq/m^3),
θ the lateral cloud spread in degrees,
and h the vertical cloud spread. This equation is probably satisfactory for most general applications.

A term common in atmospheric diffusion work is χ/Q, expressed in s/m^3, which is merely the calculated or predicted dilution under a given set of meteorological conditions at a point downwind. This term is converted to actual concentration by multiplying by Q, the activity released. Thus, if χ/Q is given as 10^{-5} s/m^3 at some point, the concentration at that point from a release of 6 Ci/s (222 TBq/s) would simply be 6×10^{-5} Ci (2.2 $\times 10^{-5}$ Bq)/m^3.

References

1. Anspaugh, L. R., et al. 1975. "Resuspension and Redistribution of Plutonium in Soils", *Health Phys.* 29:571.
2. Church, P. E. 1949. "Dilution of Waste Stack Gases in the Atmosphere", *Indus. Eng. Chem.* 41:2753.
3. Fick, A., "Uber Diffusion", *Ann. Physik. Chem.* (2)94:59 (1855).
4. Gifford, F. A., Jr., 1982. "Turbulent Diffusion Typing Schemes—A Review", *Nucl. Safety* 17:68 (1976).
5. Hanna, S. R., G. A. Briggs, and R. P. Hosker, Jr. 1982. "Handbook on Atmospheric Diffusion", US Department of Energy Report DOE/TIC-11223.
6. Hewson, E. W. 1960. "Meteorological Measuring Techniques and Methods for Air Pollution", in *Industrial Hygiene and Toxicology*, Vol. 3, L. Silverman, Ed., Interscience, New York.
7. Huschke, R. E., Ed. 1975. *Glossary of Meteorology*, American Meteorology Society, Boston.
8. Kathren, R. L. 1968. "Towards Interim Acceptable Surface Contamination Levels for Environmental PuO_2", in *Radiological Protection of the Public in a Mass Disaster,* Proceedings of a Symposium of the Fachverband fur Strahlenschutz, Interlaken, May 27-June 1, 1968 U.S. Atomic Energy Commission Report CONF-68007; also published as Battelle, Pacific Northwest Laboratories Report BNWL-SA-1510.
9. Pasquill, F. 1961. "The Estimation of the Dispersion of Windborne Material", *Meteorol. Mag.* 90:33.
10. Pasquill, F. 1974. *Atmospheric Diffusion*, Second Ed., John Wiley and Sons, New York.
11. Powell, D. C., H. L. Wegley, and T. D. Fox. 1977. "MESODIF-II: A Variable Trajectory Plume Segment Model to Assess Ground-Level Air Concentrations and Deposition of Routine Effluent Releases from Nuclear Power Facilities", U.S. Department of Energy Report PNL-2419.

12. Richardson, L. F. 1926. "Atmospheric Diffusion Shown on a Distance-Neighbour Graph", *Proc. Royal Soc. (London)*, Series A, 110:709.
13. Schmidt, W. 1925. *Der Massenaustausch in Freier Luft and Verwante Erscheinungen*, Volume 7 of *Probleme Kosmischen Physik*, Verlag von Henri Grand.
14. Semel, G. A. 1980. "Particle Resuspension: A Review", *Environ. Int.*, 4:107.
15. Slade, D. H. (Ed.) 1968. "Meteorology and Atomic Energy 1968", US Atomic Energy Commission Report TID 24190 (July 1968).
16. Smith, W. J., F. W. Whicker, and H. R. Meyer. 1982. "Review and Categorization of Saltation, Suspension, and Resuspension Models", *Nucl. Safety*, 23:685.
17. Strenge, D. L., W. E. Kennedy, Jr., and J. P. Corley. 1982. "Environmental Dose Assessment Methods for Normal Operations at DOENuclear Sites", U.S. Department of Energy Report PNL-4410.
18. Taylor, G. I. 1921. "Diffusion by Continuous Movements", *Proc. London Math. Soc.*, (2)20:196.
19. Turner, D. B. 1967. "Workbook of Atmospheric Diffusion Estimates" US Public Health Service Publication 999-AP-26.
20. U.S. Nuclear Regulatory Commission (USNRC). 1980. *Final Generic Environmental Impact Statement. Uranium Milling Project M-25. Volume III, Appendices G-V*, U.S. Nuclear Regulatory Commission Report NUREG-0706, Vol. III, App. G-V.

CHAPTER 10

Aquatic and Terrestrial Transport

Introduction

Aquatic dispersion is analagous to diffusion by the atmosphere in that both air and water are fluids, but dispersion of radioactivity in the environment by both aquatic and terrestrial transport mechanisms is more complex and difficult to characterize than atmospheric diffusion. Waterborne transport is affected by many factors (Table 10–1), including the geometry (i.e size, shape, depth) and temperature profile of the water body, influences of ground waters, weather, tides, salinity, and whether the water is impounded or free flowing. Weather may have a large effect on aquatic transport; winds, for example can lead to enhanced surface mixing, precipitation can result in increased dilution perhaps in the form of spring runoff from melting snows, and low winter temperature can produce frozen surfaces on rivers, ponds, and other water bodies which greatly affect the transport and dispersion of radioactivity.

The general aquatic circulation follows the water or hydrologic cycle of the earth. Water from the ocean undergoes solar evaporation into the atmosphere resulting in the formation of clouds which are moved by wind action over land where the moisture precipitates out as rain or snow. Subsequent runoff may pass through subsurface aquifers or as surface runoff via rivers and streams back to the ocean. About 200,000 cubic kilometers (80,000 mi³) are evaporated from the oceans annually, with another 40,000 km³ (15,000 mi³) evaporating from lakes, rivers, and other fresh water sources including the land surface itself, and transpiration by plants. The time for the cycle to be completed is highly variable and may range from a few hours to many years; indeed, some water deposited as snow and ice in high mountain ranges or that exists deep in lakes or within the ground may not return to the oceans for millions of years.

The deep waters of Crater Lake are an excellent example of water that has been trapped in one location for thousands of years. The tritium con-

TABLE 10-1   Factors Influencing Aquatic Transport

Physical Factors
  Currents
  Tides
  Winds
  Precipitation
  Turbulence
  Convection
  Gravitational Settling
  Thermal Gradients
  Resuspension
  Radioactive decay
Chemical Factors
  Dissolution of Solids
  Precipitation
  Oxidation-reduction reactions
  Chemical combination
  Ion exchange
  Sorption
  Photochemical Reactions
Biological Factors
  Uptake and concentration by biota
  Excretion
  Transport by mobile species

centration in these waters is predictably quite low, since there is little mixing with surface waters nor any other way for the tritium content to be maintained. Thus, within a relatively short period of time, perhaps a short as a century, tritium levels will have fallen to undetectable levels. The short half-life of tritium (12.26 y) coupled with the constant level of tritium production by natural processes in the atmosphere also permits determination of snowfall and other information regarding the snowpack in the mountains by analysis of the tritium content of the snow at various depths, assuming suitable corrections are made for tritium from anthropogenic sources such as weapons tests and nuclear reactor operations.

The simplified water cycle is affected to a great extent by weather. The amount of evaporation is largely determined by insolation, and wind, air temperature, and the presence of condensation nuclei further affect the distribution of radioactivity.

The Marine Environment

The oceans cover an area of about 3.6×10^8 km^2 (1.4×10^8 mi^2), or about three-fourths of the surface of the earth, and occupy a volume of about 1.4×10^9 km^3 (3.3×10^8 mi^3). The mean depth is about 3.8 km (2.5 miles), with the maximum depth of about 10.5 km (6.5 mi) occurring in the Pacific Ocean just east of the Philippine Islands. Portions of the continents extend outward into the ocean, gradually sloping downward to a depth of about 200 m (600 feet); these are the continental shelves. Typically, the continental shelf extends out to about 50 km (30 miles), but shows some variability and may be almost non-existent as is the case off the west coast of South America, or several hundred miles long as occurs off the coast of China.

The oceans constitute a great natural reservoir of radioactivity, and are estimated to contain in excess of 4×10^9 curies (1.5×10^{22} Bq) of ^{40}K, 4×10^8 curies (1.5×10^{21} Bq) of ^{87}Rb, and 1×10^8 Ci (3.7×10^{18} Bq) of ^{226}Ra (Osterberg 1982). In addition, the oceans contain significant amounts of uranium, thorium, and their daughters, plus various radionuclides of anthropogenic origin mostly associated with nuclear detonations. By far, however, the greatest actvity is from ^{40}K, and ocean water typically shows a concentration of about 340 pCi/l (12,500 Bq per cubic meter).

A key characteristic of the ocean is salinity. The mean salinity is about 3.5 weight per cent, about three-fourths of which is sodium chloride, and the remainder largely from salts of magnesium, calcium, and potassium. By contrast, an isotonic solution is only 0.9%. Salinity in the deep layers is slightly greater than in the surface layers. Sea water is highly buffered and shows a constant alkalinity, having a pH of 8.2.

The oceans are also characterized by relatively cold temperatures, near the freezing point of water or even slightly below except near the surface. Thus, the mean ocean temperature is 3.52 °C, with the median temperature 2.1°C (Montgomery 1958). There is relatively little vertical mixing which results in a stratified ocean with three major layers. The surface layer may extend to a few hundred meters but typically has a mean depth of 75 meters, which is about the range of penetration of sunlight. It is in this layer that much of the marine life is found.

The surface layer is characterized by much movement and mixing from tides and currents, which are largely the result of wind action. Thus, in the surface zone or layer vertical mixing is essentially complete and rapid, resulting in near uniform temperature, salinity, and density. The intermediate zone is a layer about 1000 meters thick which is generally known as the thermocline or pycnocline. In this zone, temperature decreases and density increases rapidly with depth. There is little vertical movement or mixing in this intermediate layer or in the deep ocean lying below it, although there may be streams of horizontal motion sometimes resulting from temperature gradients. Radioactive tracer studies have shown horiozontal spread in this layer, but essentially none vertically. In one such study, a tracer released below the thermocline was found to spread over a horizontal area of 100 km^2 (18 mi^2) while remaining in a band less than one meter thick (Pritchard et al. 1971).

Mixing and diffusion processes in the oceans are difficult to model or to generalize. Surface mixing and diffusion is largely related to winds which produce currents that may carry radioactivity large distances fairly rapidly. The Gulf Stream, for example, travels at a rate of 7 km/h (4 mph) and thus may transport materials more than 150 km (100 miles) per day. In the Pacific, the Japanese current travels at about half this rate. Physical mixing in the surface layer also depends on the depth of the water, the bottom and shoreline configuration, tides, and the temperature and depth at which the radioactivity is introduced. Transport and mixing may also be significantly affected by biological processes which may remove and concentrate specific radionuclides.

Horizontal diffusion in the ocean cannot be characterized by Fickian type diffusion equations but is generally described by an exponential model, with several such models having been put forth (Pritchard et al. 1971). One group of models contains a time dependent diffusion velocity parameter while another is characterized by the rate of turbulent energy transfer. No single adequate model, however, has been developed for open ocean horizontal diffusion, although the problem has been extensively studied. In general, currents are the major cause of horizontal motion in the surface layers, with relatively little vertical movement of activity. Thus, at or near the surface, horizontal diffusion may be at the rate of several km/h. In the deep ocean, horizontal or lateral diffusion is considerably less, with movement thought to be on the order 10^{-3}m/sec (Pritchard 1971). Vertical diffusion is considerably slower, generally on the order of a few meters per year.

Radionuclides from fallout or otherwise introduced into the surface layers of the sea thus tend to remain in the first 30–60 meters of depth, although the horizontal dispersion may be very great and cover many square kilometers (Koczy 1960). Particulates gradually settle until they reach the thermocline, where there appears to be a holdup. After a lengthy period, the particulates may reach the bottom where they tend to remain and be buried by sediments, especially those of biological origin. This model is, of course, greatly simplified, for open ocean diffusion is highly complex and does not lend itself to easy mathematical description.

Coastal Waters and Estuaries

Diffusion in coastal waters is sometimes assumed to be Fickian except close in to the shore, permitting the maximum concentration C_m along the axis of the current at any distance x to be described by (NAS 1959)

$$C_m = \frac{Q}{2D(2\pi KUx)^{0.5}} \qquad (10\text{--}1)$$

in which C_m is the maximum concentration of the radionuclide along the
    axis of the current
    Q is the activity released per unit time
    D is the depth
    K is the diffusivity constant, typically with a value of $6 \times 10^{-3} m^2/min$
    U is the velocity of the current
    and x is the distance of travel.

This equation ignores deposition or settling.

Generally, the same diffusion models that hold for coastal waters hold for large lakes and small landlocked seas. Such models usually assume a Gaussian distribution of activity, and are similar in form to the corresponding atmospheric diffusion equations, as illustrated by Equation 10–2 which calculates the concentration from a continuous release of activity (USNRC 1977)

$$\chi = \frac{W}{2\pi u \sigma_y \sigma_z} f(\sigma_z, z, z_s, d) f(\sigma_y, y, y_s) \qquad (10\text{--}2)$$

The graphical solution to this equation is shown in Figure 10–1. Note that despite certain simplifying assumptions such as the constant depth of water and straight shoreline used in this model, aquatic diffusion cal-

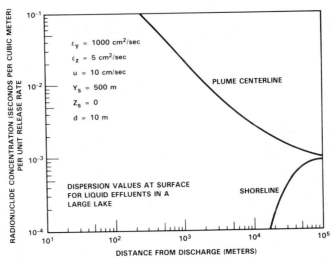

FIGURE 10–1   Graphical solution to equation for radionuclide dispersion in a lake (USNRC 1977).

culations are far more complex than the corresponding atmospheric diffusion equations. For example, one of the two diffusion functions at the end of Equation 10–2 is a summed exponential of several variables whose solution is not easily obtained. Thus, aquatic diffusion in large bodies of water is not easily predicted.

In the near-shore zone, or strip of water lying adjacent to the shoreline, movement and mixing of radionuclides is largely governed by local meteorological conditions, the freshwater runoff from the land, and the physiognomy of the coast (Pritchard et al. 1971). Radioactivity introduced into the near-shore zone will be transported from the point of introduction by the existing currents at the time of discharge, and will undergo turbulent mixing. If discharges are continuous or made over a long period of time, there will be a buildup of a low level concentration field in the vicinity of the discharge location. Plume movement and plume concentrations will vary greatly, and does not lend itself to mathematical modelling.

Estuaries are semi-enclosed coastal water bodies with free connection with the ocean in which seawater is measurably diluted with freshwater from land drainage (Pritchard 1967a). Tidal action enhances the mixing of fresh and saline water producing a transition zone between fresh water and marine biological habitats. Estuaries are classified into four types based on their mixing patterns (Pritchard 1967b):

1. Type A, in which seawater flows upstream in the form of a salt water wedge underlying the fresh water. The surfaces of the water bodies are clearly defined and there is little mixing.

2. Type B, in which the seawater flows into the head of the estuary under the fresh water, but in which there is sufficient mixing of the two to blur the clear definition of each layer, and to create a vertical salinity gradient with relatively little horizontal mixing.

3. Type C, in which vertical mixing is nearly complete resulting in the disappearance of the vertical salinity gradient. In this situation, a horizontal gradient develops, with the salinity increasing towards the left hand bank of the estuary when facing downstream in the Northern hemisphere, and the right hand bank in the Southern hemisphere.

4. Type D, in which turbulent mixing is so great that the salinity gradients in both the vertical and horizontal directions are eliminated; in other words, mixing is virtually complete.

An estuary can fall into more than one category depending on circumstances such as tides and fresh water flows, but in general, most fall into the category of partially mixed estuaries, i. e. Type B or Type C.

Despite extensive study, estuarine mixing is incompletely understood, and the great differences in estuaries does not permit the development of a simple model to describe diffusion and mixing. A one dimensional model has been developed (USNRC 1976) based the technique of tidally averaged approximation which has been successfully applied in several estuaries. The model assumes a one dimensional estuary with constant cross-section which results in the following partial differential equation

$$\frac{\partial C}{\partial t} + u_f \frac{\partial C}{\partial x} = E_L \frac{\partial^2 C}{\partial x^2} - \lambda C \qquad (10\text{--}3)$$

which can be solved for an instantaneous release of activity to yield

$$C = \frac{Q}{A\sqrt{4\pi E_L t}} \exp - \left[\frac{(x - u_f t)^2}{4E_L t} + \lambda t \right] \qquad (10\text{--}4)$$

in which C is the concentration of radioactivity across the cross-section A of the estuary at a distance x and time t from the point of release, E_L

is the longitudinal diffusion coefficient, Q the quantity of activity released, λ the decay constant of the radionuclide, and U_f the net downstream freshwater velocity. For continuous discharge, Equation 10-4 can be integrated by means of the convolution integral, but the resulting equation is quite complex.

Rivers, Streams, and Lakes

Dilution and dispersion of radioactive materials in rivers and streams is to a large extent governed by the basic principles of hydraulic flow. In an open channel such as a river or stream, the average flow or mean velocity of radioactive discharge is simply expressed by the quotient of the volume of the discharge divided by the cross-sectional areas of the channel. In channels or rivers with uniform or near uniform cross-section, the maximum velocity of flow occurs aprroximately midway between the two banks and a some distance—generally on the order of one-third of the depth—below the surface. In very shallow streams, however, the maximum velocity may occur at or near the surface, while in deep rivers it may occur midway down. In general, for a given surface velocity, wide and deep streams have greater mean and bottom velocities than do shallow streams.

Prediction of diffusion and downstream concentrations can be estimated by classical diffusion theories and is related to the mean velocity or flow rate of the river or stream. One of the simplest cases assumes a continuous discharge into a small turbulent stream with rapid and essentially complete mixing. The mean cross-sectional concentration of radioactivity, C, in Ci or Bq/m^3, immediately downstream of the point of discharge is given by

$$C = \frac{Q}{F_s + F_d} \qquad (10-5)$$

in which Q is the discharge rate in Ci or Bq/sec, and F_s and F_d are the flow rates of the stream at the point of discharge and of the discharge, respectively, in m^3/s. The stream flow rate can be calculated from.

$$F_s = yzu \qquad (10-6)$$

in which y and z are the width and depth of the stream in meters, and u is the mean velocity of the stream. Combining Equations 10–5 and 10–6 gives

$$C = \frac{Q}{yzu + F_d} \qquad (10\text{--}7)$$

Note that the release rate Q can be expressed as the product of the discharge concentration, C_d, and the discharge flow, which, when substituted in Equation 10–7 gives

$$C = \frac{C_d F_d}{yzu + F_d} \qquad (10\text{--}8)$$

Equation 10–8 is highly simplified in that it not only assumes instantaneous and complete mixing and a constant stream flow and cross-section, but does not take into account the various removal processes, dilution from tributaries, and other factors that affect concentration such as the variability in stream velocity already discussed above. The important removal processes include biological uptake, adsorption, settling and deposition. Concentration may also be reduced by dilution from flow from tributaries or other discharge points, or from precipitation. The effect of the latter is usually minor as the contribution from rain or snow is only a small fraction of the total volume of the stream. Increased concentration may result from evaporation, seepage, and resuspension of settled solids.

In general, removal processes are exponential, and can be expressed by.

$$C_x = Ce^{-kx} \qquad (10\text{--}9)$$

in which C_x is the concentration at some distance x downstream, in meters, and k is the removal coefficient expressed in reciprocal meters. Dilution can be expressed by the term D and evaporative loss by E and are simply additive algebraically to the stream flow term in the denominator of Equation 10–8. The contribution from resuspension of settled material is a function of many factors, but can be generalized as the quantity or activity Q reintroduced into the stream per unit length of the stream bed times the distance downstream, and is directly additive to the quantity term in the numerator of Equation 10–8. Including these terms and Equation 10–9 yields

$$C_x = \frac{C_d \, F_d + Q_x}{yzu + F_d + D - E} \, e^{-kx} \qquad (10\text{--}10)$$

which provides a generalized if somewhat ideal model for computing downstream concentration.

In actual practice, the downstream concentration from a point of discharge will be governed by complex diffusion principles more or less analagous to those for atmospheric dispersion. Thus, the concentrations will be found to follow the stream flow rates, with the highest concentration occurring at the point of maximum velocity for any given downstream distance. Waterborne transport is highly specific to the transporting body, and modelling should be done on an individual basis (Strenge, Kennedy, and Corley 1982). For large rivers, however, uniform concentrations will occur at downstream distances greater than several times the width of the river and can be estimated by

$$C = \frac{Q}{F_s} \qquad (10\text{--}11)$$

in which Q is the activity discharged and F_s. Note that the flow rate in Equation 10–11 is simply the inverse of the dilution factor and is expressed in units of volume per time. Equation 10–11 also holds for a continuous or extended discharge if Q is expressed in units of activity per unit time.

The above equations modelling dispersion of radioactivity in rivers and streams, although simplified, provide a reasonably accurate picture of what actually occurs. More sophisticated models have been developed for both the Columbia River in Washington State and the Clinch River in Tennessee, based on empirical observations of the discharges of radioactivity from adjacent nuclear operations, viz. the production reactors at Hanford in the case of the Columbia and the Oak Ridge National Laboratory in the case of the Clinch (Onishi 1977a, 1977b). These finite element models require the use of a sophisticated computer code for solution, but provide a far more accurate and complete characterization of the diffusion and sedimentation process in these two rivers.

A more generalized study of radionuclide dispersion into rivers and streams showed that the principal mechanism of transport is hydrodynamic, although influenced by a variety of factors such as the amount and kind of suspended material in the waterway, the chemical character of bottom sediments and the river or stream itself, and the removal by biota (Yousef, Kudo, and Gloyna 1970). Existing models, however, even of the highly simplified type derived above, provide reasonably satisfactory estimates.

For a lake or other impoundment, the following equation can be used to obtain the concentration, $C_{(x,t)}$, at distance x from the release point and time t after the discharge:

$$C_{(x,t)} = \frac{Q(1 - e^{-kt})}{x} \qquad (10\text{-}12)$$

in which k is a rate constant for the rate of removal from the lake. Removal from the lake is by radioactive decay, deposition in sediments or on the shore, lake drainage to a river or other water body, and biological mechanisms. For physical processes the rate constant k assumes the form.

$$k = \lambda + a + FV^{-1} \qquad (10\text{-}13)$$

in which λ is the radioactive decay constant, a is the coefficient of excess heat loss (which may be significant if the discharge is heated), F the flow rate through the lake, and V the volume of the lake. When Equations 10-12 and 10-13 are combined, the resultant shows that concentration is, as expected, directly proportional to the quantity released and inversely proportional to the distance from the point of release, and also greater with decreasing flow or lake drainage, and with a smaller lake volume.

Subsurface Water Movement

Radioactivity released to the environment may get into subsurface or ground waters in several ways. The radioactivity may be directly discharged to the ground water through wells, or may seep into the ground water through stream or lake beds, or through the soil if deposited on the surface or above the water table. The subsurface movement of water is relatively well understood. The Darcy equation provides the rate of seepage, Q, through a permeable, non-absorbing soil medium:

$$Q = uiA \qquad (10\text{-}16)$$

In Equation 10-16, Q is ordinarily given in gallons or cubic feet per minute, u is the permeability of the soil medium, i is the hydraulic gradient, and A and the area of the land through which the water is seeping. Typically, the hydraulic gradient is near unity, and the permeability on the order of 10^{-6} to 10^{-4} ft/min (5×10^{-9} to 5×10^{-7} m/s).

The Darcy equation is particularly applicable to seepage from lakes or other impoundments and can be adapted to an absorbant soil medium and corrected for radioactive decay by the addition of the appropriate terms to yield.

$$Q = uiAe^{-kd\lambda t} \tag{10-17}$$

in which k is the coefficient of absorption or removal from the soil, d is the depth of the soil, λ the decay constant, and t the time. The coefficient of removal from the soil is a complex term, and includes removal by both ion exchange and adsorption as well as biological uptake by plant roots. Thus, this term is highly variable and dependent not only on the type of soil but the vegetation as well.

The Darcy equation only describes seepage or movement of radionuclides down into the ground water, but does not describe movement through the ground water. Transport of radionuclides via underground movement is difficult to determine and best estimated from monitoring data at specific locations (Strenge, Kennedy, and Corley 1982). Movement through groundwater to rivers and streams and then to the ocean can be relatively rapid, and can result in the dispersion of radioactivity from a specific source over a wide area.

Removal of radioactivity from subsurface water may proceed by several mechanisms. Biological activity may be quite important, with radionuclides being removed from the ground water by burrowing animals or microorganisms as well as by plants. Organic complexing agents may modify the chemical form or the radioactivity, altering its solubility or ionic character. Ruthenium, for example, forms many complexes and has been widely studied in ground water. Soil sorption is, however, the most significant mechanism and can result in complete removal of radionuclides from ground water, thus obviating further movement.

Terrestrial Transport

Terrestrial transport is to a great extent determined by the following interrelated factors:

1. Climate and weather.
2. Terrain and topography.
3. Soil type and characteristics.
4. Vegetation and animal life.

The most important physicochemical factors are related to soil type and characteristics, which determine the availability of the radionuclides for uptake by plants and animals.

Soils are basically a mixture of organic and inorganic materials that serve as the basis for life. Soils are highly variable in characteristics and thickness and are usually found in the form of strata or layers known as horizons of differing physical characteristics. The uppermost layer of the soil is known as the topsoil or surface layer and is commonly about a half meter (1 to 2 feet) thick. It is in this layer that most life processes take place, and hence it is rich in organic materials. Underlying the topsoil is the subsoil horizon, usually about 30 cm (1 foot) thick, and consisting mostly of inorganic material. These two layers together are known as the solum and lie on top of a rubble or pervious rock layer a 30–60 cm (1–2 feet) thick from which the upper layers will eventually be formed by weathering. The lowest layer is the parent layer or bedrock, which may be impervious to water.

Surface soils are categorized on the basis of texture or fineness of particle size as sands, silt, and clay. Sands range in size from 20 to 2000 μm in diameter, and have a relatively small surface area per unit mass (Table 10–2). Because of their large particle size, they are relatively pervious to the flow of water. Chemically, many sands are SiO_2 and hence nonreactive and insoluble. Radionuclides transported by water are thus likely to rapidly travel through a sandy soil until they encounter ground water or reach the impervious bed rock layer.

TABLE 10–2   Physical Data of Soils

Soil Type	Diameter (μm)	Particles/g	Surface Area (cm^2/g)
Sand	20–2000	10^4	23
Silt	2–20	6×10^6	454
Clay	<2	9×10^{10}	8×10^6

Adapted from Whicker and Schultz (1982), p. 34.

Silts lie intermediate between sands and clays with particle sizes in the range 2–20 μm and are also relatively pervious to water. Because of their small size, they have significantly greater surface area per unit mass than do sandy soils, and are sufficiently small to be transported by flowing surface water, including runoff from rainfall. Hence, silts may be involved in the rapid and long range movement of radionuclides through the environment.

The clays have the smallest particle size, <2 μm in diameter, and hence the greatest available surface area per unit mass. Clays are the so called heavy soils, typified by adobe and thick muds. Thus, they are resistant to water movement through them. Also, clay particles may carry a predominently negative surface charge making them attractive to cations, which sorb onto the surface of the clay particles and thus act as an ion exchange medium. Radioactive materials passing through a clay soil will thus be removed and held by the soil particles and are available for uptake by plants.

Sorption by soil is an important mechanism by which radionuclides may be removed from a fluid medium such as air or water. Sorption is a surface phenomenon, and may result from molecular or electrostatic attraction, chemical bonding, or capillary forces. Molecular attraction results from van der Waals forces and is relatively weak. Electrostatic attraction, on the other hand, is relatively strong and accounts for the ordinary ion exchange properties of soils. Chemical reactions may also produce tight bonding between a radioactive nuclide and the surface, but reactions may also be reversible. Capillary forces produce surface binding though surface tension between a solution (e.g. soil moisture containing a radioactive species) and solid (e.g. soil particle) phase.

Sorption is frequently described in terms of the distribution coefficient, K_d, which is simply the ratio of the activity in the solid phase to that in the liquid phase, or

$$K_d = \frac{\text{Concentration in solid phase}}{\text{Concentration in liquid phase}} \qquad (10\text{--}18)$$

The concentration in the solid phase is ordinarily expressed in units of activity (usually μCi or Bq) per gram or, for surfaces, activity per unit area. Liquid phase concentration is ordinarily expressed in units of activity per ml, although in the SI system, the units are Bq per cubic meter. Thus, the distribution coefficient has units of ml/g or ml/unit area. Values of K_d are quite variable and are determined by the physical and chemical properties of the soil, the liquid and the specific radionuclide of interest. For clays, typical values of K_d are in the range 10^3–10^5 for radio-nuclides of cobalt, cesium, and other anionic fission products with similar values reported for marine sediments (Sorathesn et al. 1960, Duursma and Gross 1971) and somewhat lower values for actinides (Serne et al. 1977).

The distribution coefficient is inversely related to the velocity of the radionuclide leaching through the soil and is expressed by (Essington and Fowler 1976)

$$V_r = \frac{V_w \theta}{[(1 - \theta)\rho k_d + \theta]} \qquad (10\text{--}19)$$

in which V_r is the velocity of the radionuclide in soil, V_w the velocity of the solution containing the radionuclide, θ the fractional porosity of the soil, ρ the grain density or, alternatively for surface activity, the grain area-volume ratio, and, of course, K_d the distribution coefficient. Equation 10–19 mathematically demonstrates that the greater the solubility of the ion, the more rapidly it migrates through the soil, and, as $K_d \rightarrow 0$, the velocity of the ion may approach the limiting value of the velocity of the leaching water.

The ion exchange properties of soils and sediments are inversely related to particle size because smaller particles have a higher surface area to volume ratio. The elemental composition of the soil or sediment is likewise very important as appropriate ions must be present in sufficient quantity to permit the exchange to occur. The sequence of sorption from least to greatest of radionuclides from sea water by sediments is (Duursma 1969):

$$^{45}Ca < {}^{90}Sr < U, Pu < {}^{137}Cs < {}^{86}Rb < {}^{65}Zn$$
$$< {}^{59}Fe < {}^{95}Zr\text{-}Nb, {}^{54}Mn < {}^{106}Ru < {}^{147}Pm$$

In soils, the sequence is (Whicker and Schultz 1982):

$$H < Sr < Ra < Ca < Mg < Cs < Rb < K < NH_4 < Na < Li$$

Other factors significantly affecting soil sorption include pH, rainfall or other precipitation, which can drastically alter the pH in addition to producing dilution and washout effects, and the presence of organic material, which improves the ion exchange capability of the soil. In general, most of the factors favoring ion exchange are present to a greater degree in clay soils and loams, and least in sands. Cation absorption capability

is commonly expressed as milliequivalents of cation necessary to neu-
tralize the surface charge of 100 grams of soil at a pH of 7, and typically
ranges from a few mEq/100 g to a few hundred mEq/100g.

Movement of radioactivity through the soil can either be vertical or
horizontal and is usually relatively slow, seldom exceeding a few centimeters per day. Vertical movement is most rapid in sandy soils and slow-
est in clay, and is accelerated by rapid water movement. Thus, penetration into the soil is enhanced by precipitation and is greatest in areas with
heavy rainfall. Depending on the depth of the soil, radionuclides may not
reach the water table for months or even years. Animal life in the soil
enhances downward movement, while vegetation tends to hinder it
because of the uptake by plants which results in removal from the soil
with subsequent return to the surface when the plant dies, or the animals
feeding on the plant excrete the radionuclides or themselves die.

The degree or depth of penetration of a specific radionuclide is influenced by many factors including the concentration of its stable isotopes
in the soil, soil type, precipitation, pH, and the valence of the radionuclide; divalent cations tend to be more tightly bound than monovalent
cations. Thus, the two important fallout fission products ^{90}Sr and ^{137}Cs
are generally found tighly bound in the upper few centimeters of soil,
with the latter more tightly bound (Alexander 1967). Similarly, tran-
suaranic elements tend to remain in the upper layers of the soil; in one
study in the vicinity of the Rocky Flats weapons plant, more than 90%
of the plutonium deposited between 6 and 18 years previously was found
in the top 10 cm (Krey 1976).

The process by which radionuclides deposited on the surface of the
ground penetrate into the soil or are removed from the site of deposition
by water, wind, or other natural mechanical disturbances is known as
weathering. A month after deposition, the concentration or distribution
of fission products as a function of soil depth appears to be exponential,
with a relaxation length (i.e. decrease by a factor of e) of 3 cm (Beck
1966). Similar observations have been made with transuranic elements,
with most environmental plutonium existing in a strongly adsorbed state
on surface soils (Watters et al. 1980).

Penetration of sorbable solutes through a permeable bed of soil can be
expressed in terms of the following second order partial differential
equation

$$D \left(\frac{\partial^2 C}{\partial yz} \right) - u_y \frac{\partial C}{\partial y} = \frac{\partial C}{\partial t} + k(K_m C - m) \qquad (10\text{--}20)$$

in which D is the diffusivity of the solute through the interstices of the soil, u_y is the vertical convective velocity, K_m is the distribution coefficient, C the concentration of radionuclide in the liquid at a given time and depth, and m the sorbed concentration at the same time and depth. If the surface is saturated and absorption has not occurred, Equation 10–20 simplifies to.

$$C = C_o e^{-py} \qquad (10\text{--}21)$$

in which C is the concentration in the soil at depth y, C_o the concentration at the surface, and p the linear uptake coefficient.

Vegetation and Resuspension

Radionuclides deposited on the surface of the ground or on vegetation are prone to both horizontal and vertical transfer. Soil and vegetation can be either contaminated externally by deposition of airborne radioactivity or by water runoff or precipitation that contains radioactivity. In addition, vegetation can take up and perhaps even concentrate radionuclides from the soil. If eaten by animals, externally contaminated plants represent a means by which radioactivity can bypass the soil and be directly introduced into the food chain.

Deposition from air is ordinarily expressed in terms of the product of the deposition velocity and the air concentration, and thus has units of activity per unit area per unit time, typically $\mu Ci/cm^2$-day. Deposition velocity, V_d is a function of both wind speed, atmospheric stability, particle size, and surface roughness (Healy 1974; Sehmel 1975), but can be approximated in units of cm/s by

$$V_d = 0.0003 u d_p \qquad (10\text{--}22)$$

in which u is the wind speed in m/sec and d_p the particle diameter in μm. Equation 10-22 was empirically developed with plutonium particles at the Nevada Test Site, and applies over the range of wind speeds of 2–12 m/s and particle diameters of 1–100 μm (Sehmel 1973).

Foliar deposition can be removed from plants by several mechanisms:

1. Radioactive decay.
2. Evaporation.

3. Washoff by rain.
4. Other weathering as by wind, etc.
5. Death of the plant.
6. Eating or mechanical removal by animals.

In addition, the surface concentration or activity per unit area of foliage may be diluted by the growth of the plant, or may be reduced in the aggregate by leaves or other portions of the plant dying and dropping to the ground. Evaporation is a particularly important removal mechanism for volatile nuclides such as the radioiodines, returning them to the atmosphere where diffusion is more likely to occur. The rate of removal of a radionuclide from the surfaces of plants appears to be a power function of time, with initially rapidly decreasing levels that become asymptotic. Studies of ^{89}Sr and ^{131}I in fallout revealed an initial weathering half-time of 1.4 days for during the first five days after deposition, increasing to 20 days during the next ten days, and to 30 days for the period 15 to 30 days after deposition, and 130 days beyond 30 days (Martin and Bloom 1980). Other studies have shown an exponential decrease of about 5%/day from foliar surfaces for fission products (Chamberlain 1970). Earlier studies with fallout radioiodine revealed half-times for loss from foliage due to physical processes other than decay of 6 and 18 days which, when combined with physical decay, produced effective removal half-times of 3.5 and 5.5 days, respectively for ^{131}I (Menzel 1967). For calculation of the doses to man from airborne reactor effluents, the U.S. Nuclear Regulatory Commission recommends a value of 0.2 for the fraction of deposited activity retained on crops or other vegetation (USNRC 1977).

Removal of foliar deposition by natural processes is thus quite variable and complex. To a great extent, removal is governed by particle size, with large particles more readily removed than smaller ones (Menzel 1967). Broad leaved plants are less likely to retain deposited radionuclides, while cereals and grains, because of their physical configuration, are more likely to show high foliar retention levels.

The concentration of a radionuclide taken up from the soil by a plant may be highly variable, and in general is species dependent. For a given species, uptake tends to lessen as the plant grows, in part because the roots draw from deeper and hence less contaminated volumes of soil as they grow down into the soil. Growing plants thus may greatly soil concentrations of radionuclides, and may produce seasonal variation by removing significant quantities from the soil during the growing season and returning the radionuclides to the soil during the winter when the plant dies or drops its leaves. Modelling of this terrestrial transport mechanism is difficult, and is highly species and weather dependent.

Resuspension of material deposited on the surface of the soil represents an important and often overlooked horizontal terrestrial transport mechanism. The activity may have originally been deposited as fallout from the atmosphere, but it could also have been transported to the location by water which later evaporated or soaked into the soil, or from biological sources such as dead plant material or animal scats. Resuspension results from wind or other mechanical disturbance and is usually described in terms of the resuspension factor (RF) or resuspension rate. The resuspension factor is defined as the ratio of the activity per unit volume in air at a reference height above the ground (usually 1 meter) to the activity per unit area on the surface of the ground. Thus, the RF has units of reciprocal length, and is usually expressed in units of m^{-1}. The resuspension factor is highly variable, depending on the freshness of the deposition, particle size, weather and atmospheric variables including precipitation, humidity, wind speed, turbulence, and temperature, and the ground roughness, moisture content, and other physical factors. Values ranging over 11 orders of magnitude from 10^{-13} to 10^{-2} have been reported in the literature (Kathren 1968) but the value of $10^{-6}/m$ appears to be suitable for general use outdoors for freshly deposited material (Brodsky 1980).

The resuspension rate, R, is the fraction of the radionuclide present on the ground that is resuspended per unit time by winds or mechanical disturbance (Healy 1980). Thus it is similar to the resuspension factor, but has the dimensions of reciprocal time, usually day^{-1}. It has been used to calculate the velocity of resuspension by multiplying it by the ratio of the activity per unit surface area and the activity per unit volume in the soil (Slinn 1978), e.g.

$$m/d = Rx \frac{\mu Ci/m^2}{\mu Ci/m^3} \qquad (10\text{--}23)$$

Resuspension is also sometimes described by mass loading, which involves measurement of the mass of soil particles in the air and thus directly relates the measured soil concentrations to those in air. The air concentration is simply the product of the activity per unit mass of soil and the mass of soil per unit volume of air. Although the RF is the most widely used, the mass loading approach appears to be superior for some applications such as generic studies (Healy 1980). Typical mass loading values are on the order of 100 $\mu g/m^3$.

Mathematical models of resuspension have been developed based on wind erosion models and as a function of wind speed. Air concentration, C, can be approximated with the following model (Martin and Bloom 1980):

$$C_{air} = k(u - u_T) \frac{C_{soil}}{u} \qquad (10\text{--}24)$$

in which C_{air} is the air concentration in $\mu Ci/m^3$,
　　　　u is the wind speed in m/s
　　　　u_T is the threshold wind speed for particle resuspension
and C_{soil} is the surface contamination level in $\mu Ci/m^2$.

Resuspension has also been expressed in terms of the following power function model

$$C_{air} = Ku^n \qquad (10\text{--}25)$$

in which u is the wind speed and K and n are constants. Values of n have been found to be about 2.1 for various soils, nuclides, and climates (Martin and Bloom 1980).

It should be noted that the resuspension factor is subject to change with the age of the deposition, generally getting much smaller with time. Based on empirical data, an exponential decrease with time has been proposed, with a half-time of 45 days (Kathren 1968). Other more complex models have also been used to express the change in resuspension factor with time, but generally these are not much better at prediction than the simple exponential model.

References

1. Alexander, L. T. 1967. "Depth of Penetration of the Radioisotopes Strontium-90 and Cesium-137", U.S. Atomic Energy Commission Report HASL-183.
2. Beck, H. L. 1966. "Environmental Gamma Radiation from Deposited Fission Products", *Health Phys.* 12:313.
3. Brodsky, A. 1980. "Resuspension Factors and Probabilities of Intake of Material in Process (or "Is 10^{-6} a Magic Number in Health Physics?)", *Health Phys.* 39:992.

4. Chamberlain, A. C. 1970. "Interception and Retention of Radioactive Aerosols by Vegetation", *Atmos. Environ.* 4:57.

5. Duursma, E. K. 1969. "Chemistry of Sediments", in *Annual Report of the International Laboratory on Marine Radioactivity, Monaco*, International Atomic Energy Agency Technical Report Series 98, pp. 92–97.

6. Duursma, E. K. and M. G.Gross. 1971. "Marine Sediments and Radioactivity", in *Radioactivity in the Marine Environment*, National Academy of Sciences, Washington, pp. 147–160.

7. Essington, E. H. and E. B. Fowler. 1976. "Distribution of Transuranic Nuclides in Soil: A Review", Los Alamos Scientific Laboratory Report LA-UR-77-385.

8. Healy, J. W. 1974. "A Proposed Interim Standard for Plutonium in Soils". U. S. Atomic Energy Commission Report LA–5483.

9. Healy, J. W. 1980. "Review of Resuspension Models", in *Transuranic Elements in the Environment*, W. C. Hanson, Ed., U.S. Department of Energy Technical Information Center Report DOE/TIC-22800, pp 209–235.

10. Kathren, R. L. 1968. "Towards Interim Acceptable Contamination Levels for Environmental PuO_2" in *Radiological Protection of the Public in a Nuclear Mass Disaster*, Proceedings of a Symposium of the Fachverband for Strahlenschutz, Interlaken, May 27-June 1, 1968, U.S. Atomic Energy Commission Report CONF–68007; also published as Battelle-Northwest Laboratories Report BNWL-SA-1510.

11. "Koczy, F. F. 1960. "The Distribution of Elements in the Sea", in *Proceedings of a Symposium on the Disposal of Radioactive Wastes*, International Atomic Energy Agency, Vienna.

12. Krey, P., et al. 1976. "Plutonium and Americium Contamination in Rocky Flats Soil 1973", U. S. Energy Research and Development Administration Report HASL-304.

13. Martin, W. E., and S. G. Bloom. 1980. "Nevada Applied Ecology Group Model for Estimating Plutonium Tansport and Dose to Man", in *Transuranic Elements in the Environment*, W. C. Hanson, Ed., U.S. Department of Energy Technical Information Center Report DOE/TIC 22800, pp. 459–512.

14. Menzel, R. G. 1967. "Airborne Radionuclides and Plants", in *Agriculture and the Quality of Our Environment*, N. C. Brady, Ed., American Association for the Advancement of Science, Washington, D. C., pp. 57–75.

15. Montgomery, R. B. 1958. "Water Characteristics of Atlantic Ocean and of World Ocean", *Deep-Sea Res.* 5:134.

16. National Academy of Sciences-National Research Council (NAS). 1959. *Radioactive Waste Disposal Into Atlantic and Gulf Coastal Waters, A Report from a Working Group of the Committe on Oceanography*, National Academy of Sciences-National Research Council Publication 655, Washington D. C.

17. Onishi, Y. 1977a. Finite Model Elements for Sediment and Contaminant Transport in Surface Waters", Battelle Pacific Northwest Laboratory Report BNWL-2227.

18. Onishi, Y. 1977b. "Mathematical Simulation of Sediment and Radionuclide Transport in the Columbia River", Battelle Pacific Northwest Laboratory Report BNWL-2228.

19. Osterberg, C. L. 1982. "Why Not in the Ocean", *IAEA Bulletin* 24(2):30.

20. Pritchard, D. W. 1967a. "What is an Estuary: Physical Viewpoint", in *Estuaries*, American Association for the Advancement of Science Publication 83, Washington, pp. 3–5.

21. Pritchard, D. W. 1967b. "Observations of Circulation in Coastal PLain Estuaries", *op. cit.*, pp. 37–44.

22. Pritchard, D. W. et al. 1971 "Physical Processes of Water Movement and Mixing" in *Radioactivity in the Marine Environment*, National Academy of Sciences, Washington, pp. 90–136.

23. Sehmel, G. A. 1973. "Particle Resuspension from an Asphalt Road Caused by Car and Truck Traffic", *Atmos. Environ.* 7:291.

24. Sehmel, G. A., 1975. "Experimental Measurements and Prediction of Particle Deposition and Resuspension Rates", Battelle Pacific Northwest Laboratory Report BNWL-SA-5228.

25. Serne, R. J., et al. 1977. "Batch K_d Measurements of Nuclides to Estimate Migration Potential at the Proposed Waste Isolation Pilot Plant in New Mexico", Battelle Pacific Northwest Laboratory Report PNL-2248.

26. Slinn, W. G. N. 1978. "Parameterizations for Resuspension and for Wet and Dry Deposition of Particles and Gases for Use in Radiation Dose Calculations", *Nuclear Safety* 19:205.

27. Sorathesn, A., et al. 1960 "Mineral and Sediment Affinity for Radionuclides", U.S. Atomic Energy Commission Report CF–60–6–93 (1960).

28. Strenge, D. L., W. E. Kennedy, and J. P. Corley. 1982. "Environmental Dose Assessment Methods for Normal Operations at DOE Nuclear Sites", Battelle, Pacific Northwest Laboratory Report PNL-4410.

29. U. S. Nuclear Regulatory Commission (USNRC). 1976. "Draft Liquid Pathway Generic Study", Report No. NUREG–140.

30. U.S. Nuclear Regulatory Commision (USNRC). 1977. "Calculation of Annual Doses to Man from Routine Releases of Reactor Effluents for the Purpose of Evaluating Compliance with 10CFR Part 50, Appendix I", Regulatory Guide 1.109, Rev. 1.

31. Watters, R. L., et al. 1980. "Synthesis of the Research Literature", in *Transuranic Elements in the Environment*, W. C. Hanson, Ed, U. S. Department of Energy Technical Information Center DOE/TIC-22800, Washington, D. C., pp. 1–44.

32. Whicker, F. W. and V. Schultz. 1982. *Radioecology: Nuclear Energy and the Environment*, Vol. II, CRC Press, Boca Raton, pp. 33–34.

33. Yousef, Y. A., A. Kudo, and E. F. Gloyna, "Radioactivity Transport in Water". 1970. University of Texas Center for Research in Water Resources Technical Report 20, CRWR-53.

CHAPTER 11

Biological Transport and Pathway Analysis

Introduction

More than 300 individual radionuclides currently reside in the biosphere. About 200 of these representing 35 different elements have been introduced into the environment by nuclear weapons testing and nuclear reactors. To these must be added the naturally occurring radionuclides, some of which have been extracted from the earth and concentrated as has been the case with uranium, radium, and thorium. Although in the aggregate the number appear formidable, only about a dozen or so nuclides are present in sufficient concentration and suitably biologically available to be significant from the standpoint of dose to man or other living things; these include tritium, ^{14}C, ^{40}K, $^{58,60}Co$, ^{85}Kr, $^{89,90}Sr$, $^{129,131}I$, $^{134,137}Cs$, ^{239}Pu, plus radium and uranium.

Even so, the problem of biological transport is quite complex. Radionuclides can be transported through the food chain to man, with both the rate and amount transported determined by many factors, both physical and biological, that go far beyond the consideration of simple ingestion or inhalation. In both physical and biological systems, radionuclides behave very much if not identically to their stable homologues, and generally follow the same biochemical or metabolic pathways. Thus, ^{131}I will behave exactly like stable iodine, and tritium like ordinary hydrogen. This makes the task of evaluating and understanding biological transport of radionuclides far simpler; it also enables an understanding of the behaviour of the naturally occurring stable elements by studying the more easily and convenient radioactive isotopes present in the environment.

Organisms tend to accumulate and retain radionuclides, and the radioactivity concentration of an organism is customarily described or characterized in terms of the specific activity. As used in radioecology, the

243

specific activity has several definitions or meanings. The most precise but not necessarily most common usage is as the ratio of the number of radioactive atoms of the element in question to the total number of atoms of that element in the organism. Another and perhaps more commonly used usage of the term is as the activity, either gross or from a specific radioelement or radionuclide, per unit mass of the organism, or per unit mass of the stable element. Specific activity is also sometimes used to mean the activity per unit or total mass of an organ or the entire organism, particularly if the nuclide is well distributed throughout. Each of these may be numerically quite different and therefore specification of the units or context of use is extremely important. In no case should the biological specification of specific activity be confused with the physical (radiological) term, which is an expression of the activity per unit mass of the radionuclide—i.e. the number of curies or becquerels per gram.

Stable elements can be used as chemical analogues for studying and understanding the behaviour, movement, and distribution of radioactivity through the environment if the stable element chemistry of the organisms constituting the links in the biological pathways is known. In addition, the specific activity ratio of the element at the source of access to the food chain must be known. The specific activity ratio, SAR, is defined as

$$\text{SAR} = \frac{\begin{array}{c}\text{Radionuclide}\\\text{in food}\end{array}}{\begin{array}{c}\text{Total element}\\\text{in food}\end{array}} = \frac{\begin{array}{c}\text{Radionuclide concentration}\\\text{in consumer}\end{array}}{\text{Total element in organism}} \qquad (11\text{--}1)$$

and is commonly expressed in $\mu\text{Ci}/\text{g}$. The SAR is valid only if there is complete environmental mixing, no biological discrimination among isotopes, and an adequate knowledge of the chemistry of the ecosystem. Only the second condition—i.e. no biological discrimination among isotopes of an element—is essentially always met. Complete environmental mixing is seldom achieved, even in small areas, and knowledge of the chemistry of the ecosystem may or may not be sufficient. Thus the SAR may be more of a theoretical concept in many instances than an apt model or empirical tool.

Food chain kinetics and element distribution require knowledge of many bioenvironmental factors, including the geochemical characteristics of the ecosystem, the dietary levels of various nutrients in foodstuffs, and the physiological demands of the consumers. For man, the cultural characteristics also need to be considered. Since knowledge of these factors is usually quite limited, it is difficult to draw hard and fast conclusions regarding the bioenvironmental transport and distribution of radionuclides.

Biological Concentration

The uptake of radioactive elements by plants or animals puts these substances into the ecological food chain. The term food chain is used by ecologists to signify the transfer of food energy from its original source, usually in plants, through the successive groups of animals that eat the plants and in turn are eaten by other animals. A typical food chain might have three to five successive links or trophic levels (Whittaker 1975). The first or lowest level in the typical food chain is known as the producer level. Producers are photosynthetic plants, plankton, or similar organisms that takes up radioactivity directly from non-living portions of the environment such as the soil or water.

The next level is that of the primary consumer, or the first animal to feed on the plant food. This is followed by the first carnivore or secondary consumer level, which is characterized by an animal feeding on an herbivore. The fourth trophic level is that of the secondary carnivore or tertiary consumer, and the fifth level that of the tertiary carnivore. Obviously, the food chain can have fewer than five or even three levels, as would be the case of an omnivore such as a bear or man eating the plant producer directly. Similarly, a food chain could have many more than five trophic levels, this being determined by the number of carnivores in the chain.

Although the boundaries of the various trophic levels are not sharply defined, man is ordinarily at the end of the chain or at the highest trophic level. The radioactivity thus passes through one or more other organisms before it reaches man, and it may be concentrated by one or more of the organisms in the chain before it reaches the final level. Thus, the highest trophic level will usually have the greatest concentration of radionuclides, and hence incur the greatest dose. The highest trophic level, usually man, is thus considered to be the indicator organism.

Concentration may occur in varying degrees and at any trophic level. The concentration ratio, CR, is also known as the concentration factor and is usually defined as the ratio of the element in the consumer or a specific tissue, organ, or compartment of that consumer to that in what is consumed, or to what is in the environment, viz.

$$CR = \frac{\text{Concentration in consumer}}{\text{Concentration in environment}} \qquad (11\text{--}2)$$

The concentration is usually expressed in terms of activity per unit mass (e.g. $\mu Ci/g$ or Bq/g) or, for aquatic media, activity per unit volume (e.g. uCi/ml or Bq/m^3). Thus for terrestrial systems the CF will be dimensionless while in aquatic systems it may assume the dimensions of volume per mass. Note that the concentration ratio may refer to many things such as the ratio of the concentration of a specific nuclide in the whole animal relative to that in the general environment, or to the concentration of a specific nuclide in a particular tissue or organ relative to than in the environment or a portion of the environment such as a major foodstuff of the organism. However, if unspecified, the concentration ratio is generally assumed to have plants as its base in the case of terrestrial ecosystems, and water in the case of aquatic systems.

Concentration is also known as bioconcentration or bioaccumulation, terms strictly applicable only if the concentration ratio exceeds unity. All three terms, are essentially identical in meaning. Concentration ratios are empirically determined and may be highly useful for prediction of routes and rates of transfer of radionuclides through the food chain to man, and are necessary for the calculation of radiation doses.

Concentration ratios are affected by many factors, including the chemical and physical form of the radionuclide in the environment, the biological need for the specific element or its homologues by the plant or animal, season and other climatic factors such as temperature, the age of the organism, the specific tissue or organs involved, and the specific characteristics of the local ecosystem. The presence of high concentrations of stable isotopes of a given radioelement will result in reduced concentrations in an organism, as implied in Equation 11–1. Similarly, the specific need of the animal for an element at a particular age in its life cycle may result in an increase or reduction in the concentration ratio; for example, a young animal laying down new calcium based bone or shell might be expected to have a greater concentration of ^{90}Sr, a calcium homologue, than an older animal whose bone structure may have been essentially complete prior to the introduction of the radiostrontium into the environment.

Among terrestrial animals, concentration ratios typically range from less than 1 to several hundred. Typically, values for tritium and ^{14}C are unity. Among insects and other invertebrates, the concentration ratio for radiostrontiums may be as low as 0.1, or as great as 100 for the so-called "calcium-sink" or shelled invertebrates such as snails, millipedes, and isopods. [137]Cs, which is virtually ubiquitous in the environment as a result of nuclear testing, tends to concentrate in the muscle tissue of the higher trophic levels, with a concentration factor of 2–3 noted at each step in the food chain (Reichle, Dunaway, and Nelson 1970).

For plants, concentration ratios for most elements are unity. For herbivorous mammals feeding on these plants, bioconcentration factors of less than unity are common for many elements, including essential ones such as cobalt, iron, and iodine, which usually have CR values of about 0.5. Elements such as strontium and cesium have a wider range of values for the CR, generally over an order of magnitude from about 0.3 to 5. Radium is well discriminated against, the CR being on the order of 0.01 (Reichle, Dunaway, and Nelson 1970). In invertebrates, the CR for most elements is <1 except for certain essential elements such as sodium and phosphorous which have CR's in the range of 10 to 30, and potassium with a range about an order of magnitude less.

Bioconcentration in aquatic biota shows much greater diversity than in terrestrial species. The chemical content of the water and the elemental needs of the organism are of prime importance. Organisms growing in nutrient poor (oligotrophic) systems show the greatest concentration of radionuclides as they must utilize a greater proportion of the available element to sustain life. In nutrient rich (eutrophic) waters, concentration ratios are very much smaller.

Specific organisms or classes of organisms may exhibit special characteristics with regard to elemental uptake and concentration. Strontium and magnesium tend to accumulate in the shells of clams and snails, and strontium in the calcareous skeletons of various marine organisms and in the bones of vertebrates. Phosphorus is generally highly accumulated in aquatic species at all trophic levels, while cesium and potassium are concentrated to a relatively small degree (Reichle, Dunaway, and Nelson 1970). There is also considerable difference between concentration factors in the marine and fresh water environment; in general, the mean CR for most elements studied is greater in freshwater than in marine organisms although the ranges of values observed may overlap. For example, the mean concentration ratio for cesium in freshwater organisms has been reported as about 900, with a range of 80–4,000, as compared with a mean of 51 and a range of 17–240 in marine plants (Eisenbud 1973). For cobalt, the corresponding mean values of CR are 6,760 in freshwater plants and 553 in marine plants. Similar effects are noted with higher trophic levels. The effect, however, is not consistant; for example, iodine is more heavily concentrated by marine biota than by freshwater organisms.

TABLE 11-1   Typical Plant-Soil Concentration Ratios.

Element*	CR	Element*	CR
H	4.8	Mo	0.12
C	5.5	Tc	0.25
Na	0.052	Ru	0.05
P	1.1	Rh	13
Cr	0.00025	Ag	0.15
Mn	0.029	Te	1.3
Fe	0.00066	I	0.02
Co	0.0094	Cs	0.01
Ni	0.019	Ba	0.005
Cu	0.12	La	0.0025
Zn	0.4	Ce	0.0025
Rb	0.13	Pr	0.0025
Sr	0.017	Nd	0.0024
Y	0.0026	W	0.018
Zr	0.00017	Np	0.025
Nb	0.0094		

*Listed in order of increasing atomic number.
Source: USNRC 1977.

The great diversity of concentration factors observed for various elements under various conditions makes it very difficult to generalize, even as has already been done to some extent. However, it is essential to have some knowledge of the concentration ratio for the purpose of evaluating biological transport of environmental radioactivity and for environmental radiological impact evaluations and dosimetry. Representative bioconcentration data are given in Tables 11-1 and 11-2 which are based on recommendations of the U.S. Nuclear Regulatory Commission. While these tables provide a reasonably good characterization of the relative concentration of elements by biota, the values cited are probably on the average conservative because of their intended application to environmental dose estimation from nuclear power plant operations. Again, it should be emphasized that concentration factors for any given element may vary from one to several orders of magnitude in a single trophic level or even species depending on the specific conditions in the ecosystem and therefore the best course of action is to empirically determine the appropriate CR's for the radionuclides and ecosystems of concern.

A useful adjunct to the concentration factor or ratio is the discrimination factor, which characterizes the relative uptake of two elements by

TABLE 11–2   Typical Concentration Ratios for Aquatic Organisms.

Element*	Freshwater Fish	Invertebrate	Saltwater Fish	Invertebrate
H	0.9	0.9	0.9	0.93
C	46,000	91,000	18,000	14,000
Na	100	200	0.067	0.19
P	100,000	20,000	29,000	30,000
Cr	200	2,000	400	2,000
Mn	400	900	550	400
Fe	100	3,200	3,000	20,000
Co	50	200	100	1,000
Ni	100	100	100	250
Cu	50	400	670	1,700
Zn	2,000	10,000	2,000	50,000
Br	420	330	0.15	3.1
Rb	2,000	1,000	8.3	17
Sr	30	100	2	20
Y	25	1,000	25	1,000
Zr	3.3	6.7	200	80
Nb	30,000	100	30,000	100
Mo	10	10	10	10
Tc	15	5	10	50
Ru	10	300	3	1,000
Rh	10	300	10	2,000
Te	400	6,100	10	100
I	15	5	10	50
Cs	2,000	1,000	40	25
Ba	4	200	10	100
La	25	1,000	25	1,000
Ce	1	1,000	10	600
Pr	25	1,000	25	1,000
Nd	25	1,000	25	1,000
W	1,200	10	30	30
Np	10	400	10	10

*Listed in order of increasing atomic number.
Source: USNRC 1977.

an organism (Polikarpov 1966). The discrimination factor, DF, is simply the quotient or ratio of the two concentration ratios, or, mathematically

$$DF = \frac{CR_1}{CR_2} \qquad (11\text{–}3)$$

The discrimination factor is frequently used to describe the relative uptake of elements from the water by aquatic organisms, and is particularly useful in comparing elements with similar chemistry such as strontium and calcium. One application of the DF was to express the discrimination between the uptake of ^{90}Sr and elemental calcium by aquatic orgainisms in terms of the so-called strontium unit. The strontium unit is obtained by combining Equations 11–2 and 11–3 to yield

$$DR = \frac{C_{Sr}}{C'_{Sr}} : \frac{C_{Ca}}{C'_{Ca}} \qquad (11-4)$$

in which C_{Sr} and C_{Ca} are the concentrations of ^{90}Sr and calcium in the organism, and C'_{Sr} and C'_{Ca} the concentrations in the aquatic medium. If the indicated arithmetic operations are carried out, Equation 11–4 becomes

$$DR_{1,2} = \frac{C_{Sr}C'_{Ca}}{C_{Ca}C'_{Sr}} \qquad (11-5)$$

The numerator and denominator in Equation 5 are the stronium content in the organism and the water, respectively, expressed in strontium units which were popularly used during the heyday of atmospheric weapons testing in the 1950's and appear widely in the scientific literature of that time.

The concentration ratio can also be used to determine the fraction, f, of a radionuclide concentrated from the environment by aquatic organisms (Polikarpov 1966). Since the concentration ratio is expressed as

$$CR = \frac{C}{C'} = \frac{fW'}{f'W} \qquad (11-6)$$

in which C and C′ are the concentrations of the radionuclide in the organism and the water, respectively, W and W′ the weights of the organism and the water, and f and f′ the fraction activity in the organism and water, respectively. Equation 11–6 can be solved for f to yield

$$f = \frac{CR}{CR + W'/W} \qquad (11-7)$$

Equation 11–7 is also applicable to the terrestrial situation.

Biological Turnover

The concentration of a radionuclide in an organism is directly related to the balance between the intake of the radionuclide and its elimination from the organism by excretion or radioactive decay. Intake can be via food or water or from respiration of airborne radioactivity. Mathematically, the equilibrium radioactivity concentration at time t, C(t), can be represented by

$$C(t) = \frac{IC_o}{\lambda_2 - \lambda_1} (e^{-\lambda_1 t} - e^{-\lambda_2 t}) + C_o e^{-\lambda_2 t} \qquad (11-8)$$

in which I is the intake rate (expressed in grams of food per gram of consumer per day)

   A is the concentration of the radionuclide in the food, in terms of $\mu Ci/g$ or other suitable units of activity per unit mass

   C_o the original concentration of the radionuclide in the organism $(\mu Ci/g)$

   λ_1 the removal coefficient from the food (days^{-1})

   λ_2 the biological turnover or excretion coefficient (days^{-1})

and   t the time in days.

For the inhalation case, I becomes the breathing or respiration rate expressed in units of volume per time and A the concentration of the radionuclide in the air expressed in units of activity per unit volume. Note that Equation 11–8 assumes that all the radionuclide taken into the organism will be assimilated or absorbed into the tissues; such may not be the case, especially with respect to intake via respiration. Hence correction must be made for the fractional uptake, and this is usually done by multiplying the right hand side of Equation 11–7 by a factor f which is equal to the fraction of the radionuclide assimilted by the organism.

If the organism was initially free of the radionuclide, the concentration will increase with time until a limiting or equilibrium value will be reached. The time at which this occurs, t_{eq}, can be obtained from Equation 11–8 by putting $C_o = 0$ and differentiating with respect to time to yield

$$\frac{dC}{dt} = \frac{\lambda_2 t IC_o}{\lambda_2 - \lambda_1} e^{-\lambda_2 t} - \frac{\lambda_1 t IC_o}{\lambda_2 - \lambda_1} e^{-\lambda_1 t} \qquad (11-9)$$

which can be solved for t_{eq} by setting the differential equal to 0,

$$\frac{\lambda_2 tIC_o}{\lambda_2 - \lambda_1} e^{-\lambda_2 t} = \frac{\lambda_1 tIC_o}{\lambda_2 - \lambda_1} e^{-\lambda t}$$

and solving for t

$$t_{eq} = \frac{1}{\lambda_2 - \lambda_1} \ln \frac{\lambda_2}{\lambda_1} \tag{11-10}$$

Organisms with short life spans may not live long enough to accumulate equilibrium or near equilibrium concentrations of even very short-lived radionuclides; hence their dose will also be relatively small. Long lived organisms, however, may accumulate appreciable doses from the continuous equilibrium concentrations of various environmental radionuclides in their bodies.

For initially uncontaminated animals feeding on a contaminated food source, the intake of a radionuclide, r, neglecting decay, can be expressed by

$$r = \frac{\lambda_b Q_t}{f(1 - e^{-\lambda_b t})} \tag{11-11}$$

in which λ_b is the biological excretion constant,
 Q_t is the body burden at time t,
 and f the fraction absorbed from the gastrointestinal tract.

If a burden of radionuclide existed previously, it must be added to any new intake. For continuous intake, assuming a constant level in the environment, Equation 11-11 reduces to

$$r = \frac{\lambda_b Q_{eq}}{f} \tag{11-12}$$

Assimilation, represented by f, is not equal for all nuclides but is highly variable and depends on many factors including

- chemical form and solubility
- needs of the animal
- blocking by stable isotopes
- residence time of food in gut

- type of food
- age of animal
- source of the food.

Removal of a radionuclide from the organism is dictated by two separate processes. One is radioactive decay, which can be characterized by the radioactive decay constant λ_r, which is simply the probability of an atom undergoing radioactive decay in a given time period. The biological turnover or excretion coefficient, λ_b, is the biological analogue of the radioactive decay constant, and hence is the probability that an atom will be removed from the organism by biological processes. The total probability is thus the sum of the two, and is known as the effective removal coefficient and symbolized by λ_{eff}.

As the biological turnover coefficient is analogous to the physical or radioactive decay constant, the biological half-life, T_b, can be derived in the same manner as the radioactive half-life, and is thus equal to $0.693/\lambda_b$. Similarly, the effective half-life, T_{eff}, is equal to $0.693/\lambda_b$. Since the effective removal coefficient is equal to the sum of the radioactive decay and biological removal coefficients, it holds that

$$\frac{0.693}{T_{eff}} = \frac{0.693}{T_r} + \frac{0.693}{T_b}$$

which can be solved for the effective half-life, T_{eff}, to yield

$$T_{eff} = \frac{T_b T_r}{T_b \times T_r} \tag{11-13}$$

in which T_b and T_r are the biological and radiological half-lives, respectively. Note that if either the biological or radiological half-life is very much greater than the other, the effective half-life will approach and for all practical purposes equal the value of the smaller.

In practice, Equation 11-13 finds considerable application in biological research studies to determine the biological half-life of an element. The effective half-life can easily be determined by direct observation, and the radiological half-life, if unknown, can relatively easily be determined. Thus, simply by using a radioactive isotope of an element, it is possible to easily determine the biological turnover time within an organism.

A general relationship between body weight and biological half-life has been reported for radiocesium and several other radioelements, with the biological half-life increasing as a power function of body weight (Reichle, Dunaway and Nelson 1970). A specific power function can be fitted to various classes of animals, viz. insects, other inverebrates, cold-blooded vertebrates, and warm-blooded vertebrates. Independent evaluation of the data, however, leads to the conclusion that any relationship, except perhaps in the warm-blooded vertebrates, is tenuous (Pisarcik and Kathren 1983).

Biological uptake and turnover is associated with metabolism which is species dependent as well as a function of body size, age, sex, physical activity, physical condition, metabolic rate, and deposition of the radio-nuclide in biologic sinks. A biologic sink is a compartment within the organism that receives the radionuclide but does not release it. For example, the turnover of elements in bone is very slow, and hence bone acts as a sink for calcium and its chemical analogues including the radiostrontiums and radium.

Bioaccumulation of ^{3}H, ^{14}C, and ^{32}P

Although each radionuclide is unique with regard to biological uptake and concentration, it is nonetheless possible to generally characterize the behaviour of certain specific nuclides and chemical groups. Tritium, a heavy radioactive isotope of hydrogen, is widespread throughout the biosphere largely in the form of tritiated water (HTO), with smaller amounts as a gas (HT) or as various tritiated carbon compounds. Uptake is thus largely in the form of water. Bioconcentration is nil, and the concentration ratio is ordinarily assumed to be equal to 1 (Bard 1977). However, as the mass of the tritium atom is threefold greater than that of ordinary hydrogen, chemical reaction rates and diffusion may be significantly slower, and this may result in a concentration factor of less than unity as has been noted in some organisms lying near the bottom of the food chain. Organically bound tritium may produce concentration ratios slightly greater than unity in some biota (Koranda and Martin 1969).

Carbon-14 is also widespread in the environment and in biological materials. The relative mass difference between ^{14}C and the predominant stable isotope ^{12}C is only 7/6, not nearly so great as with tritium and ^{1}H, and hence the differences in reaction kinetics or diffusion between the two carbon isotopes is relatively small. Chemical bonding of ^{14}C is stronger than its stable isotope, but this does not appear to produce any significant effect on biochemical processes. Thus, the behaviour of ^{14}C is virtually identical to that of ^{12}C, and the bioconcentration ratio for this radionuclide is essentially identical to that of stable carbon.

^{32}P, with a 14.3 day half-life is widely used as a tracer and in other biomedical applications and is frequently released to the environment. Phosphorus is an element essential to living things but has a relatively low abundance in the biosphere; hence, ^{32}P tends have a relatively large concentration ratios and slow biological turnover in some biota particularly in aquatic species at the low end of the food chain.

As a group, the three light radionuclides ^{3}H, ^{14}C, and ^{32}P are characterized by a high degree of food chain transport and assimilation, along with bioconcentration factors of unity, with few exceptions such as that noted for phosphorous above. All are essential to life, with ^{3}H and ^{14}C found more or less uniformly distributed throughout living things, while phosphorus tends to accumulate in bone. Effective half-lives are relatively short, typically on the order of days.

The Alkaline Metals

The alkali metals—Li, Na, K, Rb, Cs, and Fr—comprise Group IA of the periodic chart, and exhibit quite similar biochemical behaviour. K, Rb, and Cs all have important radioisotopes. In general, the ratio of ^{40}K to stable potassium is constant throughout biological systems, there being no selective fractionation or uptake. Concentration ratios of potassium are quite variable, and can range to several thousand among freshwater biota (Eisenbud 1973). The behaviour of cesium generally follows that of potassium, but tends to be concentrated to a greater degree than potassium, increasing with trophic level (Pendleton et al. 1965). Freshwater fish at the higher trophic levels have been found to have CR's of several thousand, but most of the bioconcentration occurred at the lowest (i.e plankton) trophic level (NCRP 1977). In general, CR's for cesium in terrestrial systems are in the range of 0.1 to 0.5 in invertebrates, and up to 7 in carnivores (Reichle, Dunaway, and Nelson 1970). This may be due to the longer biological half-life of cesium in most organisms (Whicker and Schultz 1982). Concentration of cesium and rubidium will be enhanced in potassium poor environments.

The Group IA metals tend to have a high degree of food chain transport, with relatively low concentration factors. Concentration factors are close to unity, and may be less in some instances. Assimilation by biota is high, and these metals tend to be more or less uniformly distributed throughout the organism. Biological half-lives tend to be on the order of weeks.

Group IIA of the periodic chart is known as the as the alkaline earth metals and includes calcium, strontium, barium, and radium. Chemically, these elements exhibit similar biological behaviour in that they tend to accumulate in bone and in the calcareous skeletal tissues. The biological half-life of these elements is long in the calcareous biota, and hence the skeleton acts as a biological sink for radioisotopes of strontium and barium, and for radium.

Uptake and hence the bioconcentration factors are quite different for Ca, Sr, and Ra. In general, the concentration ratio of radium in most biota is considerably less than one. A value of 0.01 has been cited for herbivores (Reichle, Dunaway and Nelson 1970), and the assimilation factor in man is only 0.3 (ICRP 1980).

The uptake and concentration of strontium is of special significance because of the large amount of its long lived isotope ^{90}Sr which has been injected into the environment from nuclear weapons tests, and because of its biological similarity to calcium. A special unit known as the *observed ratio* (OR) is commonly used. The OR is defined as the Sr:Ca ratio in an organism relative to the Sr:Ca in the diet or, in the case of plants, the soil. In plants, the OR is usually approximately unity, suggesting that there is little discrimination between strontium and calcium (Comar 1965). Uptake and concentration of both strontium and calcium by plants is very small, on the order of 0.2 (USNRC 1977). In animals the observed ratio is commonly less than one, suggestive of discrimination against strontium; concentration ratios are in the vicinity of unity except for invertebrates, where the CR is typically 0.1 (Reichle, Dunaway, and Nelson 1970).

Despite the low OR values, appreciable concentration ratios of radiostrontiums have been observed in freshwater organisms; although the CR usually falls in the range of one or less, values to perhaps as much as several hundred have been reported (Polikarpov 1966). Concentration in marine biota is significantly less than in freshwater species; for example, in freshwater plants, the mean concentration ratio is 200 as compared with 21 in saltwater species (Eisenbud 1973). Unlike cesium, concentration of strontium does not increase with increasingly higher trophic level.

The alkaline earth metals of Group IIA show a moderate to high degree of food chain tansport, along with successive trophic level concentration ratios of less than unity. Biological half-lives tend to be relatively long. Radium, the heaviest member of the group, shows the most extreme behaviour, having the longest biological half-life, but smallest degree of uptake.

The Actinides

Elements with atomic numbers beginning with 89, actinium, are known as the actinides. This class of elements includes all the artificially produced elements with $Z > 92$, including neptunium ($Z = 93$), plutonium ($Z = 94$), americium ($Z = 95$), and curium ($Z = 96$), as well as naturally occurring uranium ($Z = 92$) and thorium ($Z = 90$). Actinides have a similar although very complex chemistry, and can exist in several oxidation states. They higher actinides are of particular interest environmentally, not only because they are not found in nature but because of their very high radiotoxicity. Uranium and thorium are of special interest as these elements are the parents of naturally occurring chains discussed in Chapter 3.

As a class, actinides readily tend to form insoluble oxides, and therefore are not biologically mobile. Unlike most of the lighter elements, assimilation is very poor, particulary if intake is in the form of an oxide. Concentration ratios, however, tend to be highly variable, and are usually greater in aquatic than in terrestrial systems. Plutonium has been the most widely studied of the transuranics, largely because of its widespread environmental distribution and use. Most field studies have been conducted in relatively arid locations in the Western continental United States in the vicinity of sites where plutonium is used or processed, such as the Nevada Test Site near Las Vegas, the Rocky Flats facility near Denver, and the Hanford Reservation in south-central Washington state, and the Los Alamos National Laboratory in New Mexico. However, some studies have been carried out at the Oak Ridge National Laboratory near Knoxville, Tennessee, and in South Carolina near the Savannah River Plant.

In terrestrial systems, concentration ratios of 10^{-6}–10^{-3} have been reported for plutonium on the basis of laboratory studies (Price 1973). Uptake is facilitated by lowering the pH of the soil or the addition of chelating agents. For plutonium, the soil uptake ratio appears to increase exponentially with time. CR values were one to two orders of magnitude greater in field studies conducted in the vicinity of the Oak Ridge

National Laboratory in eastern Tennessee, but those in the humid climate near the Savannah River Plant were similar to values obtained in the more arid Western regions. Studies with americium uptake by rice, soybeans, and other plants showed CR's ranging from one to two orders of magnitude than those obtained with plutonium, but still well below unity (Adriano, Wallace, and Romney 1980). Both field and laboratory studies of the uptake of transuranic elements by terrestrial plants from soil shows the following hierarchy: $Np > Cm$, $Am > Pu$ (Price, 1972; Schreckhise and Cline 1980). In one series of experiments, the relative uptake of Np was 2,200 to 45,000 times that of plutonium, while Am and Cm were only 10–20 times as great (Schreckhise and Cline 1980).

Although transuranic uptake by plants has been extensively studied, relatively little information is available on uptake and assimilation by terrestrial animals higher up in the food chain. However, long term studies have been conducted in arctic tundra ecosystems, revealing caribou:lichen concentration ratios of 10^{-4}–10^{-3} (Hanson 1980). Because of the low assimilation of plutonium and other actinides by animals, CR values are expected to be low.

In aquatic systems, significant CR values have been observed for plutonium. Concentration ratios of several thousand have been observed in phtoplankton from both marine and freshwater environments; concentration by organisms higher in the food chain appears more variable, with values for fish ranging from 1 to several hundred (Noshkin 1972, Watters et al. 1980).

Fewer studies have been conducted with uranium and thorium, and the available data indicate concentration ratios and biological uptake and distribution data similar to that of the transuranic actinides. Concentration ratios of 10 to 400 have been reported for uranium in marine biota (Polikarpov 1966). Concentrations tend to diminish with increasing trophic level in both aquatic and terrestrial systems, suggesting reduction in concentration factor among the higher animals (Kovalsky, Vorotnitskaya, and Lekarev 1967). Commonly, a CR of unity is assumed for uranium.

Other Elements

There are numerous other elements with radioactive isotopes in the environment; most are of minor significance, either because of low concentration or short half-life. Radiocobalt is of significance in that it is released directly to the environment from the operation of nuclear power plants. Cobalt is readily concentrated by both terrestrial and aquatic organisms; CR's of 100 to 100,000 have been observed in freshwater organisms, and 10 to 1000 in marine organisms (Polikarpov 1966). In terrestrial systems, plant:soil concentration ratios are usually in the range of one or less, although a range of 0.01–2 has been reported (Grummitt 1976); uptake and concentration by animals probably falls in the same range.

^{65}Zn has been widely studied as it was a predominant waste product associated with the operation of the single pass production reactors on the Columbia River at Hanford. Zinc is an essential biological trace element and is readily accumulated by biota and transported through food chains. Concentration factors in the range of 1000 to 10,000 have been reported for aquatic biota, and a plant:soil ratio of 0.1 is usually used for environmental radiological hazards evaluations (Polikarpov 1966; USNRC 1977).

The importance of the rare earth elements or lanthanides lies in the fact that these middle weight elements include a great many radioactive fission products. Half-lives are usually short, however, and the biological significance of these elements is further reduced as they are not essential to life and thus are poorly assimilated. Concentration ratios for these elements are typically less than unity. Cerium, for example, has been the subject of considerable study; the range of plant:soil concentration ratios for this element range are generally well below unity, ranging from a high value of perhaps 0.25 for a few plants grown in sandy soils, to more typical values of 0.001 to 0.0001 (NCRP 1978).

A final group of elements to be considered is the inert gases, Group VIII of the periodic table. As these elements are non-reactive chemically, uptake is minimal, and concentration therefore negligible.

References

1. Adriano, D. C., A. Wallace, and E. M. Romney. 1980. "Uptake of Transuranic Nuclides from Soil by Plants Grown Under Controlled Environmental Conditions", in *Transuranic Elements in the Environment*, W. C. Hanson, Ed., U.S. Department of Energy Publication DOE/TIC-22800. pp. 336–360.
2. Bard, S. T. 1977. "Environmental Deposition of Tritiated Water Vapor", *Dissertation Abstracts* 38:3073-B.

3. Comar, C. L. 1965. "Movement of Fallout Radionuclides through the Biosphere and Man", *Annual Review of Nuclear Science* 15:175.
4. Eisenbud, M. 1973. *Environmental Radioactivity*, Second Edition, Academic Press, New York, pp. 153-5.
5. Grummitt, W. E. 1976. "Transfer of Cobalt-60 to Plants from Soils Treated with Sewage Sludge", in *Radioecology and Energy Resources*, C. E. Cushing, Jr., et al., Eds., Dowden, Hutchinson and Ross, Stroudsburg, PA, p. 331.
6. Hanson, W. C. 1980. "Transuranic Elements in Arctic Tundra Systems", in *Transuranic Elements in the Environment*, W. C. Hanson, Ed., U.S. Department of Energy Publication DOE/TIC-22800, pp. 441–458.
7. International Commission on Radiological Protection (ICRP). 1979. "Limits for Intakes of Radionuclides by Workers", ICRP Publication 30, Supplement to Part 1, *Annals of the ICRP* 3:1.
8. Koranda, J. J. and J. R. Martin. 1969. "Persistance of Radionuclides at Sites of Nuclear Detonations", in *Biological Implications of the Atomic Age*, A. M. Goulden, Ed., U.S. Atomic Energy Symposium Series 16, Washington, D.C., pp. 159–168.
9. Kovalsky, V. V., I. E. Vorotnitskaya, and V. S. Lekarev. 1967. "Biogeochemical Food Chains of Uranium in aquatic and Terraneous Organisms", in *Radioecological Concentration Processes*, B. Aberg and F. P. Hungate, Eds., Pergamon Press, New York, pp. 329–338.
10. National Council on Radiation Protection and Measurements (NCRP). 1977. "Cesium-137 from the Environment to Man: Metabolism and Dose", NCRP Report No. 52, Washington, D.C.
11. National Council on Radiation Protection and Measurements (NCRP). 1978. "Physical, Chemical, and Biological Properties of Radiocerium Relevant to Protection Guidelines", NCRP Report No. 60, Washington, D.C.
12. Noshkin, V. E. 1972. "Ecological Aspects of Plutonium Dissemination in Aquatic Environments", *Health Phys.* 22:537.
13. Pendleton, R. C., C. W. Mays, R. D. Lloyd, and B. W. Church. 1965. "A Trophic Level Effect on ^{137}Cs Concentration", *Health Phys.* 11:1503.
14. Pisarcik, D. and R. L. Kathren. 1983. "The Relationship of Biological Half-Life to Body Weight in Mammals", *Health Phys.* 45:202.
15. Polikarpov, G. G. 1966. *Radioecology of Aquatic Organisms*, Reinhold Book Division, New York.
16. Price, K. R. 1972. "Uptake of ^{237}Np, ^{239}Pu, ^{241}Am and ^{244}Cm from Soil by Tumbleweed and Cheatgrass", U.S. Atomic Energy Commission Report BNWL-1688.
17. Price, K. R. 1973. "A Review of Transuranic Elements in Soils, Plants, and Animals", *J. Environ. Quality* 2:62.
18. Reichle, D. E., P. B. Dunaway, and D. J. Nelson. 1970. "Turnover and Concentration of Radionuclides in Food Chains", *Nuclear Safety* 11:43.
19. Schreckhise, R. G., and J. F. Cline. 1980. "Comparative Uptake and Distribution of Plutonium, Americium, Curium, and Neptunium in four Plant Species", in *Transuranic Elements in the Environment*, W. C. Hanson, Ed., U.S. Depratment of Energy Publication DOE/TIC-22800, pp. 361–370.
20. U. S. Energy Research and Development Administration (USERDA). 1976. "Workshop on Environmental Research for Transuranic Elements", Report ERDA–76/134, Washington, D.C.

21. U.S. Nuclear Regulatory Commission (USNRC). 1977. "Calculation of Annual Doses to Man from Routine Releases of Reactor Effluents for the Purpose of Evaluating Compliance with 10 CFR Part 50, Appendix I", Regulatory Guide 1.109, Washington, D.C.

22. Watters, R. L., et al. 1980. "Synthesis of the Research Literature", in *Transuranic Elements in the Environment*, W. C. Hanson, Ed., U.S. Department of Energy Publication DOE?TIC-22800, pp. 1–44.

23. Whicker, F. W., and V. Schultz. 1982. *Radioecology: Nuclear Energy and the Environment*, Volume 1, CRC Press, Boca Raton.

24. Whittaker, R. H. 1975. *Communities and Ecosystems*, Second Edition, Macmillan Publishing Co, New York, p. 214.

CHAPTER 12

Pathway Analysis and Environmental Dose Assessment

Generalized Pathways

Environmental transport pathways describe the movement of radionuclides through the biosphere to the highest trophic levels, usually man, and are of prime importance in evaluating the biological impact of radioactivity in the environment. There are two general categories of pathways; the liquid pathway (Figure 12–1), which involves transport or entry into the food chain via water, and the gaseous pathway in which the radionuclide is initially introduced to the biosphere via the atmosphere (Figure 12–2). Both include the introduction of the radioactivity in any physical form—liquid, solid, or gas—but ordinarily substances moving through the aquatic pathway are introduced in the form of liquids, while those moving through the gaseous pathway are in the form of particulates, vapors, or gases as from stack effluents.

The gaseous pathway is sometimes also known as the terrestrial pathway, as its transfer mechanisms are principally land based. In this case, the introduction of radioactivity into aqueous systems via fallout or washout of airborne radionuclides would be considered a liquid rather than gaseous pathway, despite the original mode of introduction into the biosphere.

Pathways are also sometimes classified as direct and indirect. Indirect contamination occurs when the radioactive material enters the soil, (or, in the case of an aquatic system, an intermediate species), before passing on into the first trophic level or the plants. Indirect pathways are governed by the total quantity of radionuclide available in the soil, and thus are sometimes described as "cumulative dependent" (Comar 1965). Because of the time needed for uptake by plants, short-lived radionuclides do not contribute significantly to the indirect pathways; however, longer lived material deposited prior to the development of intermediate species is still available to developing plants or animals intermediate in

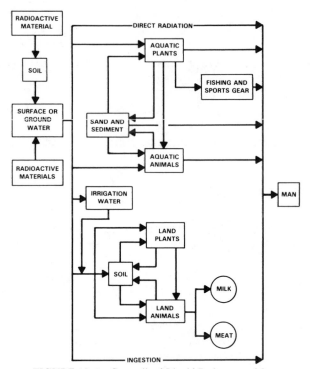

FIGURE 12-1   Generalized Liquid Pathways to Man.

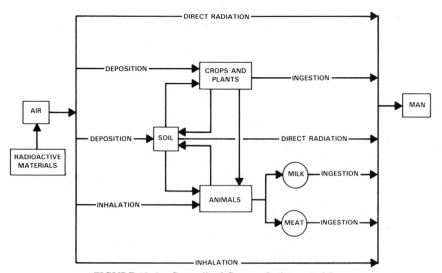

FIGURE 12-2   Generalized Gaseous Pathways to Man.

the chain. Also, radionuclides in soil or aquatic systems can be diluted or concentrated, or rendered unavailable to the food chain by various physicochemical processes.

In direct pathways, passage through the soil or intermediate species is by-passed. Thus direct pathways are usually reflective of recent events and provide a more immediate dose to the higher trophic levels. Radioactive half-life is thus of reduced or even minimal importance. Unlike indirect pathways, direct pathways are thus rate dependent in that they are affected by the rate at which the radionuclides are released into the environment, be it the rate of deposition or fallout, or the rate of release via a gaseous or liquid effluent.

As seen in Figure 12–2, gaseous pathways have basically five routes of importance:

1) direct radiation (i.e. external exposure)

2) deposition on plants

3) deposition on soil

4) inhalation by animals

5) inhalation by man.

Other pathways, such as direct ingestion by animals or man are of lesser importance. The direct radiation pathway is an external dose situation resulting largely from the immersion of an individual to a plume or cloud containing radioactivity. The exposure may result from either beta or photon radiations (alpha particles do not have sufficient energy to penetrate the dead layers of the skin) and is usually associated with the noble gases released from operating nuclear power plants, although it could occur from any source of airborne radioactivity, including particulate matter deposited on the ground, buildings, or other surfaces as radioactive fallout. For exposure to the noble gases, the radiation dose is calculated on the basis of immersion in an assumed infinite hemispherical cloud. Dose factors for key radionuclides are shown in Table 12–1. The legally allowable maximum permissible concentrations of the noble gases in air are based on the immersion dose, as are the Derived Air Concentrations (DAC's) put forth by the International Commisssion on Radiological Protection (ICRP 1979a, USNRC 1983).

Deposition or fallout of radioactivity on plants or on the soil or on buildings or other surfaces can also result in direct radiation exposure to man. The total dose delivered will be determined by many factors including the type and energy of the radiation, half-life of the radionuclide, the uniformity of its distribution, the shielding factors afforded by buildings (or even by the air), terrain, and even the height of the individual being

TABLE 12-1   Immersion Dose Rates from Selected Noble Gases.

Nuclide	Half-life	Energy (MeV)		Concentration* to deliver 1 mrad/h
		Beta	Photon (yield)	
^{85}Kr	10.8y	0.67	0.514(41%)	4×10^{-6}
^{133}Xe	5.27d	0.346	0.081(37%)	4×10^{-6}
^{135}Xe	9.14h	0.920	0.250(91%)	1×10^{-6}
			0.610(3%)	

*μCi/cm^3

exposed. This variability makes accurate modelling difficult. Simplified models based on the dose produced by a uniform infinite field one meter above the surface of the ground are thus most frequently used, with exposure rates commonly expressed in units of μrad/h per mCi/km^2. Frequently, the beta component is ignored, not so much because it may be insignificant, but because of the difficulties associated with modelling. For most gamma emitting fission and activation products, exposure rates one meter above the surface of uniformly contaminated range from about 10^{-4} to 10^{-2} μrad/h per mCi/km^2 (2.7 $\times$ 10^{-17}–2.7 x10$<^{-15}$ μGy/h per Bq/km^2) (Beck 1980).

Deposition on crops and other plants may result in entry of the radionuclide into the food chain in several ways. The potential pathways and their interrelationships are many and complex. One pathway is by direct absorption into the plant through the foliage; another is by ingestion of the contaminated plant by animals or man. The deposited material can also be resuspended from the plant and inhaled by animals, or it can be deposited on the soil and taken up by plants. Radionuclides deposited on plants may also be washed off and enter the ground where they can be taken up by plants, or may enter aquatic systems in this manner.

The concentration of radioactivity in and on vegetation, C_{veg} can be readily calculated by (IAEA 1982)

$$C_{veg} = \left[\frac{dR(1 - e^{-\lambda_1 t_e})}{Y\lambda_1} + C_S B \right] e^{-\lambda t_h} \qquad (12-1)$$

in which d is the deposition rate onto the soil from all processes, (activity per area per time)

R is the fraction of the activity interecpted by the vegetation (dimensionless)

λ_1 is the effective rate constant for the reduction of the activity from the vegetation (reciprocal time)

λ the radioactive decay constant (reciprocal time)

C_s is the concentration of the radionuclide in dry soil (activity per mass)

t_e is the time that the plants are exposed to contamination during the growing season

Y is the productivity or yield of the of the edible portion of the vegetation (mass per area)

B_v the soil to plant concentration factor of the radionuclide from the soil by edible portions of plants (dimensionless)

and t_h the holdup time between harvest and consumption of the food.

The concentration of the radionuclide in dry soil, C_s, can be calculated by

$$C_s = \frac{d[1 - e^{-\lambda_2 t_b}]}{P\lambda_2} \qquad (12\text{--}2)$$

in which λ_2 is the effective rate constant for reduction of the activity concentration from the root zone of the plant (reciprocal time)

P is the effective surface density for the root zone (mass per area)

and t_b the period of long deposition for soil activity (time).

Equations 12–1 and 12–2 illustrate the interrelationships of the various factors that affect the concentration of a radionuclide in and on vegetation. The deposition factor, d, is derivable from other terms, and is basically the deposition rate of the radionuclide by both wet and dry deposition mechanisms. The fraction of deposition activity intercepted by crops is based on the fractional area covered by the vegetation and must be modified to include retention also, giving due consideration to translocation and other removal processes such as washoff from the plant surfaces. For cesium on grain, a value of 0.5 is suggested (Aakrog 1971); for other nuclides, and for cesium on other food crops, 0.2 (Moore et al. 1979).

The productivity, Y, is a function of the ability of the land to produce vegetation and the amount of the radionuclide in the edible portions of the plant. For agricultural crops, productivity or yield is in the neighbourhood of 0.6 kg/m^2 (Baes and Orton 1979). The soil to plant concentration factor, B_v, is dependent not only upon which element is involved, but also on a host of other factors including solubility, the amount of stable isotopes in the soil, and the needs of the plant. Typical values are given in Table 11–1.

The effective removal constant for activity from the crops, λ_1, is actually the sum of the radioactive decay constant and the removal coefficients for environmental processes. The removal coefficients for environmental processes will be highly variable, and related not only to the particular element, but to weather, the type of plant, and numerous other factors. A reasonable estimate can be obtained from empirical observations of the half-life of various radionuclides on plants, recalling that the removal coefficient can be calculated by dividing this half-life into 0.693. Typical values of the removal half-life on vegetation are 10 days for iodine and 15 days for particulates (Miller and Hoffman 1979).

The effective surface soil density, P, varies not only with type of soil but with depth. For peats, the values range from 10 for shallow rooted plants (<2 cm) to 150 kg/m^2 for plants with a rooting zone depth to about a foot (30 cm). Values for other soils are 30 to 400 (IAEA 1982).

The inhalation pathways provide a mechanism for direct introduction of radionuclides into the food chain at a high trophic level. This pathway assumes greatest significance for soluble radionuclides, and hence, as will be shown, is of great importance for radioiodines, radiocesiums, and radiostrontiums.

For certain radioelements such as the radiostrontiums, radioiodines, and radiocesiums, the milk and meat pathways are of major significance (Figure 12–2). The concentration of a radionuclide in milk, C_{milk} is a direct function of the contamination level of the food and water ingested, and can be calculated by.

$$C_{milk} = F_{milk} I e^{-\lambda t} \qquad (12-3)$$

in which F_{milk} is the transfer coefficient to milk (reciprocal volume)
    I the daily intake (activity)
    λ the radioactive decay constant (reciprocal time)
and t_f the mean time of transport from the feed to the receptor.

The intake, I, is the product of the concentration in the food (or water) and the amount consumed. The mean time of transport is commonly assumed to be four days in the absence of site specific data. The transfer coefficient to milk is expressed in units of reciprocal volume, and is simply the fraction of the daily intake that is excreted in a liter of milk, and

TABLE 12-2   Transfer Coefficients to Beef.

Element	Coefficient (kg^{-1})	Element	Coefficient (kg^{-1})
Na	0.02	Cs	0.02
P	0.08	Ba	0.0002
S	0.1	La	0.002
Cr	0.02	Ce	0.002
Mn	0.001	Pm	0.002
Fe	0.003	Sm	0.002
Co	0.003	Eu	0.002
Ni	0.005	Pb	0.0008
Zn	0.02	Bi	0.02
Sr	0.0006	Po	0.003
Y	0.002	Ra	0.0005
Zr	0.02	Ac	0.00002
Nb	0.3	Th	0.0001
Tc	0.01	Pa	0.001
Ru	0.002	U	0.03
Ag	0.005	Np	0.001
Sb	0.001	Pu	0.00001
Te	0.02	Am	0.00002
I	0.01	Cm	0.00002

Source: IAEA 1982, p. 70.

ranges from a high of 0.02 for phosphorus, through 0.01 for iodine, and zinc, 0.001 for strontium, and values lower by one to two orders of magnitude for most other elements (Ng et al. 1977).

A similar calculation can be made to determine the concentration of radioactivity in meat, C_{meat}:

$$C_{meat} = F_{meat} I e^{-\lambda t_f} \qquad (12-4)$$

in which the terms are analogous to those in Equation 12-3. Values for the meat transfer coefficient are given in Table 12-2; the tranfer time is taken to be 20 days in the absence of site specific data (IAEA 1982).

Although the liquid pathways (Figure 12-3) appear more complex than the gaseous, they are simpler from a modelling standpoint and can typically be expressed by the simple linear equation.

$$C_{aquatic} = C_{HOH} CR \qquad (12-5)$$

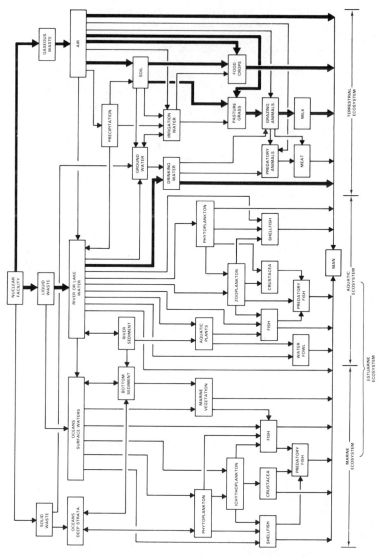

FIGURE 12–3 ^{90}Sr Pathways to Man. The important pathways are illustrated with heavy lines.

270

in which $C_{aquatic}$ is the concentration of the radionuclide in the aquatic species, C_{HOH} the concentration in the water, and CR the concentration ratio or bioaccumulation factor (Table 11–2).

Specific Pathways for Important Nuclides

Although radionuclides can reach man by a number of pathways, some are more important than others. For planning purposes, the concern is the critical pathway, which is defined as the pathway that will provide, for a given radionuclide, the greatest dose to a population, or to a specific segment of the population. This segment of the population is known as the critical group, and may differ depending on the radionuclides under consideration, age, dietary, or other cultural factors. The dose may be delivered to the whole body or to a specific organ; the organ receiving the greatest fraction of the permissible does level is known as the critical organ.

Potential critical pathways for various internally deposited radionuclides have already been touched upon. For the noble gases, the critical pathway is direct external exposure from immersion in a cloud. The immersion dose can be readily calculated if the concentration in the cloud is known. For calculation of the dose rate from beta radiation, a semi-infinite cloud—i.e. one with radius greater than the range of the betas—is assumed and gives rise to the following equation

$$\frac{dR}{dt} = \frac{k_1 k_2 E_{av} \chi}{\rho k_3} \tag{12–6}$$

in which dR/dt is the dose rate in rad/h

E_{av} the average energy of the beta particle in MeV

χ the concentration of the radionuclide in Ci/m^3

ρ the density of the air ($= 1293 \ g/m^3$)

k_1 a constant equal to 1.6×10^{-6} erg/MeV

k_2 a constant equal to 1.33×10^{14} disintegrations per hour-Ci

and k_3 a constant equal to 100 erg/g-rad.

Putting in the constants, Equation 12–6 becomes

$$\frac{dR}{dt} = 27.6 E_{av} \chi \tag{12–7}$$

which can be integrated with respect to time to give the total dose, R, in rad.

$$R = 27.6E_{av}\chi t$$

in which t refers to the time, in hours, of exposure. Hence for a cloud or plume, t refers to the time of passage in hours. The calculation can also be done in SI units, yielding

$$R = 1.3 \times 10^{-17} E_{av} \chi t \qquad (12\text{--}8a)$$

with R in gray, χ in Bq/m^3 and t in seconds.

The dose from x and gamma radiation is not so easily determined because of the greater penetrating power of the photons. However, if the cloud is assumed to be hemispherical, the photon dose rate dR/dt in rad/h can be obtained from

$$\frac{dR}{dt} = \frac{0.5k_1k_2E_{av}\chi}{k_3} \qquad (12\text{--}9)$$

which, when putting in the constants becomes

$$\frac{dR}{dt} = 13.8E_{av}\chi \qquad (12\text{--}10)$$

or, for total dose in rad, can be integrated with respect to time to yield

$$R = 13.8E_{ave}\chi\, dt = 13.8E_{av}\chi t \qquad (12\text{--}11)$$

or, for

$$R = 8 \times 10^{-11} E_{av} \chi t$$

where R is in gray, χ in Bq/m^3, and t in seconds. In Equations 12–8 through 12–10, symbols are as defined for Equation 12–6 above with the exception of E_{av}, which is the mean photon energy absorbed. The equations for photon dose rate and dose neglect backscattering from the ground; air attenuation is included in the E_{av} term.

Tritium has several important major pathways. For airborne releases, these are 1) direct inhalation and immersion, including absorption through the unbroken skin, 2) passage through food crops to man, and 3) passage into drinking water via precipitation. For liquid releases, the important pathway is through drinking water to man. Note that these are by no means the only pathways for tritium; indeed they represent only a few of the potential tritium pathways although they are the most likely.

Pathways for ^{14}C follow the carbon cycle (Figure 2–4). Thus, the major pathway for ^{14}C is the air—food crop via photosynthesis—man, with secondary important pathways via direct inhalation, and to a lesser extent through grazing animals eaten for meat. The activity of carbon reflects the specific activity of ^{14}C in carbon dioxide. There is a lag time of about a year from the time of production in the atmosphere to the appearance in plants. Prior to nuclear testing, the ^{14}C content of the atmosphere had stabilized at 4×10^{28} atoms, rising to a peak of about 7×10^{28} atoms about 1964 from the weapons testing and dropping back to near normal by the late 1970's. Thus, plant tissues laid down during the peak will have a somewhat greater activity of ^{14}C than those prior or subsequent to the years of intensive testing, and this can be used to determine the age of certain plant tissues.

The transition elements include elements with atomic numbers 21–28, 39–46, and 72–78. These elements are centrally located in the Periodic Chart and differ principally in the number of d-electrons. Thus, they are chemically similar. The group includes several important corrosion activation products associated with operation of nuclear reactors, notably ^{65}Zn, ^{58,60}Co, ^{55,59}Fe, and ^{54}Mn, as well as important fission radioisotopes of yttrium, zirconium, niobium, molybdenum, technecium, ruthenium, rhodium, and palladium. Other than ^{95}Zr-Nb with a half-life of 65.5 days for the longer lived parent, the fission radionuclides have very short half-lives and hence little environmental persistance; hence, they are not of significance from the standpoint of transport to man via food chains.

For the transition elements with sufficiently long half-lives, there are several important pathways to man. Direct radiation exposure pathways via soil to man and river or lake sediments to man may both be significant. Values for the distribution factor, which is the ratio of the concentration in the sediment to that in water, are on the order of 10^4 or greater for most transition elements (IAEA 1982). Of even greater significance are the aquatic pathways via plankton and shellfish to man. High concentration ratios have been noted for the transition elements in shellfish and plankton, particularly for zinc and cobalt, making ingestion of shellfish the major aquatic pathway for reactor corrosion products (Thompson et al. 1972).

For ^{89,90}Sr, the two radiostrontiums of environmental significance, the major pathways to man as are shown in Figure 12–3. The pathway via milk is of particular importance in the case of infants and children, not only because they are more likely to drink large quantities of milk but also because they are laying down bone and teeth, where the strontium, a calcium analogue, tends to accumulate. The food crop pathway is also important largely because the downward movement of strontium in soils in relatively slow; even in soils with low clay and humus content, through which movement is fastest, most of the strontium will remain in the upper few centimeters several years after deposition. Low calcium content of the soil furthers strontium uptake by plants, as does low pH. Treatment of soil with lime to increase the pH has been suggested as a means of reducitng plant uptake of radiostrontiums from soil.

Because strontium is chemically similar to calcium, the observed ratio is useful to describe the relative uptake of strontium by plants. For the above ground portions of plants relative to soil, the OR is typically 1 (Comar 1965). For fish muscle relative to water, the OR is usually in the range 0.2 to 0.4. For milk relative to the diet, the OR is approximately 0.1; in human muscle tissue relative to diet, the OR is about 0.25, and may approach 1 in infants (Comar 1965).

For radioiodines, virtually all pathways are important. However, because of the short half-lives of most radioiodines, the direct inhalation pathway along with ingestion of unwashed fruits and vegetables, and drinking of cow and goat milk are probably the most significant. Because of its extremely low specific activity, ^{129}I is not of particular environmental significance.

The milk pathway is highly significant for ^{131}I, which has a radiological half-life of 8.05 days. Once deposited on vegetation, this nuclide has an effective half-time of 5 days in grazing animals, which concentrate the iodine in their milk. For cow milk, the bioconcentration factor is 700; for goat milk, 3000 (Voilleque and Pelletier 1974; Hoffman 1975).

The behaviour or radiocesium has been extensively studied, particularly in Arctic ecosystems (NCRP) 1977. Two radiocesiums—134 with a 2.07 year half life and 137 with a 30 year half-life—are environmentally important. Although cesium and potassium are biologically similar, they are not bioloigically interdependent and hence are not as easily tracked by the observed ratio as are strontium and calcium. There are five primary pathways for cesium. The first is the direct external radiation pathway via freshwater sediments, which is of concern primarily for individuals who reside or spend a great deal of recreational time on a river or lake shore. The other two aquatic pathways of importance are the direct drinking water pathway, and the pathway through freshwater fish, especially botton feeders such as carp and sucker.

Two air pathways are important, and round out the total of five. these are the air—pasture grass—grazing animal—milk pathway, and the air—pasture grass—grazing animal—meat pathway. Cesium is efficiently absorbed from the gut of animals for transport to muscle and milk, although its biological half-life is relatively short. Pathways through plant foods are relatively unimportant as cesium is poorly absorbed by the plants from the soil, and is relatively uniformly distributed throughout all portions of the plant and hence does not tend to concentrate in the edible portions. Grains, however, do tend to have relatively high concentrations although fruits and root vegetables, which have a high water content, tend to have low concentrations of cesium.

Radiocerium pathways have also been extensively studied (NCRP 1978). Three cerium radioisotopes have half-lives greater than a month—139 (140 days), 141 (32.5 days), and 144 (284 days); the latter has the greatest environmental significance because of its longer half-life. The major pathways to man for radiocerium are via aquatic crustacea and molluscs, which show concentration factors to 6,500 (Polikarpov 1966). Food crops also represent an important if secondary pathway. Cerium absorption through the gastrointestinal tract is relatively low, however, which limits the uptake of this radioelement by people.

Radium and daughters released to the environment via uranium mill tailings or other front end fuel cycle wastes are most likely to reach man by the direct inhalation pathway from windblown particulates, or through the ground water or drinking water pathway due to seepage. For plutonium, the major pathway is the direct inhalation; uptake by plants and through the gastrointestinal tracts of humans and food animals is small (ICRP 1979a; Kathren 1968; Martin and Bloom 1980).

In summary, milk constitutes a significant pathway for the following radioelements: iodine, strontium, and cesium. Drinking water is of particular import for tritium, strontium, cesium, and radium. The transition elements cobalt, zinc, and manganese and radiocerium are most likely to reach man via shellfish, or may provide external exposure from sediments or ground deposition. The sediment—ground deposition pathway is also important for radiocesiums. The direct external dose pathway is significant for the noble gases and ^{14}C, and to a lesser extent for tritium. Finally, the meat pathway is significant for radiocesiums.

Environmental Dose Assessment

Environmental dose assessment involves the determination of the dose or dose equivalent incurred as a result of radioactivity or radiations in the environment. Normally, dose assessment is limited to man, as man is the organism of concern. In fact, man makes a good indicator organism because of the well documented and large body of knowledge regarding radiological effects and the fact that man is ordinarily the highest and frequently final trophic level.

Two methods are generally used to determine radiation doses and dose equivalents from environmental sources:1) comparison with existing concentration guides or other published values for intake of radionuclides in food and water, or for immersion, and 2) direct calculation using various known metabolic and pathway data. Fortunately, only a relatively small number of pathways are of significance, producing most of the dose to man from specific radionuclides of importance. However, when determining the environmental dose, it is necessary to consider all pathways and all exposed tissues, summing the dose from each to provide the total dose commitment for each organ as well as the total body from a given radionuclide in the environment. Also, as has been shown above, breathing air and drinking water may not be the most significant pathways for certain nuclides; thus, the application of existing concentration guides, which deal exclusively with air and water, may not provide an adequate assessment of the dose incurred.

Dose calculations involve determining the energy absorbed per unit mass in the various tissues—Thus, for radionuclides taken into the body, the metabolism of the radionuclide in the body must be known along with the physical characteristics of the nuclide such as energies and types of radiations emitted, and half-life. For internal radiations, the dose equivalent rate, dH/dt, to the organ of reference can be calculated by the following basic equation

$$\frac{dH}{dt} = \frac{k_1 k_2 q E_{eff}}{k_3 m} e^{-\lambda_{eff} t} \qquad (12-12)$$

in which k_1 is a constant equal to 1.6×10^{-6} erg/MeV

k_2 a constant equal to 1.32×10^8 disintegrations per hour per uCi or 3600 disintegrations per hour per Bq if SI units are used

k_3 a constant equal to 100 erg/g-rad or 1 J/kg-Gy in the SI units

q the quantity of activity in the organ of reference, in uCi or Bq, as appropriate

E_{eff} is the effective absorbed energy, in units of MeV/ disintegration m the mass of the organ, in grams

λ_{eff} the effective removal constant, in reciprocal time

t the time after uptake by the organ.

If the specified constants are used in Equation 12–12, dH/dt will have units of rem/h, and simplifies to

$$\frac{dH}{dt} = \frac{2.1qE_{eff}}{m} e^{-\lambda_{eff}t} \qquad (12\text{--}13)$$

If the constants are in SI units, the numerical constant in Equation 12–13 becomes 0.0058 and dH/dt has units of Sv/h.

Dimensional analysis of Equation 12–12 suggests that the units for dH/dt are in fact dose (i.e. rad or Gy) per time rather than dose equivalent (rem or Sv) per time. This is true if the quality factor is not applied to the energies of the radiations involved. Conversion from dose to dose equivalent is accomplished by the effective absorbed energy term, which is includes the quality factor for, each radiation and energy group combination emitted by the radionuclide and is thus identical to the summation term put forth in by ICRP Committee II on Permissible Dose for Internal Radiation (ICRP 1959) and is equivalent to the specific effective energy (SEE) term of ICRP 30 (ICRP 1979a). Values for the effective absorbed energies can be obtained from the literature, excellent summaries having been published by the International Commission on Radiological Protection (ICRP 1959, 1979a).

The committed dose equivalent over a period of time can be obtained by integration of Equation 12–13 with respect to time to give

$$H = \int_0^t \frac{2.1qE_{eff}}{m} e^{-\lambda_{eff}t} \, dt \qquad (12\text{--}14)$$

which, reduces to

$$H = \frac{2.1qE_{eff}}{m\lambda_{eff}} \qquad (12\text{--}15)$$

if the upper limit of integration is infinity or very long relative to the effective half-life of the radionuclide in the organ. Recent guidance from the ICRP specifies the use of 50 year committed dose equivalent (ICRP 1979a); others including the U. S. Environmental Protection Agency have used 70 years. However, as most radionuclides other than the bone seekers (e.g. radium, actinides, ^{90}Sr) have effective half-lives of weeks or months in the body, it makes little difference in most cases which time base is used.

Equations 12–12 through 12–15 are based on the *uptake* and not the *intake* of a radionuclide. Only a fraction of the intake, or quantity taken into the body, will reach the organ of reference, and consideration must thus be given to the fractional uptake by the organ. The fractions will, of course, differ depending on route of intake, the specific element, solubility, block by stable elements, and a whole host of other factors, some physical, some biological. Compilations of fractional uptake for most radionuclides is available in the publications of the ICRP (ICRP 1959, 1979a, 1979b) and in the regulatory guidance put forth by the U.S. Nuclear Regulatory Commission (USNRC 1977). The calculations, however, have been performed for most radionuclides and organs and have been published in the form of tables of dose factors providing for both adult and child the mrem per μCi inhaled or ingested to various body organs (Soldat et al. 1974; Hoenes and Soldat 1977; USNRC 1977).

In addition to the dose or dose equivalent calculation, the assessment of environmental dose must include many other factors. A complete appraisal would therefore include a compilation of all potential sources of exposure from all radionuclides. Clearly, this can be an enormous task. For example, consider a situation involving the release of a single radionuclide—^{60}Co to a river. In this case, there is but one radionuclide of concern, but numerous exposure pathways. The dose equivalent incurred by any individual will be the sum total of the dose equivalent from each pathway. Some of the pathways may be obscure, yet be significant dose contributors; in the example cited, a fisherman standing on the bank would receive an external exposure from the ^{60}Co deposited in shoreline sediments, as well as from ingestion of fish or shellfish from the river. The river may also serve for recreational purposes, imposing the potential for external exposure from boating or swimming activities. Note that the latter produces an immersion dose, while the former is analagous to exposure from a uniformly contaminated surface. The river may also contribute to internal exposure by serving as the source of drinking water or irrigation water.

Many models have been developed to calculate the doses to individuals in the population from anthropogenic sources of environmental radioactivity. The Concentration Factor (CF) method of the ICRP is a relatively simple scheme useful for many situations involving an equilibrium situation with chronic or continuous release of radioactivity with a few predominant radionuclides and pathways (ICRP 1979b). This method basically involves determining the various compartments in the critical pathways and multiplying each donor-compartment concentration by the appropriate concentration factor, and summing over all donor compartments. The resultant concentration is then combined with the rate of intake, and the product is compared with the published values for the Annual Limit on Intake (ALI) for the nuclide(s) in question. The ALI calculations include all the necessary uptake and specific effective energy factors for each specifc organ of reference (ICRP 1979a).

A second method of modelling proposed by the ICRP is the Systems Analysis (SA) method in which the dynamic behavior of the radionuclides in a given environment with a given set of coupled compartments is modelled as a function of time by a series of differential equations. This method is not subject to the limitations of the CF method, but can be used with intermittant or instantaneous releases of any number of radionuclides and any number of pathways (ICRP 1979b). Basically, what is done is to conceptually approximate the kinetics of the actual system. This method is more complex than the CF method, but does not have the temporal variation limitations incumbent upon the latter. Thus, the SA method uses basic first order kinetics to account for the possible time dependence of transfer coefficients.

For nuclear power plants, the U. S. Nuclear Regulatory Commission has developed a comprehensive set of models along with estimates of all necessary concentration and transfer factors (USNRC 1977). The NRC includes the equations for the following pathways:

1. Liquid Pathways
   a. Potable Water
   b. Aquatic Foods
   c. Shoreline Deposits
   d. Irrigated Foods—all radionuclides except tritium
   e. Irrigated Foods—tritium
2. Atmospheric Pathways
   a. Noble Gas Immersion
      1. Gamma Dose from stacks > 80m
      2. Beta Dose + Gamma dose from all other releases
      3. Total Body Dose from stacks > 80m

    4. Annual Skin Dose from stacks > 80m
    5. Annual Total Body Dose from all other releases
    6. Annual Skin Dose from all other releases
  b. Radioiodines and Other Radionuclides
    1. Ground Deposition
    2. Inhalation
    3. Ingestion in Food.

At U.S. Department of Energy Sites, several different dose models are used with the specific analyses a function of the site characteristics and purpose (Kennedy and Mueller 1982).

In practice, environmental dose calculations are performed largely to estimate the dose incurred by the critical segment of the population exposed in the vicinity of a nuclear facility. Thus, what is usually done is to calculate the doses incurred by a so-called average or typical resident and those incurred by the hypothetical maximum individuals. In the example involving the release of ^{60}Co to the river given above, the latter might be an avid fisherman and river recreationist who grows his own vegetables in a personal garden irrigated with river water. Assumptions are made regarding the quantity of fish ingested, the amount of time spent in various recreation pursuits on and in the river, and with regard to the amount of irrigation water used as well as the types of plant foods grown. The resultant calculation produces a hypothetical maximum dose, which may in fact be manyfold greater than is incurred by other members of the population.

Frequently, environmental dose calculations result in very small values for the exposed population or hypothetical maximum exposed individuals. When compared with other sources of exposure, and especially exposure from natural background sources, these levels appear miniscule or even trivial. Since the primary purpose of most environmental dose calculations is to assist in evaluating the radiological hazards from operation of various anthropogenic sources such as nuclear power plants or other fuel cycle activities, or medical, industrial, or other applications of radionuclides, it seems pointless to consider the hazard from a dose that is a tiny fraction of the dose from other sources. The *de minimis* concept, has been borrowed from legal practice to describe these doses and has begun to be widely applied in environmental assessments. The *de minimis* level is that which provides an inconsequential or negligible additional radiation exposure; a dose equivalent rate of 1 mrem/y (10 μSv/y) to an individual member of the general population—approximately 1% of the natural background level—has been proposed for commercial and Department of Energy operations (Rodger et al. 1978; Kathren et al. 1980).

References

1. Aakrog, A. 1971. "Prediction Models for Strontium-90 and Cesium-137 Levels in the Human Food Chain", *Health Phys.* 20:297.
2. Baes, C. F., III, and T. H. Orton. 1979. "Productivity of Agricultural Crops and Forage, Y_v", in *A Statistical Analysis of Selected Parameters for Predicting Food Chain Transport and Internal Dose of Radionuclides*, F. O. Hoffman and C. F. Baes, III, Eds., Oak Ridge National Laboratory Report ORNL/NUREG/TM–282, p. 43.
3. Beck, H. L. 1980. "Exposure Rate Conversion factors for Radionuclides Deposited on the Ground", U. S. Department of Energy Report EML-378.
4. Comar, C. L. 1965. "Movement of Fallout Radionuclides through the Biosphere and Man", *Ann. Rev. Nucl. Sci.* 15:175.
5. Hoenes, G. R. and J. K. Soldat. 1977. "Age-Specific Radiation Dose Commitment Factors for a One Year Chronic Intake", U.S. Nuclear Regulatory Commission Report NUREG/CR-1072.
6. Hoffman, , F. O. 1975. "A Reassessment of the Parameters Used to Predict the Environmental Transport of ^{131}I from Air to Milk", Institute fur Reaktorsicherheit Report IRS-W–13, Cologne, Germany.
7. International Atomic Energy Agency (IAEA). 1982. "Generic Models and Parameters for Assessing the Environmental Transfer of Radionuclides from Routine Releases", Safety Series No. 57, International Atomic Energy Agency, Vienna.
8. International Commission on Radiological Protection. 1959. "Report of Committee II on Permissible Dose for Internal Radiation", ICRP Publication 2, Pergamon Press, New York.; also *Health Phys.* 3:1 (1960).
9. International Commission on Radiological Protection (ICRP). 1979a. "Limits for Intakes of Radionuclides by Workers", ICRP Publication 30, *Annals of the ICRP* 3(3/4):1.
10. International Commission on Radiological Protection (ICRP). 1979b. "Radionuclide Release in the Environment: Assessment of Doses to Man", ICRP Publication 29, *Annals of the ICRP* 2(2):1.
11. Kathren, R. L. 1968. "Towards Interim Acceptable Contamination Levels for Environmental PuO_2", Battelle Pacific Northwest Laboratories Report BNWL-SA–1510; also in *Radiological Protection of the Public in a Nuclear Mass Disaster*, Proceedings of a Symposium of the Fachverband fur Strahlenschutz, Interlaken, Switzerland.
12. Kathren, R. L. et al. 1980. "A Guide to Reducing Radiation Exposures to As Low As Reasonably Achievable (ALARA)", U. S. Department of Energy Report DOE/EV/1830-T5.
13. Kennedy, W. E., Jr. and M. A. Mueller. 1982. "Summary of the Environmental Dose

Models Used at DOE Nuclear Sites in 1979", Battelle Pacific Northwest Laboratory Report PNL-3916.

14. Martin, W. E., and S. R. Bloom. 1980. "The Nevada Applied Ecology Group Model for Estimating Plutonium Transport and Dose to Man", in *Transuranic Elements in the Environment*, W. C. Hanson, Ed., U.S. Department of Energy Report DOE/ TIC-22800, pp. 459–512.

15. Miller, C. W. and F. O. Hoffman. 1979. "An Analysis of NRC Methods for Estimating the Effects of Dry Deposition in Environmental Radiological Assessments", *Nuclear Safety* 20:458.

16. Moore, R. E., et al. 1979. "AIRDOSE-EPA: A Computerized Methodology for Estimating Environmental Concentrations and Dose to Man from Airborne Releases of Radionuclides", Oak Ridge National Laboratory Report ORNL-5532.

17. National Council on Radiation Protection and Measurements (NCRP). 1977. "Cesium-137 from the Environment to Man: Metabolism and Dose", NCRP Report No. 52, Washington, D.C.

18. National Council on Radiation Protection and Measurements (NCRP). 1978. "Physical, Chemical, and Biological Properties of Radiocerium Relevant to Radiation Protection Guidelines", NCRP Report No. 60, Washington, D.C.

19. Polikarpov, G. G. 1966. *Radioecology of Aquatic Organisms*, Reinhold Book Division, New York.

20. Rodger, W. A., et al. 1978. "*de minimis* Concentrations of Radionuclides in Solid Wastes", Atomic Industrial Forum Report AIF/NESP-D16.

21. Soldat, J. K., et al. 1974. "Models and Computer Codes for Evaluating Radiation Doses". Battelle Pacific Northwest Laboratories Report BNWL-1754.

22. Thompson, S. E., et al. 1972. "Concentration Factors of Chemical Elements in Edible Aquatic Organisms", University of California Report UCRL-50564, Rev. 1.

23. United States Nuclear Regulatory Commission (USNRC). 1977. "Calculation of Annual Doses to Man from Routine Releases of Reactor Effluents for the Purpose of Evaluating Compliance with 10 CFR Part 50, Appendix H", Regulatory Guide 1.109, Rev. 1.

24. United States Nuclear Regulatory Commission (USNRC). 1983. "Standards for Radiation Protection", *Code of Federal Regulations*, Title 10, Part 20.

25. Voilleque, P. G., and C. A. Pelletier. 1974. "Comparison of External Irradiation and Consumption of Cow's Milk as Critical Pathways for ^{137}Cs, ^{54}Mn and ^{144}Ce-^{144}Pr Released to the Atmosphere", *Health Phys.* 27:189.

CHAPTER 13

Environmental Surveillance and Measurement Programs

Basic Concepts

Detection and quantification of environmental radiations and radioactivity may be performed for a number of reasons. The basic purpose may be a desire to satisfy basic scientific curiosity and to gain knowledge and an improved understanding of the world. A second reason may be to assess the impact of human activities involving radioactivity and radiation on the environment, a process which may require differentiation of natural and anthropogenic sources of radioactivity and lead to assessments of dose from environmental sources.

Differentiation between naturally occurring and artificially introduced radioactivity implies some sort of isotopic analysis technique, as well as a knowledge of what was present prior to the introduction of manmade sources. Existing radiation and radioactivity levels from natural sources constitute the baseline level, and requires study of the environment usually for a period of 1–3 years prior to commencement of human activities that may alter the radiological environment.

Environmental dose assessment is a sequential procedure that begins with the characterization of the radionuclides or radiation sources. This logical first step is followed by determination of the radionuclide distributions and quantification in the environment, and the detection of external radiations incident on man and the uptake and accumulation of radioactivity by man. Once these data have been determined, they serve as the basis for the calculation of doses incurred from environmental sources. Any environmental surveillance program involving dose assessment must thus be designed to consider all of the factors that may result in, or modify doses to, people.

Surveillance in the vicinity of nuclear facilities has certain pragmatic objectives with regard to protection of the public and the environment.

The most important of these is the assessment or estimation of the probable limits of the actual or potential radiological exposure of people from the operation of the facility (Denham 1982). Basically, the program is carried out to ensure that the facility is in compliance with applicable laws, regulations, license and permit conditions, which presumably provide adequate protection for people and the environment. This is accomplished not only by gross measurements of radioactivity and radiations in the environment, but by differentiation of plant effluents from naturally occurring or fallout radionuclides; determining the dose to populations and individuals adjacent to the facility; acquisition and analysis of data regarding long term radiological trends in the plant environs; and establishment and verification of the general safe operation of the facility. Positive public relations may be another factor, as may the need for protection from legal actions. It is desirable to carry out the monitoring functions at reasonable cost and with minimal disruption of the natural environment, which implies judicious choice of the number and kinds of samples and analyses.

Environmental radiological surveillance around nuclear facilities is customarily classified as preoperational, operational, or postoperational depending on facility status. The preoperational program should commence a minimum of one and preferably two or three years prior to facility operation or to the receipt of radioactive materials on site; for nuclear power plants, the Nuclear Regulatory Commission recommends a two year data collection program (USNRC 1975a). This program should identify probable exposure pathways and critical population groups, provide for the selection of sampling locations and media, verify analytical procedures, and establish baseline levels. A sample museum or bank in which unanalyzed 'pristine' samples of various environmental media such as soil, dessicated or preserved biota, and even water can be kept intact over a period of years or even the lifetime of the facility is a good scheme for later verification of preoperational or baseline analytical results. If improved analytical techniques become available at some future time, these samples may be analyzed as required to provide additional information. Museum samples may be suitable only for long lived radionuclides, however, although new analytical techniques may permit detection of stable daughters of short lived radionuclides.

The preoperational program should set the stage for a smooth transition to the operational program, which commences with the operation of the facility, or, perhaps the time that radioactivity is first brought on site. Operational surveillance programs may have several aims, including (Eichholz 1978):

1. Protection of people and the environment from radiological releases.

2. Monitoring and documentation of existing and continuing radiological conditions.

3. Compliance with regulatory requirements.

4. Detection and documentation of accidental or unanticipated environmental effects.

5. Protection of the facility operators from legal liability claims.

6. Research and development, including verification of calculational models.

The operational environmental radiological surveillance program is thus basically a check on facility operations and control, and as such should be continuous throughout the operating lifetime of the facility, suitably modified from time to time to account for changes in operations.

The operational program terminates with the cessation of radiological activities or facility operation, giving way to the postoperational program, which may be much reduced and simply performed to verify long term trends and other data developed in the operational program.

Program Development

The first step in developing an environmental radiological surveillance program is to establish the purpose of the program. If the program is to be carried out to accomplish several objectives each may have its own specific requirements for data collection and analysis. However, most environmental monitoring programs are carried out largely for regulatory compliance purposes and as such are primarily concerned with establishing that the radiological discharges from a given nuclear facility have not exceeded legal or permit limits. Data collection is thus directed towards this end, and may not be as specific or in-depth as might be the case if the program were carried on for the purpose of scientific research.

For a nuclear facility, development of an environmental radiological monitoring program is basically a six step process:

1. Facility evaluation.

2. Pathway identification.

3. Pathway selection.

4. Establishing primary (i.e critical) measurement criteria.

5. Establishing secondary measurement criteria.

6. Developing dose assessment techniques.

Evaluation of the facility as a source of both direct radiation and radio-activity involves determination of radioactivity release points and of the radionuclides involved, including concentrations and quantities, and the relevant physical and biological characteristics for each. This is the deter-mination and characterization of the source term, which should include consideration of each specific radionuclide in terms of its potential hazard as well as its potential accumulation and impact on the environment. Knowledge of the facility, including the design and operating character-istics of effluent and waste treatment systems and practices is of primary importance in assessing the release of the radionuclides involved in the operation to the environment. Experience gained with similar types of facilities may be directly applicable, but should be carefully considered and evaluated particularly with respect to site specific factors. In the case of a nuclear power plant or similar major nuclear facility, consultations with the architect-engineer early in the design phase may not only be useful from the standpoint of establishing the source term, but may also provide insight into minimizing the radiological impacts on the eviron-ment. Care should be taken to consider all potential release pathways and radionuclides. For example, in the case of a reactor, activation products produced in the core or shielding materials are sometimes neglected but may in fact be important sources of radioactivity, particularly after the reactor has been in operation for several years. The mode of release will determine sampling type and frequency. Routine grab samples may be suitable for continuous, constant or 'averaged' releases from holdup tanks, while continuous sample collection may be required for intermit-tant or variable releases.

Environmental transport pathway identification is an extremely impor-tant step in the design of an environmental sampling program, and requires firsthand knowledge of the ecological system and the microen-vironment in the facility environs. Although values for many environ-mental factors such as concentration ratios and atmospheric stability can be obtained from the literature, validation experiments are necesary to ensure that these values do in fact apply and that the predictive models selected are suitable to the site (Hoffman et al. 1977). Thus, to suitably identify pathways, baseline ecological information is necessary, including information regarding the major species and biotic communities. Ideally, bioconcentration factors in key species should be experimentally obtained; this is particularly true for aquatic species which may show seasonal or other variability. However, even in the absence of empirical data regarding the concentration ratio, it is possible to make reasonable estimates by examining the stable element content of the aquatic medium. In general, eutrophic (nutrient rich) systems produce lower concentration factors than oligotrophic or nutrient poor systems.

Also necessary are basic data regarding the site meteorology and hydrology, plus demographic and dietary data pertinent to the human population surrounding the site. It is important to know the size and distribution of the population, as well as such characteristics as age distribution, location of sources of potable water, dairies, agricultural activities including home vegetable gardens and fruit growing, and any habits or activities such as hunting or fishing which might mark significant pathways and thereby affect the ultimate dose to people. A house-to-house survey in the vicinity of the site may be required to gain the necessary information. Relevant information from other similar sites can also be of value, but again may be only partially applicable.

Detailed information on meteorological and climatological parameters is essential to the determination of critical pathways and to the development of a suitable sampling program. At least a full year of data is needed to adequately reflect seasonal changes; and as is the case with background radiological information, two to three years of data is better. The data must accurately reflect the site and its surroundings; weather even a few miles from the site may be grossly different and uncharacteristic of the site itself, rendering data from nearby weather stations of little or no value.

Meteorological data to be collected include wind velocity frequency distributions, which are of prime importance in locating air sampling equipment and dosimeters to determine immersion doses. Wind velocity generally shows seasonal or diurnal variations and therefore appropriate wind roses should be prepared to assist with the determination of sampling site locations. Atmospheric stability data is also important, particularly in determining values of ground level radionuclide concentration and dilution (χ/Q) at various locations around the facility. Precipitation and its forms are also important, and may determine specific pathways and sampling requirements. Site meteorological data is normally collected at two or three heights with the aid of a suitably instrumented tower. In addition, there may be several satellite locations at which certain measurements such as rainfall may be made.

The topography of the site and adjacent lands can prove of great value in establishing monitoring locations. Hills and other high points are likely locations for air sampling as they are more likely to be in the plume rather than beneath it. Features such as channelling, valley-slope circulation, and sea breezes also influence the choice of air sampling locations.

Hydrological information of importance to the development of an environmental monitoring program includes general information on nearby water bodies, including size, depth, flow characteristics, and tidal reversals. If possible, the turnover rate should be experimentally determined, especially for quiescent bodies such as ponds and lakes. The characteristics of the sediments in these water bodies is also of importance, as these may sorb or retain various radioelements and thus serve as a reservoir for radioactivity in a food chain or pathway. An obvious but occasionally overlooked aspect of water bodies is their usage which may lead to drinking water, irrigation water or recreational use pathways of significance. Ground water conditions need also to be evaluated, and in particular any usage such as wells for drinking or irrigation.

Once the basic environmental and population data is obtained, the critical pathways—i.e. those that will provide the greatest doses to man—can be determined. In this determination, care must be taken so that the critical population group is not omitted. For example, nearby institutions with resident nonmobile populations such as hospitals, nursing homes, or jails may be of special significance. Similarly schools with their aggregation of radiosensitive children may serve as the location of a critical population group and hence determine certain critical exposure pathways such as from immersion or drinking water.

After the critical pathways have been determined, the next step is to establish the measurement criteria necessary for evaluation of the dose to the critical segment of the population during both normal and abnormal operating conditions. The various environmental media and foodstuffs to be sampled must be selected, with due consideration given to availablility (which may be seasonal), impact on the environment, sample size, numbers and relevance, and sampling technique and frequency. This process must provide for a representative sample, both in relation to the pathway and to the sampling techniques; with regard to the latter, it must be ascertained that the correct medium is being sampled, and the samples are appropriately mixed and not altered by the sampling procedure. In addition, appropriate analytical techniques need to be selected. Following establishment of the critical measurement pathways, appropriate secondary pathways and measurement criteria should be determined. These include all pathways that may result in doses or dose equivalents above the *de minimis* level, as well as pathways of temporary or seasonal significance. Depending on the site and facility characteristics, it may well be that there are none, but nonetheless the analysis needs to be made and documented.

The final step is to develop a schema for environmental dose assessment. External radiation doses are generally determined in a straightforward manner by measurement with appropriate dosimeters at key locations, although levels may be so low as to render measurements of this type difficult or even impossible (Chapter 14). The dosimeters may be designed to provide the non-penetrating or skin dose as would result from immersion in a cloud or plume of beta-emitting noble gases as well as the penetrating whole body dose equivalent. If direct measurement is not possible or impractical, external doses may be calculated on the basis of effluent release data.

Internal doses are calculated from estimates of dietary and inhalation intake of specific radionuclides based on radionuclide concentrations measured in air, water, and foodstuffs. Various calculational models are available in the literature; for nuclear power plants, the Nuclear Regulatory Commssion has published extensive regulatory guides that provide detailed techniques for environmental dose estimates from nuclear power plants, uranium mills, and fuel fabrication plants (USNRC 1977a, 1977b, 1978, 1982). In addition a variety of other methods and models are in use at Department of Energy Sites (Kennedy and Mueller 1982; Strenge, Kennedy, and Corley 1982). A key feature in all these methods is their site specificity, although many are generally adaptable in addition several other comprehensive guidance documents are available (Corley and Corbit 1983; Watson 1980; USEPA 1972; Denham 1982).

The generalized dose assessment calculation can be represented by:

$$H = kIC \qquad (13-1)$$

in which H refers to the dose equivalent, k is a specific dose factor, I the intake of the material containing the radionuclide, and C the concentration of the radionuclide in the material. The specific dose factor, k, is usually obtained from the literature, and may be dependent on time, age, sex, body weight, or individual characteristics such as pregnancy or lactation or unusually large or small organ size. In genral, however, it is taken to be constant.

The concentration may refer to the amount of radionuclide in air that is breathed, water that is drunk, or food that is ingested. If atmospheric radioactivity is being evaluated, C may be represented by

$$C = Q(\chi/Q)V_d t_a T_r \qquad (13-2)$$

in which Q is the quantity of activity released (μCi/time or μCi),

χ/Q the average dilution (m^3/time),
V_d the deposition velocity (m/time),
t_a the residence time in the environment,
and T_r the transfer coefficient.

The transfer coefficient may be the product of several other coefficients—e.g. air to soil; soil to plant, plant to animal; animal to milk; milk to man—and is clearly dependent on the radionuclide, pathway, and metabolic, chemical, and physical variables. In the case of milk, it may include the area grazed. The term t_a, residence time in the environment, may be quite variable, and should include correction for radioactive decay.

Regional vs. Local Programs

In the development of any environmental surveillance program, consideration needs to be given to the geographical area to be covered. In general, local monitoring programs that encompass a few tens to hundreds of square miles in the immediate vicinity of a nuclear plant are required for regulatory purposes. These programs are largely geared towards verifying that the radiological releases from the facility have been in compliance with regulatory requirements, and that the doses to people and the buildup of radioactivity in the immediate environment of the facility are within established limits. For most nuclear facilities, local programs are all that are necessary as release levels are so low that detection is a virtual impossibility at any considerable distance from the site. Thus, it is unusual to sample for radioactivity or measure ambient levels at distances greater than a few tens of miles from the facility except perhaps to establish control or baseline values.

Some programs conducted for different purposes than regulatory requirements may also be local in nature. These may include studies of the natural or anthropogenic radioactivity concentrations or ambient radiation in a given locale; such programs may be for the primary purpose of gaining an understanding of the radioactivity distribution or ambient radiation levels in a given locale, for radioecology or other biological studies, or for geological age dating or similar reasons. Such studies are, in general highly specific, both in terms of the physical area covered as well as the information to be derived.

Regional or global studies are often carried out to study various the distribution and transport of radioactivity and ambient radiations on a broad scale. These usually are concerned with atmospheric or aquatic transport of fallout from weapons tests, or the distribution of naturally occurring or anthropogenic radionuclides throughout the biosphere. Radioactivity surveillance networks covering large geopgraphic areas have been established in many countries, primarily to determine the effects of fallout from weapons tests or to detect such tests. In the United States, several such broad geographic studies have been carried out by various federal agencies such as the U.S. Public Health Service and, more recently, the Environmental Protection Agency.

One such program is the Environmental Radiation Ambient Monitoring System (ERAMS) of the EPA (Rowe, Galpin and Peterson 1975). This national program was instituted in July 1973 as a result of the consolidation of several smaller programs within the U.S. Public Health Service and turned over to the EPA Office of Radiation Programs when that agency came into being. While ERAMS predecessor programs were primarily concerned with measurement of fallout from nuclear weapons testing, the basic objective of the ERAMS program is identification of trends of accumulation of long-lived radionuclides in the environment, including acquisition of data for direct assessment of the radioactivity intake and dose to the American population, and to develop a dose model for application on a national level. Another objective is to provide an early awareness of potential radiological hazards, permitting mitigatory actions to be swiftly taken, thereby minimizing population dose.

The ERAMS surveillance program measures levels of radioactivity in air, air particulates, fallout, surface and drinking waters, and milk in the United States and its territories. Samples are collected by various governmental bodies at the federal, state, and local level, and delivered to the EPA analytical laboratory in Montgomery, Alabama, for analysis. Sample locations are determined in large measure by proximity of potential sources of anthropogenic releases and population centers. Emphasis is primarily directed towards identification of trends in the concentrations of long lived radionuclides in various environmental media along with monitoring shorter lived nuclides such as ^{131}I that have the capability to deliver high population doses.

The ERAMS program is built around 21 sampling sites for air particulates and precipitation; an additional 51 sites are held in reserve. Air particulates from the twenty-one locations are collected on a weekly or semiweekly basis and analyzed for gross beta activity, ^{234,235,238}U, and ^{238,239}Pu. Air samples from these locations are also analyzed for ^{85}Kr.

Monthly composite precipitation samples from each of the 21 sites are measured for gross beta and tritium concentrations. Drinking water is collected from 77 locations across the nation and analyzed for tritium and gross beta activity. Tritium is also monitored in surface waters downstream from several nuclear plants. Milk is sampled at 65 stations, with one or more in each state, and analyzed for ^{90}Sr, ^{131}I, and ^{137}Cs.

Although the ERAMS program operates with relatively few stations, it is nonetheless able to carry out its stated aims reasonably well. A major factor is the emphasis of the program on the final level of key pathways with respect to dose to people, and the analysis of samples only for certain key radionuclides. To provide improved reliability of results, data may be pooled; for example the network averages are used to calculate the long term trends of concentration in the environment. ERAMS data are reported quarterly in the EPA publication *Environmental Radiation Data* and annually in other EPA reports.

Another national program is conducted by the Bureau of Radiological Health of the U. S. Food and Drug Administration and measures radioactivity in various foodstuffs in the United States (Simpson et al. 1981). The program can be traced back to an earlier study designed to monitor the total amount of radioactivity in the typical teenage diet that was initiated in 1961 during the period of atmospheric weapons testing. This particular program was terminated in 1969, but the capability remained and in 1973 the program was reinstituted to monitor trends in the radioactivity in food samples.

Samples of selected foodstuffs typical of the total diet of an 18 year old male are collected from metropolitan areas in all ten FDA regions within the United States, including areas in the vicinity of eight nuclear power plants. Analyses are performed by the FDA laboratory at Winchester, Massachusetts, and are largely directed towards tritium, ^{137}Cs, and ^{90}Sr. This program has been able to detect differing levels of activity in the vicinity of the reactors, which is likely attribuable to reactor operations.

Program Design for Nuclear Facilities

Each nuclear facility is different and therefore unique with respect to its design, operation and relationship to its surroundings. Radiological monitoring in the environment is frequently conducted in the vicinity of these facilities to satisfy regulatory requirements. These programs in general are concerned with verifying that plant operations are being conducted in accordance with regulatory and license requirements, and establishing that the radiological impact in the immediate vicinity of a given nuclear facility is within acceptable limits. Such programs are ordinarily local in nature. They have as objectives detection of trends in environmental radioactivity concentrations or radiation levels that may contribute to human exposure and the development of data pertinent to dose rates and radioactivity concentrations for important exposure pathways (Conti et al. 1980).

Programs conducted for regulatory purposes may be constrained by specific requirements, such as those specified by the *Code of Federal Regulations* or in the Regulatory Guides published by the Nuclear Regulatory Commission. For nuclear power plants, the NRC has put forth detailed program elements and requirements specifying minimum numbers and types of samples and analytical procedures (USNRC 1975a, 1975b,1979). These have resulted in a degree of uniformity and quality among these programs, which are described in the Environmental Impact Statements and Technical Specifications of each specific nuclear power plant. Despite the specificity to large light water reactors, the guidance given by the NRC documents is frequently of general applicability, at least conceptually. Similar guidance has been provided for uranium mills and fuel fabrication plants (USNRC 1978, 1980). In general, the NRC regulatory guidance calls for both a preoperational or baseline program that includes monitoring of air, water, vegetation, animals, foodstuffs, soil, sediment, and direct radiation. This program is continued through the lifetime of the facility, and includes monitoring of all critical pathways.

The NRC suggested programs have been criticized on several grounds, including their rather broad extent and disregard of cost. Given sufficient resources—i.e. funds, personnel, equipment, and time, virtually any radiological surveillance program can be effected. However, more realistically, resources—particularly financial—are limited, necessitating good program design in order to assure that the desired information is obtained in an efficient and scientifically correct fashion. The basic environmental surveillance design process for nuclear facilities takes into account the source term, dispersion in the environment, and critical pathways as the basic elements for establishing what media are to be sampled, and which analyses are needed to provide the data upon which environmental dose estimates and accumulation of radionuclides can be made.

Consideration must be given to the physical and biological characteristics of the radionuclides in the source term, including physical and chemical state, half-life, and radiotoxicity. The facility itself needs to be considered also, including the possibility, magnitude, and type of potential accidental releases of radioactivity as well as routine or normal releases. The constancy or periodicity of release may also be an important factor in determining sampling location and frequency. Other important factors to be considered include the size and distribution of the human population in the vicinity of the facility, as well as the use made of potentially contaminated waters and nearby lands. Selection of sampling and analytical techniques must be accomplished in such a manner as to assure representativeness and adequacy. The major factors involved in the design of an environmental radiation surveillance program and their interrelationships are shown in Figure 13–1.

A distinction should be made between effluent monitoring and environmental monitoring or surveillance. Effluent monitoring refers to the measurement of the radioactivity at the point of release, and is relatively easily accomplished because of the controlled nature of the sampling location, which is usually in a waste stream being discharged through a stack or liquid discharge line. Analysis is also facilitated as release volumes are generally known and activity concentrations are relatively large in comparison with those in the environs. Moreover, there is unequivocal evidence that the radionuclide, if detected in the effluent stream, originated in the facility. Such is not the case with environmental monitoring data, which although providing an absolute measure of conditions outside the facility, are subject to a wide variety of influences such as natural radioactivity and radiations, fallout, and the vagaries of atmospheric and aquatic diffusion processes. Ideally, effluent and environmental monitoring results will be in good agreement, with the latter calculable from the former with the aid of various models. However, it should be recognized and emphasized that effluent monitoring, no matter how good or extensive, can substitute for actual measurement of radiological conditions in the environs.

In establishing an environmental monitoring program, every pathway that could produce a significant exposure or an exposure above the *de minimis* level should be routinely examined. Exposure pathways contributing more than 10 per cent of the total dose from the facility, if this exceeds more than 1 mrem (10 μSv) annually to the whole body or any specific organ, or 100 person-rem (1 person-Sv) annually (whole body) in a population of 1,000,000 persons within an 50 mile radius of the release point is a recommended guide (Corley and Corbit 1983; Denham 1982). The actual dose calculation need only be based on the measurements obtained from a single environmental medium for any single radio-

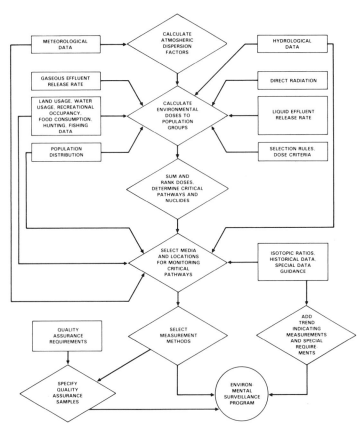

FIGURE 13-1   Environmental Surveillance Program Design Process. From Corley et al. 1981.

nuclide pathway, but should be verified by calculations based on effluent measurements. Control or background locations should be established for each pathway at locations sufficiently far from the facility such that levels detected at these points can be considered with assurance to have originated elsewhere.

Gross activity measurements, while useful as general trend indicators, should not be used to establish specific radionuclide concentrations or population doses. Dose calculations from environmental measurements should be based on measurement of specific radionuclides and should be statistically significant. Consideration should be given to seasonal and diurnal variability in environmental conditions in developing a program. Thus, sampling frequency and media may change periodically on a regular basis.

Correct location of sampling sites is essential to meeting the goals of the program. A grid pattern, with a neat layout of sampling locations with equal distances between them is usually suited to a baseline or preoperational study, or to a study of general conditions in the environment. To evaluate the radiological doses of a nuclear facility, sampling locations should be selected not on the basis of geometric regularity but on the basis of where the activity is most likely to occur. Thus, the preoperational studies of meteorology can be used to establish the locations of maximum χ/Q, and these specific locations may prove more valuable and relevant to the purpose of dose assessment and detection of facility generated environmental radioactivity. Similarly locations such as the intake of a water treatment facility or location of a major recreational facility may prove more useful than a grid or radial pattern based solely on geometry. In practice, a hybrid or mixed pattern of sampling locations often proves most suitable for an operational program.

To adequately and practically ascertain the dose to people from small concentrations of radioactivity in environmental media requires a sufficient number of samples with adequate size to ensure detection of the desired or target concentration. In general, the sample frequency and size is limited by the ability of the analytical technique; increased sensitivity can be achieved by increasing the counting time of a sample, by the purchase of new apparatus with greater sensitivity, and by collection and preparation of larger samples (Moeller et al. 1978). Evaluation of the cost-effectiveness of various sampling protocols can be made with the aid of Equation 13–3 which shows the interrelationship between the minimum detectable level of an environmental sample, MDL, the sample size or volume, V, and the lower limit of detectability, LLD:

$$MDL = \frac{k(LLD)}{Vfre} \qquad (13\text{–}3)$$

in which k is a constant for unit conversion,the sample, f the fraction of the sample analyzed, and e the counting efficiency of the detector. This equation can be rearranged to determine the sample size required to achieve a given level of detectability, inasmuch as the other parameters are either known or subject to laboratory control.

Equation 13–3 can also be used to establish reasonable laboratory counting times and to judge the sensitivity and utility of a particular analytical procedure. The LLD is simply the product of the counting time t and the minimum detectable count in that time period, represented by

C. If t(C) is substituted for LLD in Equation 13–3, solution is in terms of A, the smallest concentration that will be quantitatively measured by the procedure, or

$$A = \frac{t(C)}{Vfre} \qquad (13\text{--}4)$$

Quality Assurance

Although quality assurance (QA) is an important and necessary aspect of environmental surveillance programs, it is frequently overlooked or neglected. Quality assurance is defined as the sum total of planned and systematic activities necessary to ensure the accuracy and validity of a program by detecting and minimizing errors. QA includes quality control (QC), which is the sequence of actions which provide the means of determining and controlling the characteristics of measurement equipment and processes to meet established requirements. A systematic QA program must include good facilities and equipment, adequately trained and experienced personnel, and a knowledgeable and supportive management (USNRC 1979; Oakes, Shank, and Eldridge 1980). Its primary objective is to identify sources of uncertainty, whatever their origin, in the environmental surveillance program, and to maintain the quality of the results within accepted limits (Sanderson et al. 1980).

Major elements necessary to a suitable quality assurance program include the establishment of a suitable organization with clear responsibilities for key personnel. The program should set forth the specific quality objectives for the program and delineate the means of achieving them. The QA organization must include appropriately qualified staff with suitable delegation of responsibility and authority. QA functions should be organizationally independent of operations and should have the full support of upper management.

Procedures and records are an integral part of any formal QA program. Written procedures should encompass all programmatic activities including sample site selection and location; sample collection, handling, and analysis; calibration and standardization of measurements and equipment; and collection and analysis of data, including environmental dose calculations. A systematic program of recordskeeping and documentation should also be implemented. It is often wise to retain records for the lifetime of the operation and beyond, and this should include such primary data sources as sampling collection forms, data sheets, laboratory notebooks, and logbooks. The information contained in these documents can be of value years later in identifying and analyzing problems as well as providing legal protection.

From an operational standpoint, six necessary quality assurance components have been identified (Corley et al. 1981):

1. Equipment procurement acceptance testing.

2. Routine instrument calibration.

3. Analytical cross-check program.

4. Systematic replicate sampling.

5. Procedural audits, both routine and *ad hoc.*

6. Documentation of procedures and QA records.

Various specific techniques have been devised to assure that these components are adequately addressed, and include ensuring that instrument calibrations are relatable to the National Bureau of Standards, laboratory intercomparison programs with both unknown and spiked samples of environmental media, and independent audits of programs by outside personnel.

Finally, for a QA program to be successful, there has to be followup of findings, with an eye towards appropriate corrective actions. The range of corrective actions may be many, and all such activities, ranging from the audit or investigation that uncovered the problem through the implementation and final verification of the corrective action need to be fully and accurately documented.

References

1. Conti, E. F., et. al. 1980. Environmental Radiation Monitoring Program Objectives", in *Upgrading Environmental Radiation Data Health Physics Society Committee Report HPSR-1 (1980)*, U. S. Environmental Protection Agency Report 520/1–80–012.
2. Corley, J. P. and C. D. Corbit. 1983. *A Guide for Effluent Radiological Measurements at DOE Installations*, U. S. Department of Energy Report DOE/EP–003, Washington, D.C.
3. Corley, J. P. et al. 1981. *A Guide for Environmental Radiological Surveillance at U.S. Department of Energy Installations*, U. S. Department of Energy Report DOE/EP–0023.
4. Denham, D. H. 1982. "Planning Environmental Monitoring Programs", in *Hand-*

book of *Environmental Radiation*, A. W. Klement, Jr., Ed., CRC Press, Boca Raton, pp. 117–127.

5. Eichholz, G. G. 1978. "Planning and Validatiion of Environmental Surveillance Programs at Operating Nuclear Power Plants", *Nuclear Safety* 19:486.

6. Hoffman, F. O., et al. 1977. "Computer Codes for the Assessment of Radionuclides Released to the Environment", *Nuclear Safety* 18:339.

7. Kennedy, W. E., Jr. and M. A. Mueller. 1982. "Summary of Environmental Dose Models Used at DOE Nuclear Sites in 1979", Pacific Northwest Laboratory Report PNL-3916.

8. Moeller, D. W. et al. 1978. "Environmental Surveillance for Nuclear Facilities", *Nuclear Safety* 19:66.

9. Oakes, T. W., K. E. Shank and J. S. Eldridge. 1980. "Quality Assurance Applied to Environmental Radiological Surveillance", *Nuclear Safety* 21:217.

10. Rowe, W. D., F. L. Galpin and H. T. Peterson. 1975. "EPA's Environmental Radiation-Assessment Program", *Nuclear Safety* 16:667.

11. Sanderson, C. G., et al. 1980. "Quality Assurance for Environmental Monitoring Programs", in *Upgrading Environmental Rdiation Data Health Physics Society Committee Report HPSR-1(1980)*, U.S. Environmental Protection Agency Report 520/1–80–0012.

12. Simpson, R. E., et al. 1981. "Survey of Radionuclides in Foods, 1961–1977", *Health Phys.* 40:529.

13. Strenge, D. L., W. E. Kennedy, Jr. and J. P. Corley. 1982. "Environmental Dose Assessment Methods for Normal Operations at DOE Nuclear Sites", Pacific Northwest Laboratory Report PNL-4410.

14. U. S. Nuclear Regulatory Commission (USNRC). 1975a. "Programs for Monitoring Radioactivity in the Environs of Nuclear Power Plants", Regulatory Guide 4.1, Revision 1.

15. U. S. Nuclear Regulatory Commission (USNRC). 1975b. "Environmental Technical Specifications for Nuclear Power Plants", Regulatory Guide 4.8.

16. U. S. Nuclear Regulatory Commission (USNRC). 1977a. "Calculation of Annual Doses to Man from Routine Releases of Reactor Effluents for the Purpose of Evaluating Compliance with 10 CFR 50, Appendix I", Regulatory Guide 1.109.

17. U. S. Nuclear Regulatory Commission (USNRC). 1977b. "Methods for Estimating Atmospheric Transport and Dispersion of Gaseous Efflents in Routine Releases from Light-Water-Cooled Reactors", Regulatory Guide 1.111.

18. U. S. Nuclear Regulatory Commission (USNRC). 1978. "Measuring, Evaluating, and Reporting Radioactivity in Rleases of Radioactive Materials in Liquid and Airborne Effluents from Nuclear Fuel Reprocessing and Fabrication Plants", Regulatory Guide 4.16.

19. U. S. Nuclear Regulatory Commission (USNRC). 1979. "Quality Assurance for Radiological Monitoring Programs (Normal Operations)", Regulatory Guide 4.15, Revision 1.

20. U. S. Nuclear Regulatory Commission (USNRC). 1980. "Radiological Effluent and Environmental Monitoring at Uranium Mills", Regulatory Guide 4.14, Revision 1.

21. U. S. Nuclear Regulatory Commission (USNRC). 1982. "Calculational Models for Estimating Radiation Doses to Man Resulting from Uranium Milling Operations", Regulatory Guide 3.51.

22. Watson, D. C., Ed. 1980. *Upgrading Environmental 'Radiation Data Health Physics Society Committee Report HPSR-1 (1980)*. U.S. Environmental Protection Agency Report 520/1-80-0012.

CHAPTER 14

Ambient Radiation and Aerial Measurements

Ambient Measurements with Integrating Detectors

Ambient environmental radiation include direct radiation fields or immersion doses associated with the release of activity from a nuclear facility, as well as direct radiation fields associated with natural background. Measurement is usually accomplished with passive integrating *in situ* detectors, but a variety of active devices are also available. It is possible to measure dose or dose equivalent or their corresponding rates, energy spectra, or radioactivity concentrations at various levels, with the only limitations being expense and imagination. Photon doses or exposures are probably the most common types of ambient radiation measurement because of the high interest in the radiological impact of radioactivity releases from nuclear facilities and the relative ease with which photons can be detected and quantified. Ambient beta measurements are made in the vicinity of nuclear power plants and fuel reprocessing facilities that may discharge effluents containing the radioactive noble gases which are primarily beta emitters. In the vicinity of accelerators, ambient neutron measurements may be desired. Very low level neutron, photon, proton, and beta fluence or dose measurements may be carried out in connection with studies of the natural radiation environment.

An ambient radiation measurement device must have a high degree of sensitivity and precision in order to be able to detect background or near background levels of radiation, and permit the identification of doses of on the order of a few per cent of background as is frequently required. Direct radiation measurements in the vicinity of nuclear facilities are usually made with passive integrating dosimeters, which include photographic emulsions, radiophotoluminescent (RPL) glasses, thermoluminescent dosimeters (TLD), and integrating ionization chambers. Of these, TLD has consistantly shown the highest degree of sensitivity and precision (Gesell and de Planque 1980) and is the most widely used because of these and other advantages (Table 14–1).

Photographic films are perhaps the oldest passive integrating detector, having been used during the Manhattan District days (Parker 1980) but are now little used for environmental monitoring. The blackening of the film produced by exposure to ionizing is chemically amplified through the development process, which must be rigidly controlled with respect to temperature and time in the darkroom. The degree of radiation induced darkening is measured photometrically after development and correlated with dose. A typical photographic emulsion has a lower level of detectablitity equivalent to about 10 mrad (100 μGy) for most beta-photon radiations. At typical natural background levels of 10–20 μrad/h (0.1–0.2 μGy/h), 500 to 1000 hours—i.e. several weeks—of exposure would be required to reach this lower level of detectability. Even longer exposure times would be required to differentiate the tiny additional incremental exposure typically associated with anthropogenic sources. In addition to their inherent lack of sensitivity, accuracy and precision, photographic emulsions are susceptibile to damage or erroneous readings from exposure to heat, humidity, light, trace amounts of chemicals such as mercury, and other physical damage which greatly restricts their usefulness and exposure interval in the field. Without protection from humidity, an exposure interval as brief as a week may be all that can be tolerated in high humidity climates without loss of the latent image or even dmage to the film emulsion. The interval may be extended to a maximum of perhaps three months by sealing the film packet in a plastic envelope but even so the validity of the measurements is questionable (Kathren, Zurakowski, and Covell 1966). Also, because of their high effective Z number, photographic emulsions are highly energy dependent and the use of energy flattening filter may be required to obtain accurate results in the photon energy region below a few hundred kilovolts.

Despite these limitations, photographic emulsions may still find application in certain special environmental monitoring situations. Thick emulsions are sometimes used to detect high energy neutrons and other particles from accelerators or from cosmic rays at high altitudes if the exposure interval is sufficiently brief. However, even these applications are on the wane, being supplanted by techniques such as track etch films of plastic or mica which are not adversely affected by environmental conditions.

Radiophotoluminescent dosimeters have been sucessfully used for perimeter monitoring around nuclear facilities both in the United States and in Europe (Becker 1973). The RPL property is a permanent change

TABLE 14-1 Comparison of Photographic, Luminescent and Ionization Chamber Dosimeters for Environmental Monitoring.

Quality	Photographic Films	RPL Glass	TLD	Ionization Chambers
Precision in Field Use	+, − 25%	+, − 10%	+, − 5%	+, − 10%
Photon Energy Dependence	Very poor	Moderate	Excellent	Good
Angular Dependence	Large	Moderate	Good	Fair to Good
Lower Limit of Detection	10 mrad (100 μGy)	10 mrad (100 μGy)	<1 mrad (<10 μGy)	1 mrad (10 μGy)
Ease of Handling	Good	Good	Good	Fair
Maximum time in Field	3 months	Indefinite	Indefinite	3–6 months
Resistance to Environment	Very poor	Excellent	Very Good	Fair
Reusable	No	Yes, in some cases	Yes	Yes
Convenience	Fair	Very Good	Very Good	Fair
Cost	Low	Moderate	Moderate	Moderate

introduced into various phosphate glasses by irradiation, with the degree of change determined by measuring the ultraviolet excited light emission. Unlike photographic emulsions, RPL dosimeters are highly resistant to the environment, and can be used over a period of years to accumulate low level doses. Their level of sensitivity is similar to that of photographic emulsions—about 10 mrad (100 μGy)—as is their energy dependence, but their precision and accuracy are very much better.

Because of their excellent sensitivity, precision, and accuracy coupled with ease of handling and low cost, thermoluminescent dosimeters are widely used for integrating ambient environmental dose measurements. Thermoluminescence is a property common to all inorganic crystals, and involves the trapping of electrons within imperfections in the crystalline lattice following irradiation. Upon subsequent heating, the electrons will return to the ground state in the valence band of the crystal with the emission of visible light. The quantity of light emitted is proportional to the radiation exposure the crystal has received. The TLD may be returned to its original state of usefulness by annealing with a suitable time-temperature combination. Thus, they are quite economical in that they may be reused many times.

Use of TLD's for ambient environmental monitoring began in the 1960's, and today several phosphors are available with a lower detection level of 1 mrad (10 μGy) or less (Table 14-2). The most commonly used phosphors are LiF, CaF_2 and $CaSO_4$ with various activators such as manganese or dysprosium (Gesell and de Planque 1980). LiF, with a Z very close to that of air or soft tissue, possesses excellent energy dependence properties. In the field, TLD's need only minimal protection; as some phosphors may be light sensitive, good practice is to enclose the TL material in lighttight packaging. Packaging will also protect against dirt, which may provide spurious results, and discoloration or dissolution by precipitation or wetting, which may result in a reduced light output on subsequent readout.

Two factors that may adversely influence the accuracy and reliability of the TLD are 1) fading of the radiation induced thermoluminescence with time and 2) self-irradiation from trace amounts of radioactivity incorporated into the TLD or its packaging material. Fading is a function of the phosphor (Table 14-1) and adverse effects may be minimized by judicious choice of TLD materials and through the use of appropriate correction factors based on exposure time in the field. Self-irradiation may be significant in some phosphors, particularly if extreme sensitivity and precision is sought. These effects can be mathematically described by (NCRP 1976).

TABLE 14-2 Properties of Some TLD Phosphors Useful for Ambient Environmental Radiation Monitoring

Phosphor	Z_{eff}	Useful Range	Fading	Comments
LiF	8.2	1 mrad–10^5 rad (10 μGy–1000 Gy)	Negligible	May exhibit tribothermoluminescence
LiB_4O_7	7.4	10 mrad–1000 rad (100 μGy–10 Gy)	None	Z lower than air
CaF_2	16.3	1 mrad–10^4 rad (10 μGy–100 Gy)	5%/mo	Natural substance; self-irradiation may interfere with low level measurements
CaF_2:Dy	16.3	<0.5 mrad–10^4 rad (<0.5 μGy–100 Gy)	Large initially, then small	Requires preannealing
$CaSO_4$	15.5	<0.1 mrad–20 rad (<10 μGy–200 mGy)	Low	Various dopants used, including Mn, Dy and Tm
BeO	5.2	2 mrad–20 rad (20 μGy–200 mGy)	Low	Z lower than air; also useful as TSEE dosimeter

$$I_t = I_oe^{-\lambda t} + kx(1 - e^{-\lambda t}) \qquad (14-1)$$

in which I_t is the thermoluminescence light intensity at time t, I_o the background thermoluminescence, χ the exposure rate from ambient sources and self-irradiation, and λ the thermoluminescence fading constant.

Specific guidance with regard to the application of TLD's to environmental monitoring, including calibration and evaluation techniques, has been called out in an American National Standard (ANSI 1975) and U.S. Nuclear Regulatory Commission Regulatory Guide (USNRC 1977). Typically, TLD's are placed in the field in groups of three and positioned one meter above the ground. Care must be taken to avoid shielding effects from nearby buildings, large trees, or other structures. Suspension from thin poles or wires a distance of 50 meters from the nearest building is the usual practice for field emplacement.

In a typical monitoring program in the environs of a nuclear facility, a relatively large number of TLD measurements will be made, largely because of the ease and low cost of measurement. Thus, ambient TLD monitoring stations may be located at the same location as air and soil monitoring stations, and in addition may be placed at other locations of interest, such as along the banks of rivers or lake shorelines where radionuclides may be likely to build up in sediments. At least one such station and preferably several should be placed at a reasonably convenient location that is unlikely to be affected by the facility. This station will serve as a background or control station. For facilities such as nuclear power plants and uranium mills which may give rise to increased ambient radiation levels, the number of TLD's used for ambient environmental radiation monitoring may be relatively large. For uranium mills, the NRC recommends up to eighty located in a radial pattern outward from the mill, one dosimeter at each 150 meter interval in each of eight directions out to a distance of 1500 meters (USNRC 1980). If each package contained three dosimeters, and two processing intervals (e.g. monthly and quarterly) were used, this would require the use of 480 individual dosimeters. For nuclear power plants, the recommended practice is two or more dosimeters at each air sampling location plus 2 or more at each of three other locations of highest calculated offsite ground-level dose, plus two controls (USNRC 1975). A typical installation around a nuclear power plant might thus have 20 to 40 TLD sites. TLD's in waterproof packages has also been used for underwater measurements in rivers and other bodies of water, both in conjuction with nuclear facility monitoring and with measurements of natural radioactivity. One of the more interesting applications of the latter is a study of the depth dose profile in Crater Lake which is being carried out by the Oregon State Health Division and which, among other things, will yield the attenuation coefficient for cosmic rays in water.

TLD's are ordinarily processed on a quarterly basis, and the mean daily dose or exposure calculated from the readings. Monthly, semiannual, or annual field exposure periods are also commonly used, and it is not uncommon to place several TLD's, all with different processing frequency, at a single location to gain improved detection capability.

Integrating measurements are also made with electronic devices. Passive ionization chambers have been used with some success, although in general such devices are relatively insensitive and also subject to inaccuracies due to environmental influences. Pocket ionization chambers and similar devices have been used to a limited extent primarily for area monitoring or perimeter monitoring at some facilities, or for *ad hoc* ambient environmental monitoring purposes associated with a specific operation.

Ambient Measurements with Rate Meters

Commonly used in the field are ionization chamber rate meters which record their readings at specific intervals, and which may telemeter this information back to a central site for computer processing and storage. Most such ratemeters use a pressurized ion chamber to gain adequate sensitivity. Current flow in an air ionization chamber with a volume of one liter at atmospheric pressure is 9.3×10^{-17} amperes per μR/h. Thus, a typical environmental background level of 10 μR/h will produce a current of approximately a femtoamp, a level difficult to precisely measure without very expensive and delicate ammeters. To obviate this problem, the volume of gas in the ion chamber can be increased, either by enlarging the chamber or pressurizing the chamber. For many years, ambient environmental exposure rates were measure in the vicinity of the Hanford nuclear site with a cylindrical aluminum walled ionization chamber with a volume of 40 liters which fed its signal directly into a sensitive ammeter mounted atop the chamber.

A more common technique is to enhance sensitivity by pressurizing a more moderately sized chamber to 20 or more atmospheres and either reading out current directly or using a very high input resistance, on the order of 10^{15} to 10^{16} ohm and measure voltage. Several pressurized ion chambers designed for permanent or long term placement in the field are commercially available. These are often equipped with remote readout telemetering capability. One uses a pressurized chamber with argon as the fill gas and is capable of continuous rate measurements over the range 1 to 150 μR/h. This particular instrument will operate with either ac or dc (battery) power, and despite the necessity for a heavy wall, shows excellent energy response for photons over the range of 50 KeV to above 3 MeV. Because of the relatively thick wall, this device is essentially insensitive to beta radiation and hence cannot be used to measure the immersion dose from noble gases.

In addition to the instruments designed for field emplacement, several portable survey instruments using ionization chamber and other detectors are also available. An ideal survey instrument for ambient environmental radiation levels should have the following characteristics (NCRP 1976):

1. Be of sufficient sensitivity to permit measurement of photons to levels at least as low as 1 μR/h.

2. Have a sufficiently short time constant to permit measurements to be made rapidly and frequently.

3. Have uniform and essentially isotropic response over the energy range of interest.

4. Be rugged and unaffected by field conditions such as changes in temperature.

5. Be conveniently portable—i.e. small and lightweight.

As a practical matter, few, if any, portable survey meters exhibit all these characteristics.

Ionization chamber type portable survey meters suitable for monitoring ambient environmental radiation fields generally have a detection range of a few μR/h to perhaps a much as 1 R/h. Some are capable of detecting beta radiations as well. Typically ion chamber survey meters exhibit relatively long time constants, particularly on their lowest ranges, and in some cases several minutes are required to permit equilibration of the instrument in μR/h fields. Thus, such instruments may be impractical if a large number of measurements are to be made, or for the detection of transients. The energy and angular response of the instrument will vary according to the wall material and geometry of the detector. If the wall material is close in atomic number to that of air or soft tissue, and sufficiently thin, the photon energy dependence will be slight. Since detectors are not spherical and may be housed inside a case with a window, angular dependence or isotropicity of response is usually pronounced, particularly for photons with enegies below a few hundred kilovolts. Unsealed air ionization chambers may also require correction for changes in air density.

Geiger-Muller (G-M) counters and scintillators such as NaI(Tl) have also been used for field measurements of ambient radiation levels. These show adequate sensitivity, generally being ccapable of detecting levels down to a few $\mu R/h$. Both G-M and scintillation detectors have short time constants and hence fast response times which makes them more suited to surveys of the environment than low level ionization chambers. They can also be made very rugged and resistant to weather. However, because they are made from high Z materials, both G-M and scintillation detectors generally are quite energy dependent, with a potential overresponse to photons of an order of magnitude or greater relative to that of soft tissue or air in the region about 30 keV. To minimize or even eliminate this effect, energy flattening filters may be used with excellent results. One commercially available G-M survey meter with an internal detector has a flat response over the region of 50–2000 keV and is able to monitor photon exposure rates in the $\mu R/h$ range. Because of the need for discriminating filters, this and similar instruments are insensitive to beta radiation, although some G-M type devices are equipped with windows or sliding shields which permit an open window reading to be made.

Routine surveys with rate meters should be made in a consistant and reproducible fashion. Monitoring locations must thus be appropriately and exactly located, with care taken to select representative areas. Outcrops of rock which may be high in natural radioactivity levels, proximity to buildings, and highly nonuniform distributions of naturally occurring radionuclides may have an adverse effect on interpretation of results. A standard measurement protocol is highly recommended, with measurements made at a height of one meter above the ground. Instruments should be calibrated with sources similar in energy to that of the sources in the field; radium in equilibrium with daughters is a most suitable source.

Scintillation detectors as well as semiconductor detectors can be used for field monitoring to detect and quantify specific gamma emitting radionuclides by spectroscopy. Separation of the cosmic and terrestrial components of radiation can also be made by means of spectroscopy; in general, photons with energies above 3 Mev are of cosmic origin while those with lower energies are ordinarily of terrestrial origin. Scintillation spectroscopy can be a highly sensitive quantitative tool for specific radionuclides. Single and dual channel portable spectrometers are made that are preset to the 364 keV photon associated with ^{131}I allowing direct readout of the meter in terms of this radionuclide.

Large scale detection of environmental plutonium has been required in a few instances following crashes of aircraft containing plutonium bearing nuclear warheads. Plutonium-239 is a highly radiotoxic fissionable nuclide that decays by alpha emission with a small percentage of accompanying L x-rays with energies in the region of 17 keV from the excited ^{237}Np daughter. Field experience following the accident near Palomares, Spain, in 1966, demonstrated the impracticality of large scale area monitoring for plutonium by alpha survey techniques and led to the development of the FIDLER (Field Instrument for Detection of Low Energy Radiation) (Tinney, Koch, and Schmidt 1969). The FIDLER utilizes a five inch diameter NaI(Tl) crystal, 1/16 inch thich, to detect the 17 keV photons associated with the decay of ^{239}Pu as well as the 60 keV photons from the ^{241}Am impurity often associated with the plutonium. This hand held detector is designed to be carried such that the detector is held one foot (0.3 m) above the surface of the ground. Ordinary environmental survey meters are either incapable of detecting these low energy photons because of the thickness of their detector walls or are relatively insensitive. The FIDLER, which examines activity through a relatively large solid angle and is able to reduce background through discriminator settings, can detect 100 nCi (37Bq)/m^2 of ^{239}Pu and 10 nCi (3.7 Bq)/m^2 of ^{241}Am, levels very much greater than those of environmental concern (Kathren 1968, Healy 1974).

Large scale contamination monitoring is sometimes carried out on roads and railroad tracks with the aid of scintillation or G-M detectors mounted on trucks or rail cars. The detectors are generally shielded on the top and sides and are suspended one foot (0.3 m) above the road or rail bed surface. Monitoring can be done at speeds of about 10 miles per hour (16 km/h). Such devices must by necessity be ruggedized and are relatively sensitive if large scintillation crystals or arrays of several G-M tubes are used.

In addition to measuring exposure or dose rates, survey instruments are sometimes useful for determination of radionuclide concentrations in soil and water. Ordinarily this is done by spectroscopy, but Geiger counter and scintillation detectors have been used for years to detect and crudely quantify the quality of uranium bearing ores.

Aerial Surveys

Aerial radiation monitoring is a valid ambient monitoring technique with numerous applications including:

- nuclear facilities monitoring
- mineral exploration
- plume tracking
- soil moisture studies
- determination of the water equivalent of snow cover
- fallout measurements
- detection of lost radioactive sources
- general radiation measurements
- geochemical and geophysical surveys
- emergency planning and environmental evaluation
- special studies of natural radiation fields.

Detection is usually made with an array or single large scintillation detector. Although NaI(Tl) is usually the detector of choice, other scintillation materials and even large proportional counters have been used.

An early system was designed in 1951 for uranium exploration and used an array of six 4 inch diameter by two inch thick NaI(Tl) detectors with six photomultiplier tubes connected in parallel. The detector system and associated electronics were subsequently mounted in a light fixed wing aircraft operated by the U.S. Geological Survey to monitor fallout and natural radiation levels; this was the first Aerial Radiation Monitoring System (ARMS I) (Davis and Reinhart 1957). A second system, ARMS-II, was developed by EG&G and utilizes a single 9 inch diameter, three inch thick thallium activated sodium iodide detector (Hand, Guillou, and Borella 1962). Subsequent improved ARMS systems have been developed and ARMS monitoring aircraft are held in reserve in Nevada for lost source detection and accident applications in addition to use for routine monitoring purposes (Doyle 1974).

The national ARMS Program was initiated in mid-1958 to determine the gamma background radiation levels in various parts of the United States (Guillou 1964). A typical survey was made over flat and rolling terrain at an altitude between 300 and 900 feet (approximately 100–300 meters) over a standard ARMS square 100 miles (160 km) on a side centered on a nuclear facility. The general technique was to fly equally spaced parallel lines over the area with a spacing of one to two miles between flight lines. Later studies expanded the ARMS background studies to other portions of the United States such that a large part of the country—at least 10 per cent of the total land area—has been mapped by aerial radiological survey techniques (Doyle 1972).

Aerial monitoring systems are generally achieve their great sensitivity through use of large detectors or arrays of detectors in conjunction with on gamma ray spectroscopy. Either scanning a wide range of energies or "zeroing in" on a limited energy range or series of energies can be used, depending on the application. The detector may be shielded on the top and sides to minimize background from cosmic rays and thereby enhance sensitivity. Wide energy range detection is used for maximum sensitivity or gross measurements; at an altitude of 100 meters, the count rate from all photons with energies greater than 50 keV is approximately proportional to the exposure rate one meter above the ground (NCRP 1976). Specific peak localization or ratios of peaks may be used when prospecting for minerals or searching for lost photon emitting sources.

Typically, surveys are conducted from a fixed wing aircraft or helicopter moving at constant speed at an altitude of 150 meters (500 feet). At this altitude the cosmic ray component is minimal and the detector sees a relatively large solid angle, even if collimated by side and top shielding. This altitude is also consistant with safe aircraft operation. Of vital importance to the calculation of environmental dose rates or similar data is precise knowledge of the altitude of the aircraft, which is usually determined with a radar altimeter. The aircraft heading must also be precisely known.

Ambient cosmic and terrestrial levels can be determined readily by aerial monitoring techniques. As the terrestrial component emanates from below and the cosmic from above the detector, dual detectors or flights at two altitudes can be used to separate the two components. Coincidence techniques and corrections for airborne activity further enhance the ability of aerial monitoring to accurately measure these two components of the natural radiation field. This principle is also applied to soil moisture studies and to the determination of the water equivalent of snow cover. The technique is generally applicable to photons with energies greater than 500 keV, where the energy and angular distributions of both the airborne and ground radiations are known (Burson and Fritzsche 1973). In addition, count rates must be sufficiently great to provide good statistics and background contributions must also be well known. Upward collimation is superior for discrimination of the the respoinse due to airborne radioactivity, and shield dual detectors in a sandwich arrangement can be used with appropriate positioning and shielding to maximize detector response.

Concentrations of radionuclides both on and in the ground can be readily made with aerial survey techniques provided the photon energies emitted by the radionuclide(s) are of sufficient energy and yield, and the background is reasonably well known. In the case of radionuclides in the earth, suitable corrections must be made for photon attenuation by the earth overburden. This may be a difficult task unless the distribution is uniform with depth or well known along with the composition of the soil. Because of the altitude of the aircraft, the detector sees a relatively large area, typically on the order of several ten of thopusands of square meters. Thus, the resultant value is an average, which may or may not be a suitable representation of the ground level concentration. FIDLER type instruments using thin detectors or larger scintillators set for the 60 keV ^{241}Am peak have been tried with a modicum of success. As low altitudes are required for measurement to avoid significant air absorption of the low energy photons, helicopters are used rather than fixed wing aircraft for this purpose.

Among the more interesting and dramatic applications of aerial monitoring is the detection of lost photon emitting sources. Several such lost sources have been detected (NCRP 1976) including two 470 mCi (17.4 GBq) ^{57}Co sources emitting 122 keV photons which could not be found by ground survey techniques (Deal et al. 1971).

References

1. American National Standards Institute (ANSI). 1975. *Performance, Testing, and Procedural Sepcifications for Thermoluminescence Dosimetry: (Environmental Applications)*, American National Standard ANSI N545–1975, American National standards Institute, New York.

2. Becker, K. 1973. *Solid State Dosimetry*, Chemical Rubber Company Press, Cleveland.

3. Burson, Z.G., P. K. Boyns, and A. E. Fritzsche. 1972. "Technical Procedures for Characterizing the Terrestrial gamma Radiation Environment by Aerial Surveys", in *The Natural Radiation Environment II*, J. A. S. Adams, W. M. Lowder, and T. Gesell, Eds., U. S. Energy Research and Development Administration Report CONF-720805, p. 559.

4. Davis, F. J. and P. W. Reinhardt. 1957. "Instrumentation in Aircraft for Radiation Measurement", *Nucl. Sci. Eng.* 2:713.

5. Deal, L. J., et al. 1971. "Locating the Lost Athena Missile in Mexico by the Aerial Radiological Measuring System (ARMS)", *Health Phys.* 23:95.

6. Doyle, J. F. 1972. "The Aerial Radiological Measuring System (ARMS) Program", in *The Natural Radiation Environment II*, J. A. S. Adams, W. S. Lowder, and T. Gesell, Eds., U. S. Energy Research and Development Administration Report CONF-720805, Washington, D. C., p. 589.

7. Gesell, T. F. and G. de Planque. 1980. "Highlights of the Fourth International Inter-

comparison of Environmental Dosimeters under Field and Laboratory Conditions", in *Proceedings of Sixth International Conference on Solid State Dosimetry*, Toulouse, France, April 1–4, 1980.

8. Guillou, R. B. 1964. "The Aerial Radiological Measuring Surveys (ARMS) Program", in *The Natural Radiation Environment*, J. A. S. Adams and W. S. Lowder, Eds., The University of Chicago Press, Chicago, pp. 705–721.

9. Hand, J. E., R. B. Guillou, and H. M. Borella. 1962. "Aerial Radiological Monitoring System. II. Performance, Calibration, and Operational Checkout of the EG&G ARMS-II Revised System", U. S. Atomic Energy Commission Report CEX–59.4, Part II, pp. 66–75.

10. Healy, J. W. 1974. "A Proposed Interim Standard for Plutonium in Soils", Los Alamos National Laboratory Report LA–5483-MS.

11. Kathren, R. L. 1968. "Towards Interim Acceptable Surface Contamination Levels for Environmental PuO$_2$" in *Radiological Protection of the Public in a Nuclear Mass Disaster*, Symposium Proceedings, Interlaken, Switzerland, May 27-June 1, 1968, U. S. Atomic Energy Commission Report CONF–680507; also Battelle Pacific Northwest Laboratories Report BNWL-SA-1510.

12. Kathren, R. L., P. R. Zurakowski, and M. Covell. 1966. "Effect of Humidity and Dose on Latent Image Stability", *Amer. Ind. Hyg. Assoc. J.* 27:388.

13. National Council on Radiation Protection and Measurements (NCRP). 1976. *Environmental Radiation Measurements*, NCRP Report 50, Washington, D.C.

14. Parker, H. M. 1980. "Health Physics, Instrumentation and Radiation Protection" U.S. Atomic Energy Commission Report MDDC-783. Reprinted in *Health Phys.* 38:949.

15. Tinney, J. F., J. J. Koch, and C. T. Schmidt. 1969. "Plutonium Survey with an X-Ray Sensitive Detector", University of California radiation Laboratory Report UCRL-71362.

16. U. S. Nuclear Regulatory Commission (USNRC). 1975. "Environmental Technical Specifications for Nuclear Power Plants", Regulatory Guide 4.8.

17. U.S. Nuclear Regulatory Commission (USNRC). 1977. "Performance, Testing, and Procedural Specifications for Thermoluminescence Dosimetry: Environmental Applications", Regulatory Guide 4.13, Revision 1.

18. U. S. Nuclear Regulatory Commission (USNRC). 1980. "Radiological Effluent and Environmental Monitoring at Uranium Mills", Regulatory Guide 4.14, Rev. 1.

CHAPTER 15

Airborne Radioactivity Sampling and Analysis

General Principles of Sampling

Levels of environmental radioactivity are determined in various environmental media by a systematic program of sample collection followed by laboratory analysis. The type of sample collected may place certain limitations on or even determine the type of analysis that may be performed. In general, it is desirable to select samples that will require little or no preparation, are of small size and convenient to obtain, but which also lend themselves to sensitive low level analyses for the desired radioactive constituents at minimum cost. In all cases, however, the sample must contain sufficient total radioactivity to permit detection and quantification by the analytical technique to be used. This consideration may be overriding in determining the sample volume or other parameters that determine volume such as sampling rate.

It is important to make the distinction between sampling and monitoring as applied to environmental radioactivity. Sampling is the process by which a portion of an environmental medium is removed for subsequent analysis; the term monitoring is applied in both a generic and specific sense. In the generic sense, it refers to an ongoing program of environmental radiological measurement. In the specific sense, the term monitoring is applied to *in situ* field measurements on a continuous basis. In either case, sample collection is implied, although may not always be necessary as some measurements may be made directly.

Environmental samples are generally classified by medium, pathway, radionuclide, or method of collection. Classification by the method of collection is often based on temporal factors. Grab or single samples are samples collected all at once over a short or essentially instantaneous time period. Thus grab samples are not ordinarily indicative of mean or average conditions, but are useful for determining concentrations at a specific point in time or for sampling transitory events such as plume passage or fallout deposition following a release of activity. This type of sampling is therefore of value in evaluating environmental emergencies or for non-routine samples, particularly where a rapid measurement is desired. Grab or single sampling may be the preferred or only practical method for many environmental samples, such as seasonal growing plants, animals, milk, sediments, and soils. In these cases, grab samples actually represent temporally averaged conditions, as what is obtained has been been growing and developing over a period of time. If repeated periodically, say on a seasonal or annual frequency, grab samples lose the character of single samples and become useful for trend analysis of radionuclide concentrations or accumulation in various environmental media.

Extended time proportional samples are often described as continuous samples. Continuous sampling may be either sampling without interruption for a relatively long period of time or a series of single samples repeated at close intervals and composited for analysis. In the latter case, sampling may be accomplished via a periodic sampling regimen; for example, a sampler may be programmed such that it collect samples for the first five minutes out of each hour, or for a set time period each day. Continual sampling is thus basically the collection of a number of grab samples; continuous uninterrupted sampling may be thought of a one huge grab sample, or, more correctly, as an infinite number of grab samples. Continuous uninterupted sampling is frequently applied to effluent streams, both liquid and gaseous, and to general sampling of environmental air and bodies of water.

Clearly, there are applications for both grab and extended time samples. In general, a large number of properly spaced grab samples provides a more accurate and comprehensive picture of the environment than does a single composited sample and thus is useful for evaluating short term changes or trends. Continuous sampling thus provides the best picture. However, continuous sampling is basically an averaging mechanism; analysis of continuous samples provides the amount of activity collected over the entire sampling period, and which when divided by the total volume or mass of the sample is the mean concentration during the sampling period. To gain the temporal distribution of radioactivity concentration in a medium such as air or water, a process of continuous or continual sampling and analysis may be accomplished. Numerous air and water monitoring instruments have been designed for this purpose and are commercially available (LBL 1983). Whatever the type of sample or the sample collection procedure used, the sample must be representative of the medium under consideration. Thus, the specific sampling location should be selected such that samples containing nonrepresentative concentrations of activity are avoided unless, of course, one of the specific sampling criteria is to obtain such a sample. Great care must be taken to avoid cross contamination or loss of radioactivity during or subsequent to sampling; for example, it may be necessary to freeze or add preservatives to samples to prevent loss of volatile radioelements such as iodine. Care must also be taken to prevent plate out of radioactivity or reaction of the sample with the sample collector or container. Chemical reaction with the sample collection container or similar equipment usually removes radioactive constituents from the sample, hence reducing the concentration, although radioactive substances in the container matrix may also be freed by chemical reaction and dissolve into the sample, giving it an artificially high concentration.

The radioactive half-lives of the radionuclides being evaluated need to be considered, as do the chemical and physical form. A regretable practice all too common in some programs has been to hold samples following collection for a period of several days, during which time short-lived radionuclides may decay to concentrations below the minimum detectable level provided by the laboratory analysis. The concentration of radionuclides such as ^{131}I, with an 8.05 day half-life, may be significantly depleted in only a few days; hence, such samples should be transported to the laboratory and analyzed as soon after collection as practicable. Decay corrections should not be applied indiscriminately. Again considering a sample containing ^{131}I, two months after collection, approximately eight half lives will have passed requiring a decay correction factor of nearly 250. Or, looking at the situation another way, less than one half of one per cent of the activity originally present remains. In general, measurements of radionuclide concentrations arrived at by application of a decay correction factor greater than 2 should be treated as suspect; those with a decay correction factor greater than 6 are considered invalid in many programs.

In any environmental surveillance program, sample collection should be standardized to minimize sample variability and bias introduced by nonstandard or *ad hoc* procedures. Thorough and appropriate mixing is desirable for average value samples to minimize errors that may be introduced by uneven distribution of the radioactivity in the sampling medium or area. For certain sampling procedures, particularly those involving filtration or similar techniques to remove particulates, the efficiency of removal may be less than 100%. This may be a relatively small effect; a sampler with a collection efficiency of 90% only biases the final result by 10%, which may be small in comparison to other sources of error. However, efficiency can be improved by using several successive samplers, or trains. If each sampler has the same efficiency, E, the fraction of the total removed by the nth stage in a train of samplers, F_n is given by (NCRP 1976).

$$E_{total} = E(1 - E)^{n-1} \qquad (15\text{--}1)$$

and the efficiency can be computed by measuring the relative amounts removed by pairs of successive collectors as shown in Equation 15-2.

$$E = 1 - \frac{F_2}{F_1} \qquad (15\text{--}2)$$

Sampling trains may be quite useful for radionuclide separation if the collection efficiencies differ. In this case, the samples from the later stages of a train will proportionately contain more activity from the nuclide with the poorer collection or removal efficiency, although the total activity will be less since so much was removed in the first stages.

Consideration also needs to be given to the chemical and physical state of the radionuclide(s) being sampled. For example, a radionuclide can be present in both ionic and nonionic forms. If a water sample were collected with an ion exchange column, only the ionic forms would be retained. Multiple stages of ion exchange columns would not appreciable improve the sampling collection efficiency. A different sampling technique is needed to capture the nonionic forms, and this may be connected in series with the ion exchange column, permitting the concentration of each as well as the total to be determined.

The time of sampling may also be important, as the radionuclide concentrations may not be temporally constant throughout the sampling period. This is particularly true of radionuclide concentrations in air and water in the vicinity of nuclear facilities which make periodic planned discharges of activity. If time-averaged values are acceptable, then continuous sampling may be appropriate. However, if peak concentrations or avaraged concentrations during the time of release are desired, then the sampling period will have to be adjusted accordingly. An adequate sample size, of course, must be obtained.

An important and frequently overlooked consideration relates to depletion of the sampling area of the medium being sampled. Obviously this possiblility is remote in the case of air or a large body of water, but is real and highly probable in other media. For example, periodic continuous sampling for sediment may scour a river bottom, and collection of certain biota may deplete the populations of these species in the area of interest. This not only has an adverse effect on the environmental surveillance program, but can also result in irreparable damage or irreversible alterations to the environment.

Airborne Particulate Radioactivity

Particulates in air are most commonly sampled by removal from the air by filtration through cellulose fiber, fiberglass, or molecular seive (membrane) type media. The latter are particularly suited to sampling alpha emitters, as deposition of particulate material is largely on the surface. Membrane filters are quite fragile and can only be used with low flow rates. However, the other types of filters are more sturdy and easily handled, and thus are more suited to sampling for beta-photon emitters. Cellulose-asbestos filters were at one time quite popular, but their use has been discontinued because of the hazards associated with asbestos. All the types of filters mentioned generally have high efficiencies for collection of particles with sizes in the respirable range, and with the exception of the membrane type filter, have a relatively low pressure drop, permitting high flow rates, often to several cubic feet per minute. High flow rates are desirable as they allow large volumes of air to be sampled, with the concommitant deposition of greater amounts of activity, which increases the sensitivity of the measurement. Each filter has its own specific particulate removal properties and optimum flow rates, as shown in Table 15–1. However, standard practice calls for the filter to achieve a removal of 99% of the particles of dioctyl phthalate having an aerodynamic median diameter of 0.3 um at the air velocity and pressure drop expected during use (ACGIH 1974).

In addition to the sample collection device, or filter, a typical sampling arrangement for particulate radioactivity in air includes an air moving device and frequently a flow meter, plus associated ducting. The air moving device is located downstream from the sample collector; that is, the air is first passed through the sample collector, then through the remainder of the system. The blower, pump, or other air moving system is ordinarily the final component in the system. It is not desirable to pass the air through any components prior to or upstream of the filter, for this may result in plateout or removal of the radioactivity prior to reaching the filter. Sampling lines, particularly those ahead of the sample collector, should be kept as short as possible, and should be free of sharp bends or areas of turbulence or resistance to flow. This not only reduces plateout, but also is beneficial from the standpoint of requiring less air moving capability.

TABLE 15-1   Properties of Some Common Filter Media.

Filter	Pore Size (um)	Permeability Velocity (cm/s)	Efficiency Range* (%)
Cellulose Fiber			
Whatman No. 1	—	6.1	49–99.96
Whatman No. 4	—	20.6	33–99.5
Whatman No. 5	—	0.86	93.1–99.99
Whatman No. 40	—	3.7	77–99.99
Glass Fiber			
Gelman Type A	—	11.2	99.92->99.99
Gelman Type A/E	—	15.5	99.6->99.99
Gelman			
Spectrograde	—	15.8	99.5->99.99
MSA 1106B	—	15.8	99.5->99.99
Pallflex T60	—	49.3	55–98.8
Reeve Angel 934AH	—	12.5	98.9->99.99
Whatman GF/A	—	14.5	99.0->99.99
Whatman GF/B	—	5.5	>99.99
Whatman EPM			
1000	—	13.9	99.0->99.99
Membrane Filter			
Celotate-EG	0.2	0.31	>99.95->99.99
Celotate-EH	0.5	1.07	99.989->99.999
Celotate-EA	1.0	1.98	99.99->99.99
Chemplast 75-F	1.5	3	83–99.99
Gelman Teflon	5.0	56.8	85–99.90
Ghia Teflon			
S2 37PL 02	1.0	12.9	>99.97->99.99
S2 37PJ 02	2.0	23.4	99.89->99.99
S2 37PK 02	3.0	24.2	92–98.98
Millipore MF-V.S	0.025	0.028	99.999->99.999
Millipore MF-VC	0.1	0.16	99.999->99.999
Millipore MF-PH	0.3	0.86	99.999->99.999
Millipore MF-HA	0.45	1.3	99.999->99.999
Millipore HF-AA	0.8	4.2	99.999->99.999
Millipore MF-RA	1.2	6.2	99.9->99.999
Millipore MF-SS	3.0	7.5	98.5->99.999
Mitex-LS	5.0	4.94	84->99.99
Polyvic-BD	0.6	0.86	99.94->99.999
Polyvic-VS	2.0	5.07	88->99.99
Zefluor Teflon P5PJ			
037 50	2.0	32.5	94.6–99.96

TABLE 15-1 (*Continued*)

Filter	Pore Size (um)	Permeability Velocity (cm/s)	Efficiency Range* (%)
Nuclepore			
Nuclepore			
N010	0.1	0.6	>99.9
N030	0.3	3.6	93.9->99.99
N060	0.6	2.1	53-99.5
N100	1.0	8.8	28-98.1
N500	5.0	30.7	6-90.7

*For particle diameter range 0.035-1 um.

Source: *Handbook of Air Sampling Instruments,* American Conference of Industrial Hygienists, 1983.

The air mover is an important part of the air sampling system, and must be capable of continuously maintaining the desired sampling rate, which may range from liters to cubic meters per minute. Air moving systems should not be sensitive to changes in pressure drop across the filter as the filter loads with dust. Filter loading is a variable proposition, and largely determined by the amount of dust in the air. If the flow rate during the sampling period does not remain constant, large errors may be introduced into the measurement of radioactivity concentration unless the concentration itself remains constant, a most unlikely situation. The collection and counting efficiencies may also be altered by heavily loaded dust filters due to self absorption of alpha and beta particles. To avoid these difficulties, constant displacement pumps and similar devices which sample or move air at a constant volume over a suitable range of pressure drops are recommended for general environmental monitoring.

Flow measuring devices may be either differential, measuring flow rate, or integral, measuring total volume. Differential or flow rate meters are often used in conjunction with running timers which permit the total flow to be calculated; the running timer is designed to operate only when the air mover is operational, thus obviating any questions regarding elapsed time in the event of instrument failure while out in the field. Knowledge of the volume of air sampled is also useful to determine if in fact the constant displacement pump has worked correctly.

Filters containing the dust removed from the air are most commonly analyzed for radioactivity by direct counting techniques for gross activity or by spectroscopy in the laboratory. Ordinarily, when looking for radionuclides in fallout or from releases from nuclear facilities, air particulate samples are held for a time after they are removed to allow the short-lived daughters of radon and thoron to decay. One technique for rapid determination of long-lived beta activity is to count the sample twice, once about four or more hours after collection to allow the 26.8 minute ^{214}Pb daughter of radon to decay away, and the second time about 20 to 24 hours later, or after about two half-lives of the beta emitting thoron daughter ^{212}Pb. The activity at the time of the first count, A_1 is the total of both the activity from the long lived fraction, A_{LL} and the activity from radon-thoron daughters, A_{Nat1}, or

$$A_1 = A_{LL} + A_{Nat1} \tag{15-3}$$

The activity at the time of the second count, A_2 is reduced because of radioactive decay and can be expressed by

$$A_2 = A_{LL}e^{-\lambda_2 \Delta t} + A_{Nat1}e^{-\lambda_1 \Delta t} \tag{15-4}$$

in which λ_2 and λ_1 are the decay constants for the long lived and natural radioactivity components, respectively, and Δt is the elapsed time between the two counts. As a practical matter, for normal intervals of a few hours to a few days between counts, there will be no decay of the long lived fraction, and all the decay of the natural activity will be ^{212}Pb, which has a decay constant of 0.0655 h^{-1}. Thus, Equation 15–4 reduces to

$$A_2 = A_{LL} + A_{Nat1}e^{-0.0655 \Delta t} \tag{15-5}$$

The natural short-lived activity at the time of the second count can also be expressed as

$$A_{Nat2} = A_{Nat1}e^{-0.0655 \Delta t} \tag{15-6}$$

which can be combined with Equation 15–5 to yield

$$A = A_{Nat2} + A_{LL} \tag{15-7}$$

Equations 15–4 and 15–7 can then be combined and solved simultaneously for C_{LL}, yielding

$$A_{LL} = \frac{A_2 e^{-\lambda_2 \Delta t} - A_1 e^{-\lambda_1 \Delta t}}{1 - e^{-\lambda_1 \Delta t}} \tag{15-8}$$

which reduces to

$$A_{LL} = \frac{A_2 - A_1 e^{-\lambda_1 \Delta t}}{1 - e^{-\lambda_1 \Delta t}} \qquad (15\text{-}9)$$

when the decay from the long lived component is negligible.

Another a perhaps simpler technique is to use a kinetic type sampler to discriminate against small particles which carry the bulk of the natural radioactivity. Airborne particulate radioactivity of anthropogenic origin is usally associated with larger particles, which may thus be collected and analyzed directly.

Several simple techniques are used in the laboratory to accomplish economies with single samples. A filter with a great deal of activity may be cut into pieces and a portion analyzed; similarly, a portion might be kept for historical (i.e. sample museum) purposes. Analysis might also be done separately on different portions for different constituents. Filters and may be dissolved and counted by special techniques, such as in well counters or liquid scintillators to gain increased sensitivity or spectroscopic capability. Organic membrane filters are readily soluble in organic solvents and are therefore quite useful for this purpose. Cellulose fiber filters are also readily dissolved, but may require dissolution by hot concentrated acid. Dissolution of fiberglass filter media is difficult, and usually requires treatment with concentrated hydrofluoric acid.

Constant air monitors (CAM's) are essentially air samplers that are equipped with a detector and appropriate electronic counting and readout equipment, and examine the sample as it is taken in the field. There are two basic types of CAM's: fixed filter and moving filter. In a fixed filter CAM, air is drawn through a filter located close to a detection device, ordinarily a Geiger-Muller tube with a thin window or a scintillator. As the particulates containing radioactivity are deposited on the filter, the activity associated with them is immediately counted and the count rate displayed on an appropriately calibrated count rate meter.

As particulates are removed from the air stream and deposited on the filter, a buildup of natural radioactivity from radon and thoron daughters will occur. Normally, equilibrium will be reach in four hours. Once equilibrium has been reached, changes in radioactivity concentration will be detectable, but these may be attributable to changes in the natural radioactivity concentration in the air, or to a shift in the equilibrium. To continuously monitor for specific gamma emitting radionuclides, scintillators may be used in conjunction with single channel analyzers set for the appropriate photon energy band(s) of the radionuclide(s) to be detected.

The problem of natural background is minimized by use of moving filter CAM's. In these, a continuous strip of filter paper is advanced mechanically, passing first across the air sampling inlet and then across the detector. The sampling time is kept short—on the order of a few minutes—to minimize buildup of radon daughters. Filter loading and air flow problems are much reduced with moving filter CAM's, but these gains may be offset by reduced sensitivity because of the smaller volume of air sampled, and the greater complexity and increased capital cost of the apparatus, as well as the increased cost of filter material.

Calibration of CAM's is best accomplished with a source of known activity identical or similar to that being monitored. Calibration should be accomplished with sufficient frequency to ensure confidence in the results; a weekly schedule has been suggested for instruments that operate continuously (Kathren 1974).

Electrostic precipitators are occasionally used for sampling radioactive particulates in environmental surveillance. These devices operate at high voltages and use a corona discharge to give airborne particulate a negative charge which causes them to adhere to a removable metal tube surrounding the central electrode. The dust thus collected can be washed from the precipitator and analyzed in the laboratory. Sizing of airborne dust particles may be desirable in some environmental monitoring situations, and for this purpose various impactors or similar devices may be used. Particle size studies, however, are not recommended for routine environmental surveillance programs although they may be highly useful for special studies (Corley et al. 1981).

Direct dust collection from either fallout or resuspension is sometimes included in evironmental surveillance activities. Fallout or deposition collectors may simply be flat surfaces of known area from which the dust may be washed off for radiological analysis. These are subject to large errors, however, from resuspension. Sticky paper or other coated surfaces, or various protected devices such as the dustfall jar may be used to supplement air sampling for certain types of activities, particularly if electrical power is not available (Denham 1982).

Airborne Radioiodines

Radioiodines are exclusively anthropogenic and thus monitoring for these radioisotopes is ordinarily associated with a nuclear facility or with fallout from the testing of nuclear explosives. In addition, medical use of various short-lived radioidines has been increasing rapidly. Sampling and analysis of radioiodines is complicated by the fact that radioiodines may exist in any one of a number of chemical states, including elemental (I_2), as iodide or in other ionic forms, or as methyl iodide or in combination with other organic radicals. Radioiodines can also exist in particulate form, for which sample collection by ordinary air particulate filters is suitable. However, the bulk of the iodine in air is usually in gaseous form, and travels through the filter medium. Hence, any environmental sampling and analysis program for radioiodine must include provision for removal of gaseous forms from the atmosphere.

Radioiodine in gaseous form is normally sampled by drawing the air or other gas stream through a bed of a substance on which the iodine will adsorb, and then counting the material directly in the laboratory or in a CAM. Measurement of ^{131}I is nearly always accomplished by quantitative spectroscopy of the 364 keV gamma ray (85% branching ratio) associated with this nuclide; the 15% 638 keV photon is also used, and direct beta counting much less frequently (LBL 1983). Lower detection limits are in the range of 1 to 10 pCi (0.037–0.37 Bq) with low background counting systems and counting times of 100 to 1000 minutes.

The materials most commonly used to sample gaseous radioiodines are activated charcoal and silver zeolite molecular seives. The former is by far the most common, largely due to cost considerations. The iodine adsorbing material is usually placed directly behind or in series with a particulate filter to remove potentially interferring radioactivity in the form of particles, or to also remove particulate radioiodines, which also need to be considered. Removal of radioiodines from a stream of gas by activiated charcoal is subject to many variables, not the least of which is the charcoal itself. The charcoal must be free of adsorbed iodine and other gases, and therefore should be kept sealed between the time of preparation until the time of use. Alternatively, the charcoal may be roasted or heated, generally for a period of several hours at a temperature of a few hundred degrees Celsius, to drive off adsorbed vapors. This procedure may also be followed to reclaim the charcoal after use; however, the effectiveness of the cleansing procedure is highly variable, and related to not only the time-temperature relationships, but also to the type of charcoal and its prior use.

The efficiency of collection for gaseous radioiodines is also related to the type of charcoal, size of the granules, depth of the sampling bed, and flow rate, as well as the particular chemical form of the iodine. To some extent a folklore has developed, with some knowledgeable persons expressing a preference for charcoal made from coconut husks or other specific subtances on the basis of superior collection efficiency and retention. While evidence may exist to support each claim, the efficacy of charcoal collection is dictated by a complex interrelationship of the several factors mentioned above. Granule size is extremely important, for the individual particles must be sufficiently small to allow reasonably dense packing and to present a large surface area. Suitable grain size is in the range 12–30 mesh. Channeling may occur in charcoal cannisters or cartridges, and appears to be related to granule size. The channeling problem may be mitigated by sampling through several charcoal cartridges in series, or by a baffle flow cartridge design.

Both collection efficiency and retention of radioiodine by the charcoal are reduced by sampling rates that are too great (Bellamy 1974). Optimum sampling rates are determined by the depth of the charcoal bed as well as the grain size. A typical charcoal cartridge for environmental air sampling purposes will range from about a half inch (12 mm) in diameter to about two inches (47 mm) with a bed depth in the same range. Suitable sampling rates are in the range of 1–3 cubic feet per minute (30–90 liters per minute), with the larger cartridges able to maintain capability with higher flow rates.

A standard method for sampling radioiodines recommends the use of a high efficiency filter for removal of iodine bearing particulates followed by charcoal cartridge 5/8 inch (16 mm) in diameter and 1.5 inches (38 mm) deep, containing 3 g of 12 to 30 mesh KI activated charcoal, for removal of the gaseous radioiodines (APHA 1972). At a flow rate of 0.01 m^3/min for a seven day period,, this method provides a sensitivity of 10^{-13} $\mu Ci/cm^3$ (0.0037 Bq/m^3) for ^{131}I, based on gamma spectroscopy centered around counting of the 364 keV peak associated with the decay of this nuclide in a NaI(Tl) well counter. Greater sensitivity can be achieved by longer counting times or through the use of other methods, one of which incorporates a specially prepared filter impregnated with barium hydroxide-iodide-iodate, and can reduce the lower limit of detection by an order of magnitude (Baretta, Chabot, and Donlen 1968).

Silver zeolite molecular seives have been used with considerable success to sample gaseous radiodines. Control of flow rate is important with these types of sampling media as it is with charcoal; collection efficiencies usually are somewhat lower than with charcoal, but retention appears to be good under most conditions. Caustic solutions, through with the air is drawn or 'bubbled' have also been used, but in general these do not collect any organic iodine.

Because of its low specific activity, sampling and analysis for ^{129}I poses special problems in that sensitivity is inadequate by normal radiological counting procedures. This nuclide is usually determined by neutron activation analysis following chemical separation of the iodine, or by liquid scintillation counting following solvent extraction.

Airborne Tritium

Tritium in air is usually found in the form of tritiated water vapor (HTO) or in the elemental form as HT or T_2. Occasionally, very small or trace amounts of organically bound tritium may be found. Perhaps the simplest and most direct method of measuring tritium or other gaseous radioactivity in air is with a flow through ionization chamber. This method is among the earliest used for tritium in air, having been developed in the early 1940's by W. R. Kanne at the Savannah River Plant (Hoy 1961). Air is drawn through the ionization chamber at a constant known rate after being filtered to remove the particulate radioactivity, where the disintegration of the remaining gaseous activity produces a current flow.

Although theoretically possible, the method is not suited to most environmental monitoring situations because of its inherent low sensitivity for tritium which is a weak beta emitter and hence does not produce a very large amount of current per disintegration. Large chambers are required to achieve adequate sensitivity, even with sensitive ammeters; chambers with volumes as large as 51.5 liters have been used, and have reported minimum detection capabilities of 2×10^{-6} $\mu Ci/cm^3$ (74 KBq/m^3) (Marter and Patterson 1971). Flow through ionization chambers are also costly and delicate pieces of instrumentation, which further decreases their practicability. They are non-specific in their response, so the composition of the gaseous radionuclides must be known in order to achieve a suitable calibration. Even so, they are responsive to external ambient fields, including background radiation. In general, however, flow through ionization chambers are suitable for use as effluent monitors or occupational monitoring where the concentrations of tritium or other radioactive gases of interest are higher and are well above the level of natural radon in air.

A considerable improvement over the flow through ionization chamber is a proportional counter. Being a pulse counter rather than a mean level current measuring device, electronic discrimination can be used to enhance detection capability. Special counting gases such as methane are required, which are inconvenient and add to the cost of measurement.

However, minimum detectable levels of an order of magnitude lower than those achievable with ionization chambers are readily achievable, even in gamma backgrounds of a few mR/h. One specially designed system not only can dicriminate against ambient gamma levels to 3 mR/h, but compensates also for other beta emitting gases (Block, Hodgekins and Barlow 1971). This system uses a dual proportional counter setup, with one detector insensitive to tritium. The reported minimum detection level is 10^{-6} $\mu Ci/cm^3$ (37 KBq/m^3).

Tritium is usually sampled by removal of the water from the air. As elemental hydrogen oxidizes rather rapidly in air, sampling for elemental tritium is often ignored, but may be accomplished by oxidizing the hydrogen and collecting the resultant water. A similar procedure is used for organically bound tritium. Water vapor or moisture may be removed from air by a variety of methods. Most commonly used are dessicants, with silca gels being the preferred. A recommended standard method calls for the use of a silica gel column 12 inches (300 mm) in length and 1.25 inches (32 mm) in diameter through which air is drawn at a flow rate of 0.1 to 0.15 liters per minute (APHA 1972). The moisture thus collected is reevolved in the laboratory, collected, and the tritium content determined by liquid scintillation techniques (Moghissi et al. 1969). This method, of course, excludes the collection of elemental gaseous tritium, and has a lower detection limit of $< 10^{-11}$ $\mu Ci/cm^3$ (0.37 Bq/m^3). Sampling may performed for as long as two weeks without exhausting the silca gel column. The silca gel can also also be counted directly without distillation, with a reported lower detection limit only slightly greater than the distillation method (Osloond et al. 1972).

Collection and differentiation of elemental tritium and tritium in the form of water vapor can be made by a two stage sampling procedure in which the air is first passed through a dessicant which removes the airborne tritiated water. The remaining elemental tritium in the air is then oxidized by spark or catalytic means and passed through a second column of dessicant. This method has been successfully used for concentrations of tritium in the range of 10^{-6} $\mu Ci/cm^3$ (37 KBq/m^3) (Corley et al. 1981).

Various other water absorbing media have been tried, but none has achieved the acceptance nor found as suitable as silica gel, although anhydrous $CaCl_2$ has been used with some success. Methods have also been developed in which the tritiated water is allowed to condense on cold surfaces or even frozen out of the air. Again, such methods do not sample the elemental tritium which may also be present. Uranium sponges are useful for removal of elemental tritium, which is driven out of the sponge by heating and counted with either a flow through counter or oxidized and counted by liquid scintillation techniques. Separation of tritiated water vapor from air has also been accomplished with a bubbler technique, which selectively discriminates against gaseous tritium noble gases (Osborne 1972). Other techniques based on combustion have also been developed for determination of gaseous tritium.

Liquid scintillation techniques are especially useful for the detection of tritiated water (NCRP 1978), and involve mixing of a small amount of the water to be examined with a somewhat greater amount of fluorescent material in an organic solution. Toluene solutions of PPO (2,5-diphenyloxazole) or TP (p-terphenyl) are commonly used. The resultant cocktail is placed inside a lighttight detector where the scintillations produced by the decay of the tritium are detected. Typically, a liquid scintillation cocktail will contain 4 ml of water and 11 ml of scintillation solution; a lower limit of detection of a few hundred pCi/l (about 5 KBq/m³) can be achieved with practical counting times. If greater sensitivity is desired, as may be the case in geological studies, electrolytic tritium enrichment may be performed, and lower detection limits of a few pCi/l (<100 Bq/m³) may be routinely achieved with reasonable counting times.

The Noble Gases

The radioactive noble gases include several isotopes each of radon, xenon and krypton. The latter two gases are anthropogenic and associated with the operation of nuclear reactors or the detonation of nuclear explosives, although there is a limited usage of radioxenons for medical diagnosis and [85]Kr in certain consumer products. Grab sampling techniques for these gases include collection in a previously evacuated sampling flask, or collection in plastic bags or similar devices. Low volume samples may also be taken with the aid of a hand pump, with large volume grab samples collected in metal containers under a pressure of 10–30 atmospheres (Johns 1974). The samples are usually examined by direct counting trechniques, including gamma spectroscopy. All such samples are subject to interference from the presence of natural radioactivity in air.

Continuous sampling for radioxenons and radiokrypton is accomplished by adsorption on activated charcoal or by cryogenic means. The gases can be distilled out of the collecting medium and dissolved in organic solvents containing liquid scintillation compounds, and counted directly in this manner. Direct counting techniques may be accomplished for gamma emitting isotopes of krypton and xenon. A technique for ^{85}Kr has been developed in which a sample of the gas is placed in a sealed 4 ml glass vial at slightly below atmospheric pressure and beta counted directly with a solid plastic scintillator (Sax, Denny, and Reeves 1968). The reported sensitivity of this method is 10^{-7} μCi/cm^3 (3.7 KBq/m^3), which may be insufficient for some environmental applications.

Flow through chambers are also used for continuous monitoring of the radioactive noble gases. Use for this purpose is similar to that of tritium monitoring, although sensitivity is generally greater because of the increased amount of energy deposited per disintegration. The same flow-through ionization chamber instrument used for measuring tritium in air can also be used for noble gases if suitably calibrated.

Numerous methods exist for measurement of environmental levels of radon in air, and may basically be divided into three classes:

1. Measurement of radon concentrations.
2. Measurement of concentrations of radon daughters.
3. Working level determinations.

The Lucas cell is one of the oldest and still among the most popular techniques for the measurement of radon gas concentrations (Lucas 1957). Basically, the Lucas cell is an evacuated flask internally coated with ZnS(Ag) scintillation material. The flask stopcock is opened in the atmosphere to be sampled, allowing the air to enter the evacuated flask through a filter that removes the radon daughters attached to dust. The flask is then resealed and the scintillations produced on the inner walls by the decay of radon are counted directly with the aid of a photomultiplier. Typical Lucas cell volumes are 100 to 200 ml, making these cells convenient for grab sampling in the field. Sensitivity is relatively great, and can be enhanced by extgended counting times; also, the sample need not be examined immediately if suitable corrections for the decay of ^{222}Rn are made.

Another widely used technique is the two-filter method, which utilizes a cylinder with a filter at each end through which the gas to be sampled is pumped, typically for 5–10 minutes at a flow rate of about 10 liters per minute (Thomas and LeClare 1970). The first filter removes all the particulate activity but allows the radon to pass through the cylinder to the second filter, which collects the ^{218}Po daughter produced in the cylinder by the decay of ^{222}Rn during its passage. Excellent sensitivity can be achieved with this method, with sensitivity related to the size of the cylinder, flow rate, and counting time. Typically, cylinders are 1 to 5 feet long (0.3–1.5m) and a few inches ⌐50mm in diameter.

Continuous measurement of radon concentration can also be made with flow through chambers of various types; the air is always filtered to remove particulate activity prior to entering the chamber. Activated charcoal samplers of various types are also used, and are equipped with prefilters to remove particulates. A standard method uses cooled charcoal to collect the sample, which may then be treansferred to a Lucas cell or directly counted with quantitative spectroscopy techniques for radon daughters, and the radon concentration appropriately calculated (APHA 1972). Other standard methods involve collection of radon daughters and direct alpha, beta, or simultaneous alpha-beta counting, and require knowledge or assumptions regarding the state of equilibrium throughout the time of sampling (APHA 1972).

Sampling of radon daughters in air is normally accomplished by filration, as the daughters attach to particle nuclei in air. Various counting techniques, including direct alpha, beta, and quantitative gamma spectroscopy can be used, but all have the limitation of the necessity for correction for radioactive decay, which may be difficult or impossible for some of the daughters because of their exremely short half lives. Estimates of daughter concentrations can be made with the aid of alpha to beta counting ratios, or with alpha spectroscopy.

Radon daughter concentrations in air can be measured with the Kusnetz method, which is also the standard method for working level measurement (ANSI 1972). This method was originally developed in the 1950's by Howard Kusnetz, an engineer with the U.S. Public Health Service, and basically involves obtaining a grab sample of filtered air over a duration of 5 or ten minutes at a sampling rate of 2 to 10 liters per minute. The sample is counted after a delay of 40 to 90 minutes (typically 40) and the disintegration rate used with a published scale factor to compute the WL in the original sampled air. This method is reasonably accurate, and is independent of the state of equilibrium. The method, however, is relatively insensitive, having a lower level of detection of about 0.3 WL.

A modified Kusnetz method has been developed that basically uses the same sampling method, but permits counting 5 minutes after sampling (Rolle 1972). With a 10 minute counting time, a lower level of detection of about 0.01 WL is achievable. Total uncertainty in the measurement is in the range of ± 20%. The modified Kusnetz method and later refinements serve as the basis of several "instant" working level meters and radon progeny measurements (Borak et al. 1982). Many working level meters have been developed and are commercially available, using a variety of detection methods including TLD, track etch methods, electronic detection, and various types of dust collection methods (LBL 1980). Even photographic film has been used to monitor WL and radon concentrations in air (Geiger 1967).

References

1. American Conference of Governmental Industrial Hygienists (ACGIH). 1972. *Air Sampling Instruments for Evaluation of Atmospheric Contaminants*, 4th Edition, Washington, D. C.
2. American Public Health Association (APHA). 1972. *Methods of Air Sampling and Analysis*, Intersociety Committee for a Manual of Methods for Ambient Air Sampling and Analysis, Washington, D. C.
3. American National Standards Institute (ANSI). 1973. *Radiation Protection in Uranium Mines*, American National Standard ANSI N 13.8–1973, New York.
4. Baretta, E. J., G. E. Chabot, and R. J. Donlen. 1968. "Collection and Determination of Iodine–131 in the Air", *Amer. Ind. Hyg. Assoc. J.* 29:159.
5. Bellamy, R. R. 1974. "Elemental Iodine and Methyl Iodide Adsorption on Activated Charcoal at Low Concentrations" *Nuclear Saf.* 15:711.
6. Block, S., D. Hodgekins, and O. M. Barlow. 1971. "Recent Techniques in Tritium Monitoring by Proportional Counters", Lawrence Livermore Laboratory Report UCRL-51131.
7. Borak, T. B, et al. 1982. "Evaluation of Recent Developments·in Radon Progeny Measurements", in *Radiation Hazards in Mining*, M. Gomez, Ed., Society of Mining Engineers, New York, pp. 419–425.
8. Corley, J. P. et al. 1981. "A Guide for: Environmental Radiological Surveillance at U.S. Department of Energy Installations", U. S. Department of Energy Report DOE/EP-0023.
9. Denham, D. H. 1982. "Sampling Instruments and Methods", in *Handbook of Environmental Radiation*, A. W. Klement, Jr., Ed., CRC Press, Boca Raton, FL, pp. 129–153.
10. Geiger, E. L. 1967. "Radon Film Badge", *Health Phys.* 13:407.
11. Hoy, J. E. 1961. "Operational Experience with Kanne Ionization Chambers", *Health Phys.* 6:203.
12. Johns, F. B. 1974. "Sampling and Measurement of the Noble Gases in the Environment", presented at Second Atomic Energy Commission Environmental Surveillance Workshop, cited in Corley et al., p. 4.16.
13. Kathren, R. L. 1974. "Plant and Personnel Radiation Monitoring", in *Nuclear Power Reactor Instrumentation Systems Handbook*, Vol. 2, J. M. Harrer and J. G. Beckerley, Eds., U.S. Atomic Energy Commission Technical Information Center Report TIC-25952-P2, pp. 57–74.
14. Lawrence Berkeley Laboratory (LBL). 1983. *Instrumentation for Environmental Monitoring*, Vol. 1, Radiation, Second Edition, John Wiley and Sons, New York.

15. Lucas, H. F. 1957. "Improved Low-Level Alpha Scintillation Counters for Radon", *Rev. Sci. Inst.* 28:680.
16. Marter, W. L. and C. M. Patterson. 1971. "Monitoring of Tritium in Gases, Liquids, and in the Environment", *Trans. Amer. Nucl. Soc.* 14:162.
17. Moghissi, A. A., et al. 1969. "Low Level Counting by Liquid Scintillation I. Tritium Measurement in Homogeneous Systems", *Int. J. Appl. Radiation Isotopes* 20:145.
18. National Council on Radiation Protection and Measurements (NCRP). 1976. *Environmental Radiation Measurements*, NCRP Report No. 50, Washington, D. C.
19. National Council on Radiation Protection and Measurements (NCRP). 1978. *A Handbook of Radioactivity Measurements Procedures*, NCRP Report No. 58, Washington, D. C.
20. Osborne, R. V. 1972. "Monitoring Reactor Effluents for Tritium: Problems and Possibilities", Atomic Energy of Canada Limited Report AECL-4054.
21. Osloond, J. S., et al. 1972. "A Tritium Air Sampling Method for Environmental and Nuclear Plant Monitoroing", in *Tritium*, A. A. Moghissi and M. W. Carter, Eds., U.S. Environmental Protection Agency, Washington, D. C.
22. Rolle, R. 1972. "Rapid Working Level Monitoring", *Health Phys.* 22:233.
23. Sax, N. I., J. D. Denny, and B. R. Reeves. 1968. "Modified Scintillation Counting Technique for Determination of Low Level Krypton-85", *Anal. Chem.* 40:1915.
24. Thomas, J. W., and P. C. LeClare. 1970. "A Study of the Two Filter Method for Radon-222", *Health Phys.* 18:113.

CHAPTER 16

Terrestrial and Aquatic Sampling and Analysis

Terrestrial Soils and Sediments

Evaluation of the radioactivity content of terrestrial media is useful for defining both direct and indirect pathways as well as for evaluation of the accumulation and long term trends of radioactivity in the environment. Terrestrial media of specific interest include both natural and cultivated vegetation, domestic and wild animals, animal products such as milk used for food, and soils and sediments. Foodstuffs and other terrestrial media are of particularly great value in environmental surveillance programs as they provide a direct basis for assessment of the radiation dose to people from the operation of a nuclear facility or from fallout from nuclear explosives.

Sampling and analysis of soils and sediments is of value primarily in determining long term trends and accumulation of long-lived radionuclides in the environment. Soil analysis is a poor technique for assessment of small incremental increases in radioactivity and is seldom used for this purpose except in special cases involving unexpected releases of activity or for specific research studies. Sampling is thus performed relatively infrequently, generally annually or less often; for nuclear power plants, for example, the U.S. Nuclear Regulatory Commission recommends triennial sample collection (USNRC 1975).

As soil radioactivity concentrations are indicative of the distribution and accumulation of radioactivity in the environs of a nuclear facility, samples are commonly collected from a regular pattern such as a rectangular grid or radial pattern at least close in to the facility. Further from the source of the radioactivity, emphasis is placed on selecting sites within the predominant downwind sectors, and in particularly at those distances which are expected to have the highest values of χ/Q based on atmospheric dispersion calculations. Thus, wind roses and diffusion calculations provide useful and in many cases essential guidance in selecting appropriate soil sampling locations. For most sites, soil sampling will be accomplished within a ten mile radius, but individual site parameters including meteorology may indicate sample collection at locations at considerable distances from the potential point of release. This is especially true of facilities that release activity via tall stacks, or from global or generalized releases such as those associated with weapons tests.

Both surface and core type soil samples are used. Surface samples have the limitation that the radioactivity may have leached down into the ground and hence be missed. However, they are advantageous from the standpoint of providing indication of recent deposition or chemically or physically bound materials. Surface soil or sediment samples are ordinarily collected with a "cookie cutter" or similar rigid frame that encloses a known area and can be sunk into the soil to a selected depth. Generally an area of 460–930 cm^2 (0.5 to 1 square foot) is sampled to a depth of 1 cm, providing a sample with a mass of about a kilogram; a standard method for plutonium in soil has been proposed based on a ring 12.7 cm in diameter and 2.5 cm deep (USNRC 1974). This size sample can be readily analyzed by gamma spectroscopy and is also sufficient to provide an adequate detection level with most radiochemical procedures. If the location selected for sampling is covered with vegetation, the sampling depth may be extended to improve the representativeness (NESP 1975). Typically, depths of 1, 2, 2.5 and 5 cm are used, with the deeper levels selected for vegetated soils. In sparsely vegetated soils, valid data can be obtained by sampling only the top centimeter (Corley, Robertson, and Brauer 1971).

Cores and depth profiles not only provide considerably more information than surface samples but are more reproducible as well (Harley 1972). A standard method of obtaining a core sample calls for a composite of ten or more individual plugs or cores about 8.9 cm (3.5 in) in diameter obtained to a depth of 30 cm (1 foot). The sample from the upper 5 cm of soil is usually analyzed separately, providing an indication of the surface concentration. The soil from the remaining depths is composited to produce a sample several tens of kilograms in weight, and an aliquot submitted for analysis. Depth profiles may be obtained in a similar manner, but care must be taken to ensure that the soil depth is accurate. A recommended procedure is to dig a small trench, smooth the sides, and use a "cookie cutter" to remove an appropriate sample at the depths desired. Metal jigs which permit slices to be obtained from specific depths are also used (Harley 1972).

The ease of soil or sediment sample collection, as well as its representativeness, is to a large extent dependent upon the characteristics of the soil. Sandy unvegetated soils and sediments are most easily sampled, while rocky, vegetated soils present the most difficulty. Hard pan adobes may also present physical difficulties, and may require moistening to obtain the sample. It may also be necessary to screen or seive the sample to eliminate small stones and organic debris such as twigs and small branches and roots which may bias analytical results. Samples should not be taken from the identical specific location each time, except in the case of waterborne sediment depositions, for the sample collection procedure depletes the soil and alters the concentration. Hence, it is advisable to establish a sampling plot with an area of several square meters and remove a sample from a different area of the plot each time.

For determining long term trends from soil deposition measurements, analysis of gross radioactivity levels is of questionable value (Corley et al. 1981). Analysis of soil is thus performed by gamma spectroscopy or by radiochemical procedures for individual and specific radionuclides. Sample preparation for radiochemistry usually involves drying, ashing to remove organics, and leaching. Radionuclide concentrations are generally reported in units of μCi/g of dry weight or Bq/kg in the SI system.

Terrestrial Vegetation

Vegetation samples may be obtained to complement soil and sediment samples as a means of determining the environmental accumulation of radioactivity. Both foodstuffs and inedible plant materials may be sampled. Selection is based upon the characteristics of the plant; perennials and long lived plants such as trees are best suited to this purpose. Samples of growing portions of the plant—i.e. leaves and roots—rather than the woody portions will usually provide the desired information. Natural or inedible vegetation may be good indicators of specific aspects of environmental radioactivity. Mosses and lichens integrate deposition over a period of years and thus may be particularly useful for evaluation of fallout levels (Whicker and Schultz 1982). Spanish moss seems to be a natural filter for airborne particulates and so may be a useful complement to an air sampling program.

More commonly, vegetation is analyzed to monitor pathways and to determine doses to man and animals. For these purposes, foodstuffs are sampled. Selection is governed by several factors including the specific food chains and pathways to be monitored, the dietary habits of the pop-

ulation, and the availability of the required foodstuff. With regard to the
latter, sampling is frequently done seasonally of the edible portions, usually at the time of harvest and perhaps one or two times before. There
may be significant differences in the uptake and accumulation of radionuclides in different portions of the plant or in different types of plants.
Thus, the selection of plants to be sampled should include both root crops
and leafy vegetables as well as fruits and berries. The specific foodstuff
sampled as well as the frequency will be dependent on the pathways
under study.

Vegetation samples are ordinarily collected and analyzed on a seasonal
or quarterly basis, and should be obtained from locations near the point
of predicted maximum annual ground concentrations and from areas
which use water that may be potentially contaminated. Vegetation collected for long term trend determination or similar studies is usually collected annually during the growing season. Typically, one to several kilograms are required for radiological analysis. The vegetation may be
compressed into blocks or pellets and directly examined by gamma spectroscopy, or may be treated chemically for specific isotopic analysis. Ashing is usually required prior to radiochemical treatment, and freeze
drying has also been successfully used to remove the water from wet samples. As is true of all terrestrial samples, naturally occuring radionuclides
(e.g. ^{40}K, uranium, thorium and their daughters) raise the background
counting levels and obviate the use of gross activity measurements. Long
range depositions from nuclear facilities may also be masked by fallout
from nuclear weapons testing. Activity concentrations are customarily
reported in units of pCi/g dry or wet weight or Bq/kg in the SI units.

Terrestrial Animals

Both domestic and wild animals may be sampled in any environmental
surveillance program. Food animals are usually the most significant, and
thus most commonly selected for sampling. Other animals may be collected as indicators or accumulators of activity because they concentrate
certain radioelelements or because of their particular position in a food
chain, or their ranging habits or locale. Collections are usually made
from the highest trophic levels as the animals are larger, more easily
obtained, and more meaningful for dose assessment or as indicators.
However, these animals are relatively slow growing and hence may take
a long time to accumulate radioactivity. At least in the case of wild animals, they may forage over large areas, and this factor needs to be taken
into account when evaluating the choice of animal in relation to the

desired purpose of the sampling. Care should also be taken not to select rare or endangered species. Stomach content analysis can be a useful indicator of what the animal ingested prior to death.

Collection of wildlife may be accomplished by one or more of the following methods:

- hunting or trapping
- acquisition from hunters in season
- acquisition of road kills and poached animals from state agencies

Hunting or trapping requires the greatest effort but also is a more certain method of acuiring the requisite samples, particularly if the sample collections are to be made several times during the year. Domestic food animals may be obtained from local slaughterhouses or farmers or other producers. Aquisition from retail outlets is generally not done as the origins of the foodstuff may be uncertain. Eggs, if desired, are generally obtained directly from the producer.

Large animal samples usually consist of a few kilograms of lean muscle and specific organs where various radionuclides may concentrate. A sample size of 3.5 kg of edible portion is desirable, with no less than 220 grams (NESP 1975). Thyroid is commonly collected, along with liver and bone; thyroid is most useful for evaluating radioiodine distribution as this organ acts as a rapid concentrator of radioiodines. Bone concentrates radiostrontiums more slowly but is still useful for studying long term changes in the environment. Liver concentrates several radionuclides, including cobalt and plutonium. Separation of these organs from surrounding tissues must be carefully done, and is particularly difficult in the case of the thyroid. If significant amounts of fat or other tissues are included along with the organ of interest, concentrations calculated on the basis of weight will be erroneously low. If too much of the organ is dissected away in an effort to remove adherent unwanted tissues, the remaining sample may be of insufficient size to achieve a sufficiently small lower limit of detectability.

Sampling frequency is usually quarterly or semiannually, although eggs, if included in the program, should be sampled and analyzed more frequently, perhaps monthly. Analysis of edible portions requires preparation in a manner similar to that used for human consumption, such as the removal of inedible portions. Samples should be weighed soon after collection, and appropriately preserved until the analytical procedure is begun. Concentration is reported in units of activity per unit mass, typically μCi (or Bq)/kg of wet weight for muscle, and μCi (or Bq)/g for various organs.

Small animals are of limited value in most environmental surveillance programs, as these are usually not directly eaten by man. Their utility as trend indicators and concentrators of radioactivity is minimal because of their large variability in mobility, age and diet. Small rodents such as rabbits and field mice, however, are sometimes used as indicator species. These are generally analyzed whole since separation of muscle and various organs is very difficult and the sample size so small.

Milk

Several important radionuclides, notably the radioiodines and radiostrontiums, tend to accumulate in milk making sampling and analysis of this product an important part of many environmental radioactivity surveillance programs for nuclear power plants. Short-lived nuclides are of particular importance as milk is consumed relatively soon after it is taken from the animal. Milk samples should be collected from both dairy cows and goats within a radius of several miles of the facility, and should include the nearest source of the milk as well as the source with the maximum predicted air concentration from routine releases (Corley et al. 1981). In addition, fluid milk samples should be collected from nearby dairies to provide an indication of representative conditions within the milkshed.

If short-lived radionuclides such as the radioiodines are of concern, sampling may be required on a weekly or biweekly basis. For ^{90}Sr and other long-lived nuclides, sampling may be accomplished on a less frequent basis such as quarterly and still yield satisfactory results. As an adjunct to milk sampling, collection and analysis of the animal's food may also be done to give an indication of the origin of the radionuclides. To attain adequate sensitivity for ^{131}I—a lower detection limit of 0.25 pCi/l (9.25 Bq/m^3) is recommended by the U.S. Nuclear Regulatory Commission (USNRC 1973)—with reasonable counting times, a four liter sample is required. One liter may suffice for other radionuclides.

As milk is perishable, measures must be taken to preserve the sample unless delivery to the laboratory can be made within a suitable time period. Souring or curdling of milk can render the sample useless. Refrigeration and freezing are suitable techniques but require appropriate facilities which may not be readily available or convenient in the field. Also, the bulk and inconvenience of handling of many 4-liter samples of milk may be considerable. One technique that has been successfully used is to pour the milk through an ion exchange resin at the time of collection, thereby removing the radionuclides of interest and eliminating the possibility of spoilage.

Preservatives have also been used but these are generally not recommended. Formaldehyde is probably the most commonly used preservative and interferes with subsequent ion exchange onto synthetic resin as it promotes protein binding of inorganic iodine. Protein binding can be minimized or eliminated by addition of thiouracil (2-thio-4-oxypyrimidine) and stable iodine carrier to the milk. Other commonly used preservatives are citrate and sodium ethylmercurothiosalicylate; these have the same drawback as formaldehyde. The method of collection may also adversely affect the representativeness of the sample. Clean polyethylene containers are preferred; iodine and other radionuclides may plate out or bond with the walls of metal or glass containers, and these should therefore be avoided.

Radiological analysis of milk is usually accomplished by direct gamma spectroscopy or by radiochemical separation and counting. Determination of gross activity is seldom done and of little value. For iodines and other radionuclides, ion exchange methods can be used to remove the radionuclides in place of wet chemistry; protein bound radionuclides are not removed by the resin. With radioiodines, this is not a problem less than ten per cent of the iodine in milk is protein bound (Bretthauer, Mullen and Moghissi 1972). However, the fraction may be appreciable with other elements including strontium. Activity concentrations in milk are usually reported in units of pCi/l, or Bq/m^3 in the SI units.

An important factor in the radiological analysis of milk is the determination of stable calcium concentrations, which are more or less constant at about 1 g/l. Abnormal levels of calcium may be indicative of poor sampling technique or analytical errors or an abnormal or sick animal. Calcium levels are also necessary for determination of the observed ratio. Sampling and analysis of milk products such as cheese or condensed or powdered milk are seldom performed in operational surveillance programs in the vicinity of nuclear facilities, as the data obtained from analysis of fresh fluid milk is of more utility and can be used for prediction of activity concentrations in these products.

Water and Sediment

Water represents a significant series of exposure pathways as well as an important means of widespread dispersion of radioactivity in the environment. As water sampling and analysis for microbiological and chemical pollutants has been long established and is routinely carried out throughout the world, many standard sampling procedures and equipment can easily be adapted to environmental radioactivity purposes, including those of the American Public Health Association (APHA 1982), the American Society for Testing and Materials (ASTM 1983) and the Environmental Protection Agency (EPA 1974). Less standardization exists with respect to sediment sampling, although some procedures do exist and the recommended practices for soil sampling generally apply equally well to aquatic sediments (Harley 1972; Corley et al. 1981).

Samples are generally collected from both surface and subsurface (ground) waters in the vicinity of nuclear facilities. Relatively large samples are usually required to obtain adequate mixing, although much smaller samples are required for analytical purposes. Receiving bodies for radioactive or potentially radioactive discharges from nuclear facilities should be sampled both upstream and downstream of the facility. In addition samples should be collected within the mixing zone of the discharge plume. Good practice calls for sampling at several depths in large bodies of water in addition to the collection of sediment samples.

Sample frequency is dependent on the purpose of the program as well as the operating characteristics of the individual facility, if one is involved. Ground water sampling is usually of the grab type and is accomplished on relatively infrequently, seldom oftenere than quarterly. Monitoring of surface waters in the vicinity of so-called "dry" facilities—i.e. those facilities with little or no liquid effluents—needs to be done far less frequently than around facilities with continuous or frequent discharges. For the dry facilities, monthly or even quarterly grab sampling frequencies of surface waters may be adequate. For other facilities, continuous effluent monitoring or monitoring of nearby surface waters may be necessary, with integration of the activity concentration performed to obtain the total quantity released.

Various types of continuous water monitors have been devised to take a continuous representative aliquot (NESP 1975; LBL 1983). Most commonly these use small diversions from which a small sample is periodically removed and composited for analysis. Continuous water samplers may be used in conjuction with detectors and thus become water monitors. Off-line monitors and samplers are those in which a small constant fraction is diverted from a larger stream. These are most commonly found on effluent streams rather than used for environmental water bodies.

Composite samples collect small incremental samples periodically and combine them into a single sample for analysis. Thus they are most useful for providing time-averaged concentrations over the sampling period. Composite samplers may be flow proportional, sequential, or filtered. Many nonflow proportional samplers are available but are undesirable as they do not necessarily produce representative concentration values in the sample. Some samplers pass the water directly through ion exchange columns or filters which extracts the radioactivity. These samplers have the advantage of greatly reducing the bulk of the sample, and permit larger volumes to be sampled. Anion and cation columns may be used in series, and may be preceded by a filter to remove solid material. Tritium, of course, will not be removed by this technique and must therefore be sampled for by direct means.

If continuous water sampling is impractical or unecessary, weekly sampling frequencies may prove adequate for most purposes. The sample should be collected in a clean polyethylene container to avoid loss of the radioctive contaminants by plateout. When obtaining surface water samples, the container should be rinsed with the water to be sampled a few times prior to the actual collection of the sample. Care must be taken to avoid cross contamination not only from unclean sampling bottles but also from other sources; even a tritium activiated luminous watch has been implicated in sample contamination (Denham 1982).

Dissolved and suspended solid fractions should be separated and analyzed independently. The amount and radioactivity content of the suspended solids is useful for evaluation of deposition in aquatic sediments, and the dissolved solid fraction may of considerable interest in irrigation and drinking waters. Sampling should include both raw and treated waters; the latter may be greater altered with respect to radioactivity content by the water treatment process and thus reliance on these waters alone may bias the data.

To inhibit biological growth and minimize plateout of activity on the container walls, a few ml of a concentrated acid such as H_2SO_4 may be added per liter of sample. However, addition of acid can oxidize iodide to iodine which may volatilize and result in an artificially or erroneously low analytical result; for this reason, samples should not be acidified if they are to be analyzed for radioiodines. Acidification may also result in quenching of organic scintillation compounds and hence adversely affect subsequent tritium or ^{14}C analyses. Refrigeration is also effective at minimizing biological growth. Filtration to remove suspended solids in the field shortly after collection is also effective in controlling biological growth but may alter the character of the sample. Activity concentrations in water are typically reported in units of pCi/l, or in the SI units of Bq/m^3.

Aquatic sediment sampling is accomplished with dredges or scoop samplers of various types. These most commonly are of the clamshell type which can be dropped from the deck of a boat to the bottom of the water body and close upon impact collecting a sample. Shoreline samples can be easily collected with standard soil sampling methods at low water, or with a trowel or similar device.

Aquatic Biota

Aquatic biota include fish, shellfish, and plankton. These are important constutents of the food chains to man and in addition may provide important information regarding the movenment and concentration of radionuclides in the environment. For most aquatic bodies, fish represent the highest trophic level and so are desirable from an environmental sampling standpoint. They are also relatively large and hence sufficient sample can be readily obtained. Fish collections should be representative of local species and, if used for environmental dose assessment, should be typical of the principal catch in the area. Samples may be obtained directly by netting, pole fishing, or electric shock methods, or indirectly by purchase from local markets or sports fishermen. In rivers, resident fish are preferable to anadramous fish, which because of their short residence time and limited diet while migrating may not be typical of the aquatic body from which they were obtained.

Unless short-lived nuclides are of concern, fish sampling may be accomplished relatively infrequently, typically no oftener than quarterly and perhaps as rarely as annually. Larger fish should be separated into muscle, bone and viscera for analysis; smaller fish are analyzed whole. Samples may oridinarily be composited or data combined without adverse effect.

Shellfish may be collected more frequently than fish because they are filter or bottom feeders and because of their propensity to concentrate radionuclides. However, monthly or quarterly collection frequency is usually adequate. Unlike fish, shellfish are stationary and hence provide a more specific picture of the microenvironment in which they reside. In addition to concentrating radionuclides, shellfish may have aquatic sediment in the gut or within the shell. Separate analysis is usually performed on the meat, fluid, gut, and shell.

Plankton and aquatic flora represent the lower trophic levels in the aquatic ecosystem. In general, plankton and algae are excellent indicator organisms and concentrators of such nuclides as ^{58,60}Co, ^{65}Zn, and ^{137}Cs. Algae and plankton are usually sampled by filtration methods, with samples collected on a semiannual or quarterly basis, and perhaps more frequently during periods of higher discharge, at points both upstream and downstream of the discharge point. Because of the high water content of aquatic flora and plankton, relatively large samples are required. Aquatic plant sampling is infrequently done, and samples of large sessile plants are preferred.

Sample Preparation and Analysis

Unless the sample is to be analyzed by direct radioactivity counting techniques as it comes from the field, preparation is required to concentrate and extract the radionuclides of interest and render the sample in a more suitable form for measurement. Sample preparation is also required to reduce the self-absorption of alpha and beta radiation and low energy photons (NCRP 1976). Biological samples require destruction of the organic matter which may be accomplished by dry or wet ashing techniques. Rapid dry ashing is usually not performed because of the large losses of volatile and semivolatile radionuclides at the elevated temperatures required; for example, 93% of the polonium in muscle is volatilized and lost by heating at 300°C for 24 hours. Reducing the temperature to 200° reduces the loss to 30%, while no loss occurs at 100° (Martin and Blanchard 1969).

Slow heating, also a dry ashing technique, is therefore preferred. It involves heating the sample in an open container for relatively long periods, generally 24 hours or more at temperatures ≤ 100°C. Slow heating minimizes losses from spatter and produces a relatively soluble ash. It may be carried out with a hot plate or by heating the sample from overhead with an infrared heat lamp; the latter technique is frequently used to evaporate small water samples in the planchet that will be used for counting. Volatile elements such as iodine, cesium and other nonmetals may be driven off and lost in the ashing process, producing erroneously low analytical results.

The problem of volatility is reasonably well resolved by the wet ashing procedure, which involves oxidation in solution, usually with concentrated nitric acid. This technique is most useful with small samples.

Hydrogen peroxide—ferrous ion solutions may also be used, and provide good and rapid results for large samples (Sansoni and Kracke 1971). Wet ashing techniques should be used preferentially if radioiodines are present.

Decomposition of organic material can also be accomplished by the use of electrically excited oxyen passed over the sample at low pressures (Gleit and Holland 1962). Oxidation generally occurs at temperatures below 150°. The method, however,is amenable only to small samples and is slow, costly, and requires constant attention and stirring as the reaction occurs only on the surface.

Reduction of sample volume by removal of the water from wet samples is frequently accomplished for the same reasons as digestion of organic matter. Rapid and slow heating of the sample are the most common methods, and carry with them the same limitations as with digestion of organic matter. Freeze drying represents a more desirable alternative and preserves volatile radionuclides. Generally, the dry weight is obtained after the removal of the water and is about one-tenth that of the wet weight. Dry weight is a more exact and reproducible quantity, and hence is preferred to wet weight for reporting of radionuclide concentrations. Wet weight is highly variable and a function of the sample handling procedure after collection, the relative humidity and time after collection. In some biota, the wet weight can change by as much as a factor of two between the time of collection and weighing in the laboratory prior to analysis; the dry weight will change only slightly, if at all.

The next sample preparation step is the dissolution of inorganic residues. This is readily accomplished, particularly in alkaline samples by treatment with nitric or hydrochloric acid. Siliceous samples may require treatment with hydroflouric acid. If acid dissolution is not readily accomplished, acid-fusion procedures may be required. Insoluble oxides and other insoluble metal compounds may be fused with the pyrosulfates of sodium and potassium to yield a sulfate mixture soluble in water or dilute acid. When fusions are performed, the type of container used is important as the acid may attack and dissolve the walls of the container. Platinum, although expensive, is best for this purpose, and minimizes contamination of the sample. Monel metal and stainless steel containers are also suitable. Dissolution of organic residues is generally carried out at high temperatures, and if done in open containers, volatile radioelements may be lost. Carrier is frequently added during the dissolution process.

Certain samples may require minimal preparation. Milk and other liquids, for example, may be poured directly through ion exchange columns

TABLE 16-1   Typical Tracers Used
in Environmental Radiochemistry

Nuclides	Tracer
^{89,90}Sr	^{85}Sr
^{224,226,228}Ra	^{225}Ra
^{230,232}Th	^{234}Th
^{234,235,238}U	^{233}U
^{238,239}Pu	^{236,242}Pu
^{241}Am	^{243}Am

and the resin removed for laboratory counting. Vegetation may be allowed to air dry and may be compressed into blocks for direct counting. Water may be added to liquid scintillation cocktail and counted directly. Air filters require no special treatment and may be counted directly.

Special preparation is required for some samples. The lower level of detection for tritium in water may be greatly reduced by electrolytic enrichment procedures. Soil samples typically require a lengthy and slow drying period. Dissolution of air sample filters may be desired and may require special ashing procedures to eliminate the filter while retaining the adherent dust. The treatment may also be determined by the specific nuclide(s) present, and may be as simple as adding appropriate tracers or carriers. Some of the more commonly used tracers are shown in Table 16-1.

Radiochemical Separation and Counting

Many different types of radiochemical procedures are used to prepare environmental samples for the final step of counting to determine the level of activity. The choice of procedure is not always straightforward or obvious and samples may sometimes be divided and analyzed by two or more different techniques. The basic procedures are:

1. Ion exchange
2. Microchemistry with tracers
3. Complex formation

4. Coprecipitation
5. Solvent extraction
6. Electroplating and electrodeposition

Ion exchange is among the most popular radiochemical techniques and is widely used with liquid samples. A large measure of its popularity is attributable to its simplicity and excellent removal of numerous radioactive species. Judicious selection of the ion exchange resins is required, particularly if sea water is to be treated. Microchemistry with tracers is gaining popularity, particularly for the analysis of low levels of actinides and transition elements.

Among the oldest and probably still the most widely radiochemical separation procedure is coprecipitation, which is potentially usable for any element that forms an insoluble compound. One common technique is the $BaSO_4$-$Fe(OH)_3$ method which is useful for heavy elements (Sill and Williams 1971, NCRP 1976). In the presence of potassium, $BaSO_4$ will precipitate from a sulfuric acid solution and carry with it all large trivalent and tetravalent cation and ferric hydroxide is used to carry metallic cations that form insoluble carbonates and precipitates. The resultant precipitate can be easily removed and transferred to a counting planchet or can be left in a test tube or centrifuge tube after removal of the supernatent and counted in a well scintillator or GeLi detector. This is but one of many coprecipitation techniques.

Solvent extraction techniques are most commonly used for purification rather than for initial separation. This technique is adaptable to a wide variety of elements, and is frequently based on the use of chelating agents such as thenoyltrifluoroacetone (TTA) which extract various cations into the organic phase by forming water insoluble chelates. New solvent extraction techniques make use of organic phosphate esters such as di(2-ethylhexyl)phosphoric acid (HDEHP) which separates ions of like charge. If the HDEHP solution is poured through an inert column, the different ions are removed at various depths in the column and can be identified in this manner. This technique is known as extraction chromatography.

Electroplating and electrodeposition techniques are most commonly used for the alpha emitting actinides, but can be used with other metals as well. This technique produces a thin uniform layer of a the radionuclide on a precious metal disk. Such a sample is necessary for high resolution alpha spectrometry. The technique has relatively high recovery and results in a sample with minimal self-absorption.

No single technique can be used for all environemental samples, or even for a single radionuclide in different media. Table 16–2, taken from NCRP Report No. 50 (NCRP 1976), is a summary of typical radio-chemical and counting procedures for various radionuclides in environmental samples.

The end procedure is counting with one or more detectors to determine the radioactivity content of the sample. Gross activity measurements may be made with a variety of detectors and require relatively simple instrumentation, but are of limited value in most surveillance programs. Of more importance are techniques that permit the identification and quantification of specific radionuclides by their radiological properties. Most commonly used is spectroscopy which requires more sophisticated apparatus and data analysis but permits both qualitative and quantitative measurement of both alpha and photon emitting radionuclides. Beta spectroscopy is also done, as are absorption techniques to determine the beta end point or maximum energy and provide a basis for identification. Half-life determinations may also be made to aid or verify identificiation.

The minimum detectable activity (MDA) or minimum detectable concentration (MDC) is of prime importance since most environmental samples contain very small amounts of specific radionuclides. The MDA is determined by all the factors and conditions that influence the measurement (Colle et al. 1980). Thus it includes such factors as sample size, chemical yield, and the efficiency of the detector for the particle or photon being counted. High yield radiochemical separations are thus desirable since they ensure a greater amount of activity will be present. The amount of radionuclide can also be increased by obtaining and processing larger samples, but a practical limit is soon reached. The MDA or MDC can also be extended by increasing the counting time of the sample. This produces a lower limit of detection (LLD); the term LLD, frequently used incorrectly, refers to the ability of the counting system to detect a minimum amount or number of counts in a sample, and is statistically based. Much as Archimedes could have moved the earth given a big enough lever and a fulcrum, so can increasing counting time extend the LLD. However, with short-lived nuclides, the limitations of this method are obvious.

There is much confusion in the literature regarding the proper way to determine the LLD, and the definition of LLD or MDA will vary considerably depending on the individual authority. The LLD has been defined as ". . . the smallest amount of sample activity that will yield a net count for which there is a confidence at a predetermined level that the activity is present." (Harley 1972). This approach requires that the number of counts is sufficiently large to permit the use of Gaussian sta-

TABLE 16–2 Radiochemical and Counting Procedures for Environmental Samples.

Analyte	Sample	Preparation: Dissolution	Chemical Separation	Method of Measurement	Detection Limit[a]
^{7}Be, ^{24}Na, ^{38}Cl, ^{39}Cl	Precipitation, air	Filter paper—leach with acidic carrier solution	Cl separated as AgCl; ^{24}Na measured in residue; other radionuclides removed with cation resin; Be separated as hydroxide	Two parameter gamma-ray spectrometry	10 fCi l^{-1} (precipitation) 0.05 fCi m^{-3} (air)
^{14}C	Food, air	Combust, to CO_2	CO_2 purified by expansion, condensation, drying agents	Internal gas counter, anticoincidence shielded	5 pCi g^{-1} (carbon)
^{14}C	Water	Add oxalate carrier	C oxidized to CO_2 in acid solution, $CaCO_3$ collected	Beta counter (gas flow or liquid scintillation)	10–100 pCi (depending on C content)
^{24}Na	Precipitation	Add Na carrier	Na separated by ion exchange, hydroxides scavenged, Na measured in filtrate	β-γ coincidence counter	0.02 pCi l^{-1}
^{32}P	Sea water	Add PO_4^{-3} carrier	P oxidized with $HClO_4$, NH_4 phosphomolybdate extracted into isoamyl alcohol; precipitated as $MgNH_4PO_4$	Beta counter	1 pCi l^{-1}
^{55}Fe	Various		$Fe(OH)_3$ precipitated, dissolved, extracted, and added to scintillation mixture; or hydroxide dissolved in H_3PO_4-NH_4Cl and colorless complex counted	Liquid scintillation counter	0.5 pCi g^{-1} (iron)
^{131}I	Milk	Add HCHO and I carrier	I separated on anion-exchange resin, purified by extraction; PdI_2 precipitated	Beta counter	0.25 pCi l^{-1}
^{137}Cs	Milk, bone, vegetation ash, soil, water	Fuse—Na_2CO_3 or leach—HNO_3	Cs precipitated; Cs separated on NH_4 phosphomolybdate, purified by ion exchange, precipitated as Cs_2PtCl_6	Beta counter	1 pCi per sample

350

Nuclide	Sample	Pretreatment	Procedure	Counting	Sensitivity
^{137}Cs	Milk, water	—	Ion-exchange on K cobaltihexacyanoferrate	Gamma-ray spectrometry	0.01 to 10 pCi l^{-1} (varies with volume)
^{137}Cs	Sea water	—	Cs separated on NH$_4$ phosphomolybdate, silica gel mixture; SiO$_2$ dissolved with HF	Gamma-ray spectrometry	0.1 pCi l^{-1}
^{141}Ce ^{144}Ce	Precipitation	Add Ce carrier	Ce separated by oxidation-reduction and solvent extraction cycles	Gamma-ray spectrometry with Ge(Li) detector	1 pCi per sample
^{210}Pb	Bone	Ignite – dissolve in HBr	Pb extracted into quaternary amine; PbSO$_4$ precipitated and counted	Beta counter (^{210}Bi counted)	1 pCi per sample
^{210}Pb	Water, soil, ores	Treat with acids, fuse – KF and Na$_2$S$_2$O$_7$, dissolve – HCl	Pb and Bi extracted into diethyl dithiocarbamate; ^{210}Bi separated from Pb by extraction in dithizone	Beta counter (^{210}Bi counted)	1 pCi per sample
^{210}Pb ^{210}Po	Bone, tissue	Dissolve in mineral acids	^{210}Po deposited on Ag from HCl, ^{210}Pb determined by deposition of ^{210}Po after ingrowth	Alpha counter	0.02 pCi per sample
^{210}Pb ^{210}Po	Sea water	Add carriers	Pb and Bi carriers precipitated with CaCO$_3$, separated by hydroxide precipitation, ^{210}Po deposited on Ag for counting; ^{210}Pb determined by deposition of ^{210}Po after ingrowth	Alpha counter	0.1 pCi per sample

351

TABLE 16–2. (Continued)

Analyte	Sample	Preparation: Dissolution	Chemical Separation	Method of Measurement	Detection Limit[a]
^{226}Ra ^{228}Ra	Soil vegetation	Plant—dissolve in acids; soil—exchange with NH$_4$ acetate	^{225}Ra, ^{133}Ba tracers and Ba carrier added; Ra precipitated with BaSO$_4$, separated from Ba by ion exchange; electrodeposited for counting	Alpha spectrometry	1 fCi g^{-1}
^{228}Ra ^{228}Th	Vegetation, food, soil, bone	Ash—dissolve mineral acids; soil—fuse Na$_2$CO$_3$	Th extracted by ethylhexyl phosphoric acid, ^{224}Ra daughter removed with HNO$_3$ after ingrowth and counted; Ra purified as PbSO$_4$, precipitated with PbBr$_2$, ^{228}Ac daughter extracted and counted	Alpha and beta counters	1 pCi Ra, 20 fCi Th per sample
^{232}Th	Bone	Ignite at 600°C	Irradiate with neutrons, dissolve in HCL, separate ^{233}Pa by anion exchange	Gamma-ray spectrometry of ^{233}Pa	0.1 ng g^{-1}
Th	Various	Ash, dissolve with acids and	Coprecipitate with BaSO$_4$, dissolve in DTPA	Fluorescence of Th-morin complex	10 pg ml^{-1} 10 ng g^{-1}
^{90}Sr	Bone	Ignite, dissolve in HCl	^{90}Y extracted into ethylhexyl phosphoric acid; impurities removed by amine extraction	Beta counter	1 pCi per sample
^{90}Sr	Food vegetation, tissue milk, soil, water	Evaporate, ignite, fuse with Na$_2$CO$_3$; Soil—fuse with Na$_2$CO$_3$ or leach with NaOH, HCl; ^{85}Sr and Sr carrier added	Sr separated, purified by nitrate precipitations, ^{90}Y precipitated and counted	Beta counter	0.5 pCi per sample

Nuclide	Sample	Preparation	Method	Counting	Detection limit
^{90}Sr ^{137}Cs	Sea water	Acidify—HCl; added carriers	Cs removed by NH$_4$ phosphomolybdate, purified, precipitated as Cs$_2$PtCl$_6$; Sr precipitated, ^{90}Y daughter separated and counted	Beta counter	0.5 pCi per sample
^{99}Tc	Water	—	Tc purified by precipitation, solvent extraction; electrodeposited for counting	Beta counter	0.5 pCi l^{-1}
^{108m}Ag ^{110m}Ag	Biological Samples	Grind in blender with water	Slurry treated with NaOH-AgCN solution; Ag electroplated on Pt, dissolved, AgCl precipitated	Gamma-ray spectrometry (4π coincidence counter)	1 pCi per sample
^{129}I	Milk, Water	—	Milk—I first separated by anion exchange. Milk and water—I extracted into CCl$_4$, and toluene; and decolorized with UV light in presence of 2-methyl-1-butane	Liquid scintillation counter	0.3 pCi l^{-1}
Th	Various	Dissolve as necessary	Coprecipitated with Ca oxalate, purified by ion-exchange	Colorimetric, with thorin or other dyes	1 μg
th-Cf	Soil	Treat with acids, fuse—KF, Na$_2$S$_2$O$_7$, dissolve—HCl	Actinides coprecipitated with BaSO$_4$, separated by solvent extraction after adjusting oxidation states, electrodeposited for counting	Alpha spectrometry	1 fCi g^{-1}
U	Water	—	Extraction into methyl isobutyl ketone	Fluorescence of U in fused NaF	5 ng l^{-1}

TABLE 16-2. (*Continued*)

Analyte	Sample	Preparation; Dissolution	Chemical Separation	Method of Measurement	Detection Limit*
^{237}Np ^{239}Pu	Pitchblende	Dissolve in mineral acids	Np(IV) extracted into trifluoroacetylacetone, purified by anion exchange; Pu(III) coprecipitated with LaF$_3$ and UF$_4$, purified by anion exchange and electrodeposited	Np – mass spectrometry Pu – alpha spectrometry	1 fCi Pu g^{-1} (sample) 0.4 ag Np g^{-1} (sample)
Pu	Soil, air particulates	Soil—ash, extract with hot HCl; particulates—ash, dissolve—mineral acids and fusion	Pu separated and purified by cation and anion exchange; electrodeposited for counting	Alpha spectrometry	1 fCi g^{-1}
Pu	Soil	Fuse—Na$_2$CO$_3$, dissolve—HNO$_3$	Phosphates precipitated and Pu purified by anion exchange; electrodeposited for counting	Alpha spectrometry	1 fCi g^{-1}
Pu	Soil	Leach with mineral acids	Pu purified by anion exchange; electrodeposited for counting	Alpha spectrometry	1 fCi g^{-1}
Pu	Water, air, soil, food, vegetation	Treat with mineral acids	Pu purified by anion exchange; electrodeposited for counting	Alpha spectrometry	1 fCi g^{-1}

Copied from NCRP 1976 by permission of National Council on Radiation Protection and Measurement tistics, and in practice is good down to a few total counts. The LLD is basically an *a priori* estimate of the detection capabilities of a given measurement procedure obtained from knowledge of the background count and the parameters of the procedure (e.g. detection efficiency).(Currie 1968).

An early and still widely used technique considers the LLD as the count equal to two standard deviations of the background count (NCRP 1961). This provides the LLD at the 90% confidence level or the point at which a sample containing an amount of activity equal to the LLD would be detected (statistically) 90 per cent of the time. Since the standard deviation of a count is simply its square root, this value is easily calculated by taking the square root and doubling it. The effect of counting time on LLD is easily demonstrated by this technique. Consider a counting system with a background of 100 counts per minute. If a one minute count were taken, the standard deviation would be the square root of 100, or 10, and the LLD 20 counts. For a detector with a counting efficiency of 20 per cent, this would be $20/0.2 = 100$ disintegrations per minute (approximately 45 pCi or 1.7 Bq). If the counting time were increased to 100 minutes, the total background count would be 100 counts/min $\times$ 100 min $= 10,000$, and the square root would be 100. Thus, for a 100 minute count the LLD would be 200 counts, or 2 counts per minute. Again, taking a counter efficiency of 20 per cent, this corresponds to a LLD of 2 cpm/0.2 = 20 dpm or about 9 pCi (0.33 Bq).

Other more sophisticated techniques can be used to obtain the LLD. These are generally statistically based and permit the selection of two types of errors: false detection, in which a sample that actually contains activity below the detection limit is deemed positive, and false non-detection, which is the opposite case in which a sample with activity is deemed to be below the detection limit. These are termed α and β errors, respectively. If α and β are selected to be equal, the LLD can then be determined by (NCRP 1978)

$$LLD = k^2 + k\sqrt{2}\sqrt{B} \qquad (16-1)$$

in which k is the value of the standard normal deviate that is exceeded with a probability of and B is the true background count.

The LLD can also be estimated from (Colle et al. 1980)

$$LLD = \frac{k_\alpha s_0 + k_\beta s_D}{k_1 k_2} \qquad (16-2)$$

in which k_1 is the detector efficiency

k_2 is the yield per disintegration of the radiation being counted

k_α and k_β are the upper percentiles of the standardized normal variate corresponding to the preselected chance of falsely concluding that activity is present (α) and the preselected degree of confidence for detecting its presence (1 -β)

s_0 the estimated standard deviation of the net sample count when its mean is zero

s_D the estimated standard deviation of the net sample count when its limiting mean equals the LLD

The above equation (16–3) for the LLD can be used to obtain the MDA, which when putting in the various constants is calculated by (Harley 1972)

$$\text{MDC} = \frac{4.66\sigma_b}{EVYe^{-\lambda t}} \qquad (16\text{–}3)$$

in which σ_δ is the standard deviation of the background count (equal to the square root of the count

k is the activity (= disintegrations per time = 1 if activity is in becquerels and time is in seconds, or, 3.7×10^4 Bq/μCi)

E is the efficiency of the counting system

V the sample volume or mass

Y the chemical yield

λ the decay constant

and t the time between collection and counting.

Equation 16–3 provides the MDC (or MDA) at the 95 per cent confidence level (i.e. one chance in 20 of failing to detect a sample with activity at the LLD) and is widely used as well as recommended by the U.S. Nuclear Regulatory Commission for counting environmental samples (USNRC 1973).

In practice, it is useful to be able to know the interrelationships of the various factors affecting the MDA in order to make decisions on sample size and counting time required. The latter is particularly important, for long counting times greatly limit the number of samples that can be analyzed with a given counting system. In general, counting times of 100 minutes or so are acceptable, with 1000 minutes being the practical upper limit for routine environmental surveillance work. The following equation demonstrates these interrelationships:

$$A = a \frac{tC}{VfYe} \qquad (16-4)$$

In Equation 16–4, A is the target concentration, which is defined as the smallest concentration which will be quatitatively measured by the procedure which has a counting time of t, a minimum detectable count rate of C for a counting time of t, a sample volume or mass of V with a fraction carried into the purification steps of f, a chemical yield or recovery of Y, and a detector efficiency for the radiation of interest of e. The a symbolizes the appropriate unit conversion factor.

Clearly, the variables C, V, Y and e are affected by available equipment and sampling techniques and therefore are basic to determining the cost of performing an analysis. Several trials may be needed before the optimum combination is reached. Also, it should be noted that if C is set to equal the LLD, then A becomes the MDL. The equation, however, is valid for any count rate or concentration (Waite et al. 1980)

References

1. American Public Health Association (APHA). 1983. *Standard Methods for the Examination of Water and Waste Water*, Washington, D.C.
2. American Society for Testing and Materials (ASTM). 1983. *Annual Book of ASTM Standards: Water*, Philadelphia.
3. Bretthauer, E. W., A. L. Mullen, and A. A. Moghissi. 1972. "Milk Transfer Comparisons of Different Chemical Forms of Radioiodine", *Health Phys.* 22:257.
4. Colle, R., et al. 1980. "Reporting of Environmental Radiation Measurements Data" in *Upgrading Environmental Radiation Data*, Health Physics Society Committee report HPSR-1, J. E. Watson, Chairman, U.S. Environmental Protection Agency Report EPA 520/1–80–012, Washington, D.C., pp. 6–1–6–34.
5. Corley, J. P., et al. 1981. *A Guide for Environmental Radiological Surveillance at U.S. Department of Energy Installations*, U.S. Department of Energy Report DOE/EP–0023, Washington, D.C.
6. Corley, J. P., D. M. Robertson, and F. P. Brauer. 1971. "Plutonium in Surface Soil in the Hanford Environs", in *Proceedings of the Environmental Plutonium Symposium, August 4–5, 1971*, Los Alamos Scientific Laboratory Report LA–4756, pp. 85–90.
7. Currie, L. A. 1968. "Limits for Qualitative detection and Quantitative Determination", *Anal. Chem.* 40:586.
8. Denham, D. H. 1982. "Sampling Instruments and Methods", in *Handbook of Environmental Radiation*, A. W. Klement, Jr., Ed., CRC Press, Boca Raton, FL, pp 129–153.

9. Gleit, C. E. and W. D. Holland. 1972. "Use of Electrically Excited Oxygen for the Low temperature Decomposition of Organic Substances", *Anal. Chem.* 34:1454.

10. Harley, J. H., Ed. 1972. *HASL Procedures Manual*, U. S. Atomic Energy Commission Report HASL-300, New York. Updated Annually.

11. Lawrence Berkeley Laboratory Environmental Instrumentation Survey (LBL). 1983. *Instrumentation for Environemntal Monitoring. Volume 1, Radiation*, Second Edition, Wiley-Interscience, New York; also, *Instrumentation for Environmental Monitoring*, Report LBL-1, Vols. 1–4, Berkeley (1980).

12. Martin, A. and R. L. Blanchard. 1969. "The thermal Volatilization of Caesium-137, Polonium-210, and Lead-210 from *in vivo* Labelled Species", *Analyst* 94:441.

13. National Council on Radiation Protection and Measurements (NCRP). 1961. *A Manual of Radioactivity Procedures*, NCRP Report No. 28, National Bureau of Standards Handbook 80, U.S. Department of Commerce, Washington, D. C.

14. National Council on Radiation Protection and Measurements (NCRP). 1976. *Environmental Radiation Measurements*, NCRP Report No. 50, Washington, D.C.

15. National Council on Radiation Protection and Measurements (NCRP). 1978. *A Handbook of Radioactivity Measurements Procedures*, NCRP Report No. 58, Washington, D. C.

16. National Environmental Studies Project (NESP). 1975. *Environmental Impact Monitoring of Nuclear Power Plants. Source Book of Monitoring Methods*, Volume 1, Atomic Industrial Forum Report AIF/NESP-004, New York.

17. Sansoni, B. and W. Kracke. 1971. "Rapid Determination of Low-level Biological Materials Using Wet Ashing by OH Radicals", in *Rapid Methods for Measuring Radioactivity in the Environment*, Proceedings of a Symposium, International Atomic Energy Agency, Vienna, pp. 217–223.

18. Sill, C. W. and R. L. Williams. 1971. "Rapid Identification and Determination of Alpha Emitters in Environmental Samples" in *Rapid Methods for Measuring Radioactivity in the Environment*, International Atomic Energy Agency, Vienna, pp. 201 ff.

19. U.S. Environmental Protection Agency (USEPA). 1974. *Methods for Chemical analysis of Water and Wastes*, Report EPA 625/6–74–003, Washington, D.C.

20. U. S. Nuclear Regulatory Commission (USNRC). 1973. "Measurements of Radionuclides in the Environment. Analysis of I-131 in Milk", Regulatory Guide 4.3, Washington, D.C.

21. U.S. Nuclear Regulatory Commission (USNRC). 1974. "Measurements of Radionuclides in the Environment. Sampling and Analysis of Plutonium in Soil", Regulatory Guide 4.5, Washington, D.C.

22. U.S. Nuclear Regulatory Commission (USNRC). 1975. "Programs for Monitoring Radioactivity in the Environs of Nuclear Power Plants", Regulatory Guide 4.1, Revision 1, Washington, D.C.

23. Waite, D. A., et al. 1980. "Statistical Methods for Environmental Radiation Data Interpretation", in *Upgrading Environmental Radiation Data*, Health Physics Society Committee Report HPSR-1, U.S. Environmental Protection Agency Report EPA 520/1–80–012, Washington, D.C., pp. 7–1-7–19.

22. Whicker, F. Ward and Vincent Schultz. 1982. *Radioecology: Nuclear Energy and the Environment*, Vol. 1, CRC Press, Boca Raton, FL.

CHAPTER 17

Radiologic Age Dating

The idea of radiologic age dating is nearly as old as the discovery of radioactivity. Early in the twentieth century, Rutherford proposed that the quantity of helium in a radioactive mineral could be used to determine its age (Rutherford 1906). In 1907, his friend and associate, American chemist Bertram B. Boltwood, suggested that ultimate product of the decay of primordial radionuclides—lead—could be used to determine the age of geologic materials. The following year R. J. Strutt made a number of experiments to determine if the helium:radium ratio could be used to provide an estimate of the age of various geological strata, and observed minimum ages as great as 710 million years in samples of the mineral sphene obtained from the province of Ontario (Rutherford 1913). The technique using helium was more fully developed by Fritz Paneth in 1913, but was not fully successful as some of the helium gas in the samples may have escaped over the years, giving erroneously low results.

The use of radioactive decay for absolute age determination is based on three premises:

1. The geological or other material contained an initially pure parent either free from daughter activity or in a known ratio with daughters at the time of formation.

2. The parent decays at a known and constant rate into the daughter. (This may be direct or via several successive chain decays).

3. No parent or daughter is added during the period in question.

The second of these premises is a well established principle of nuclear physics, and hence poses no problems. The third premise is usually not a problem either, as the analysis can be made on portions of the material that are unlikely to have been exposed to the environment. Thus, only the first premise may give rise to questions relating to the validity of the result.

Lead Methods

The several lead dating methods are based on the series decay of uranium and thorium to lead. The chemical lead:uranium method, also known as the lead:alpha method, is relatively simple and is accomplished by radiometric analysis for the parent uranium and spectrochemical analysis for lead; the U:Pb ratio determines the age. The technique is useful for selected minerals—i.e. those that contained some uranium but relatively little lead (or a known amount) when formed. The age t of the mineral, in millions of years, is obtained by

$$t = \frac{cPb}{\alpha} \qquad (17-1)$$

in which c is a constant equal to 2600, Pb the concentration of lead in parts per million, and α the number of alpha particles emitted per hour per milligram of sample.

Since lead is also the end product of the thorium chain, the method can also be used with thorium bearing minerals. Equation 17-1 still holds, but the constant c is 1990. By determining the U:Th ratio and appropriately adjusting the constant, the method can also be used for rocks containing mixtures.

The Pb:alpha method has been most useful in geological dating, especially for zircon and lava flows, providing the date at which each was formed. The useful range is from 100,000 to at least 200,000,000 years.

A second age dating method based on lead is the isotopic Pb:U method. This method involves the mass spectrographic determination of the isotopic abundance of stable isotopes of lead. Naturally occurring lead has four stable isotopes, three of which are known to be end products of naturally occurring series decay chains (Table 17-1). The fourth, ^{204}Pb, is not believed to be formed from the decay of any long lived element. Natural lead believed to have been present at the time of formation of the earth contains all four isotopes in the abundances shown in Table 17-1.

Isotopic ratios of naturally occurring lead vary considerably throughout the world, largely attributable to the fraction of radiogenic lead (Table 17-2). Generally uranium bearing minerals and ores contain a greater preponderance of ^{206}Pb, while thorium bearing minerals are enriched in ^{208}Pb. The calculation of age must take into account the common lead or the lead originally in the material, which is indicated by the

TABLE 17-1   Origin and Abundance of Stable Lead Isotopes.

Series	Head	Lead End Product	Natural Abundance*
—	—	Pb-204	1.37%
Uranium (4n + 2)	U-238	Pb-206	26.26%
Actinium (4n + 3)	U-235	Pb-207	20.8%
Thorium (4n)	Th-232	Pb-208	51.55%

*Source: *International Cyclopedia of Chemical Science,* D. Van Nostrand (1964), p. 673.

presence of ^{204}Pb. The basic technique is to correct for the common lead in the mineral, using the common lead from uranium or thorium bearing minerals in the area as a base. The common lead or natural lead isoptopic ratios can also be obtained from sea sediments and meteoritic fragments which are believed to come from or be similar in composition to the earth's core. The corrected values provide the radiogenic lead content which can then be used to determine the age, t, of the mineral by the the Pb:alpha equation (17–1) or by the so-called lead:lead method according to the following proportion:

$$\frac{^{207}\text{Pb}}{^{206}\text{Pb}} = \frac{^{235}\text{U}(e^{-\lambda_1 t} - 1)}{^{238}\text{U}(e^{-\lambda_2 t} - 1)} \qquad (17\text{--}2)$$

TABLE 17-2   Isotopic Composition of Natural Lead.

Location	Age	Pb-204	Pb-206	Pb-207	Pb-208
Happy Jack Mine San Juan City, UT	Early Tertiary	1.94%	61 %	13.3%	24.9%
Pacific Ocean Sediments (35.2 ppm Pb)	Quaternary	1.34%	25.4%	21.1%	24.9%
Northern Urals Russia	Post-Devonian	1.3%	26.1%	20.9%	51.8%
Henburg Meteorite		2.0%	19.0%	20.6%	58.4%

Source: Faul, *Nuclear Geology,* pp. 261–293.

in which the chemical symbols refer to the number of atoms of each of the nuclides and λ_1 and λ_2 are the decay constants of ^{235}U and ^{238}, respectively. Note that only isotopic ratios are needed to solve Equation 17–2.

The isotopic Pb:U method can measure ages to $> 10^7$ years and hence has frequently been used as the basis for age of the earth measurements. However, there are several potential sources of error in age determination by this method. Hexavalent uranium is readily leached from minerals, producing an erroneously low uranium concentration and estimate of the age by increasing both the ratios of Pb-207:Pb-206 and U-235:U-238. ^{222}Rn, a daughter of ^{238}U and a precurser of ^{206}Pb, being gaseous, may escape from the mineral, especially if it is near the surface or if the material is porous. If this occurs, the ratio of ^{207}Pb to ^{206}Pb will be raised as will the calculated age. There may also have been redistribution of the elements in the mineral over the time period involved, a potential source of error in any of the lead based age dating methods. Despite these possible sources of error as well as the potential measurement error associated with mass spectroscopy, the Pb:U method is generally considered to be superior and provide the most valid estimate of age.

Similar to the Pb:U method is the Pb:Th method. Although somewhat simpler in that only one parent isotope is involved, ages obtained with this method generally disagree with values obtained with other lead methods and hence this method is not considered very reliable. As with other lead methods, leaching or transfer of the various constituents of the thorium decay chain may occur, as may the loss of gaseous ^{220}Rn, also a thorium daughter.

The ^{210}Pb method is also similar to the Pb:U method and is based on the principle that the age of a uranium or thorium bearing mineral can be determinbed by the ratio of any one member of a decay chain to that of the stable end product. Thus, radioactive equilibrium is assumed. The age is determined by the isotopic ratio of ^{206}Pb:^{210}Pb. The 206 isotope is determined by mass spectrometry and the 210 isotope radiologically, either by direct counting of the low energy ^{210}Pb betas or by allowing the freshly separated ^{210}Pb to sit for a while and counting the more energetic betas from the ^{210}Bi daughter or the alphas from its daughter ^{210}Po.

When the earth's crust was formed, lead was frozen into various minerals along with uranium and thorium from which radiogenic lead was continuously being formed. By examining the ratios of lead isotopes in crustal materials, it is possible to obtain an apparently valid estimate of the age of the earth with the aid of Houterman's equation:

$$\frac{(^{206}Pb:^{204}Pb)_{modern} - (^{206}Pb:^{204}Pb)_{primordial}}{(^{207}Pb:^{204}Pb)_{modern} - (^{207}Pb:^{204}Pb)_{primordial}} = 137.7 \frac{e^{-\lambda_1 t}}{e^{-\lambda_2 t} - 1} \qquad (17\text{--}3)$$

In this equation, based on a method proposed independently in 1946 by the German physicist F. G. Houterman and British physicist Holmes, the chemical symbols refer to the number of atoms of each nuclide, and λ_1 and λ_2 are the decay constants of ^{235}U and ^{238}U, respectively. The constant 137.7 is the present ration of ^{238}U to ^{235}U. Using this equation and measurements he had made, C. C. Patterson determined the age of the earth to be 4.55×10^9 years. Other measurements based on this and other lead methods have yielded values in the range $2-5 \times 10^9$ years.

Helium Age Dating

The content of radiogenic helium in minerals and rocks serves as the basis for helium age dating. Although this was among the earliest, if not the first age dating method based on radioactive decay, it nonetheless is fraught with sources of error. Values obtained by the method are usually considered minima because of the large possiblity that a portion of the radiogenic helium will have escaped from the material being examined. Radiation damage in rocks may be responsible for a portion of the loss. These losses are generally incalculable and may be offset to some extent by spallation reactions in the heavier elements in the material induced by high energy cosmic rays. Helium produced in this manner is richer in ^3He, and by examining the ratio of ^3He to ^4He it is possible to gain some indication of the amount of cosmogenic helium. Helium may also have been produced by alpha emitting nuclides no longer present in the material. The principal difficulty with helium methods, however, lies in the inhomogeneous distribution of uranium and thorium in most minerals and the fact that uranium and thorium can also migrate through rocks.

Generally speaking, the quantity of helium is related to the uranium and thorium content of the mineral and to its age. Production is typically on the order of 0.001 cm^3/g per 10^9 years. The rate of production of helium atoms per unit time per mg of sample, dN/dt, is described by

$$\frac{dN}{dt} = \frac{^4He \text{ atoms/mg}}{\text{time}} = ak_1Nk_2at \qquad (17-4)$$

in which a is the number of helium atoms added per mg per million years, N is the number of helium atoms per mg, k_1 and K_2 are proportionality factors characterizing the rate of helium loss relative to the number of atoms present and the openness of the structure, and t is the age of the material in millions of years.

Equation 17–4 can be rearranged to solve for t, the age of the material in millions of years. Solution, however, is not straightforward as the proportionality factors are not ordinarily known. Empirical studies, largely with zircons, have led to equations which characterize these parameters, but the characterization is in terms of a rather complex integral that does not readily lend itself to manual solution.

The amount of ^{3}He can also be used to estimate the age of geologic materials. This isotope of helium is produced from decay of tritium and directly from spallations resulting from high energy cosmic ray interactions. ^{3}He can be readily detected and quantified by mass spectroscopy. The method has been successfully used on samples taken from the surface of meteorites to determine their exposure time in space or what has been termed the cosmic ray age. Thus far, the maximum cosmic ray age of 1.5×10^9 has been observed in an iron meteorite.

Strontium-Rubidium Method

Minerals containing rubidium may be conveniently dated by determination of the ratio of the number of atoms of ^{87}Rb to its daughter ^{87}Sr, and solving the following equation for t, the age:

$$\frac{^{87}Sr}{^{87}Rb} = e^{\lambda t} - 1 \qquad (17\text{--}5)$$

In equation 17–5, λ is the decay constant of the ^{87}Rb. Measurement of the ratio is not easily accomplishedand. The usual technique is to determine the ratio of total rubidium to total strontium and then use mass spectrometric isotopic analysis to determine the nonradiogenic stable isotopes of each element. Since ^{87}Rb has an abundance of 27.83% and that of ^{87}Sr is 7.0%, these data can then be used to correct the empirically observed abundances of the various isotopes. Note that if the age of the earth is taken as 4.5×10^9 years, the fraction of radiogenic strontium produced by decay of the 4.7×10^{10} year ^{87}Rb in the earth's crust is calculated to be approximately 1% of the total strontium present.

The Sr:Rb method is perhaps the most useful method for geologic age determination. In practice it is fairly simple to perform on minerals such as mica than contain rubidium; what is done is to obtain a mass spectrograph of the isotopic abundance of strontium in the sample and compare this with a mass spectrograph of common strontium. If the common strontium content of the mineral being examined is not too large, the radiogenic component can be readily determined with but small error. Ages calculated by this method have a high degree of accuracy even when the amount of radiogenic ^{87}Sr is as little as 5 per cent.

Potassium-Argon Dating

Potassium-argon dating is based on the K-capture decay of naturally occurring ^{40}K, with a half-life of 1.28×10^9 years, to ^{40}Ar. The branching ratio for this mode of decay is 11 per cent (Figure 3–1). The method is potentially excellent, because potassium is nearly ubiquitous in the environment and ^{40}K can be easily detected and quantified by the 1.46 MeV gamma ray emitted during its decay to ^{40}Ar. Perhaps of even greater importance is that argon, the daughter product, is a gas and hence is extremely unlikely to have been present or trapped in the material when it was formed. Thus, any argon found in the sample is clearly radiogenic in origin.

There are at least two potential significant sources of error. The first of these is the loss of argon from the sample by diffusion, which is a function of the grain size and structure of the mineral. The other is the possible leaching of potassium, which forms soluble compounds, from the matrix of the mineral. Another is variation in the isotopic ratios of argon in active minerals, which may result in error if the age measurement is based on the normal $^{40}Ar{:}^{36}Ar$ ratio.

Age can be calculated from the $^{40}K{:}^{40}Ar$ ratio by the following variation of equation 17–5:

$$\frac{^{40}Ar}{^{40}K} = 0.11(e^{\lambda t} - 1) \qquad (17-6)$$

in which the constant 0.11 is the branching ratio for K-capture and λ is the decay constant for ^{40}K.

The potassium-argon method is primarily useful with potassium bearing rocks at least 50,000 years old and has been extensively used for dating stony (as opposed to iron) meteorites. The oldest of these have been found to have an age of 4.6×10^9 years, which is about the upper limit for minerals found on earth. As astronomers generally believe that meteorites and the earth are of the same origin, this estimate is generally accepted as the age of the earth.

Radiocarbon Dating

Radiocarbon dating is perhaps the best known and most widely used of all the radioactive age dating methods, and is most useful for materials of more recent origin. The radiocarbon method was suggested in 1946 by Nobel prize winning chemist Willard Libby who suggested that the carbon contained in living matter would contain a small but definite and constant fraction or concentration of cosmogenic ^{14}C. ^{14}C in fact exists in equilibrium with the stable isotopes of carbon in the tissues of living things at a concentration of 6.95 pCi/g of carbon or 15.3 disintegrations per minute (7 Bq) per gram of carbon. After the death of the organism, the uptake and turnover of carbon would cease, and the radioactive fraction would decay away with its characteristic 5600 year half life. Thus, carbonaceous objects of archeological or other significance to man, including oil, could be dated by ascertaining the fraction of ^{14}C in a sample and comparing it with the fraction in a sample from modern times.

The method outlined above appears simple and obvious. However, there are uncertainties and difficulties. One of these is related top the burning of fossil fuels which has, over the past century or so, released large quantities of carbon dioxide into the atmosphere. As this carbon was very old, coming from coal and oil deposits started millions of years ago, it is depleted in ^{14}C, and hence has the tendency to dilute or reduce the concentration of radiocarbon in plants and animals. An opposite effect has been the result of the introduction of large quantities of ^{14}C into the biosphere by detonations of nuclear explosives in the atmosphere.

In addition, comparison of radiocarbon dates with tree ring values has shown some fluctuation in the ^{14}C concentration in the atmosphere between 1400 and 1700 B.C. Comparison of radiocarbon determined ages with ages of archeological materials accurately established by other methods have revealed that for the period from 100 B.C. to 1400, radiocarbon dating gives values that are too large; prior to 100 B.C. the radiocarbon values are too small. At about 1600 B.C., the radiocarbon values are about 175 years (5%) too small, increasing to about 300 years (6%) at 3000 B.C. The discrepancy appears to be a result of slight variations in atmospheric levels of ^{14}C which may be attributable to variations in the earth's magnetic field over the years, which would alter the cosmic ray intensities and hence ^{14}C production near the earth. Suitable corrections are available, however, and the useful range of radiocarbon dating is at least 1000 to 100,000 years; in the range of 1000 to 50,000 years, the time frame of great archeological significance, the uncertainty in the method is less than 5 per cent (Aitken 1974, Baxter and Walton 1971).

The radiocarbon method is of necessity destructive. ^{14}C decays by emission of a beta particle with a maximum energy of only 0.156 MeV; there are no associated gamma photons. Detection is thus best accomplished with a gaseous sample that is produced by oxidation of the carbon to carbon dioxide which can then be detected by an internal counter. Conversion to the gaseous form not only minimizes self-absorption but also eliminates other beta emitters such as naturally occurring ^{40}K that might be present and interfere with the counting. The basic method is to wash the sample to remove organic material of a more recent era, soak it in hydrochloric acid for several hours, wash with water, dry, and burn to obtain carbon dioxide. The combustion gases are captured in a dry ice trap after passage trough a dessicant to remove the water. Nitrogen, radon and other gaseous impurities are removed by bubbling through an ammonium hydroxide solution. The CO_2 is precipitated as $CaCO_3$ by treatment with $CaCl_2$ and the hydrochloric acid treatment repeated to obtain purified carbon dioxide which can then be counted directly by flow through techniques if the sample is sufficiently large, or reduced to carbon and thinly deposited on a planchet and directly beta counted.

The radiocarbon technique has been used to successfully date a variety of articles of significance in human history. The Dead Sea Scrolls have thus been dated at 35 B.C. $\pm$ 300 years, and the prehistoric Lascaux cave paintings in France at 14,000 B.C. $\pm$ 900 years. Charcoal from the Stonehenge site in England has been dated at 1846 B.C. $\pm$ 275 years. These dates typify the useful range of the method, which has been proposed as a means of helping to resolve the controversy surrounding the Shroud of Turin by definitively establishing the age of the fabric.

Fission Track Methods

When uranium or other heavy elements undergo fission, either spontaneously or neutron induced, the resultant heavy ionized fission fragments leave behind them a microscopic damage track. If the fission fragment track penetrates the surface, the polished surface of the material can be treated with strong acid or alkali producing a preferential etching of the damage track. The etched track appears as a pit or hole which can be detected by an ordinary light microscope or by a spark counter. The number of tracks per unit area is related to the number of fissions which in turn is a function of the uranium content of the material, the spontaneous fission half-life, the cosmic ray neutron flux, and the time. Ordinarily, a calibration can be made by exposing the mineral to a known neutron flux in a reactor.

The method is more useful for pottery dating than for geological materials. When pottery is fired, the existing fission fragment tracks are annealed out, providing a zero basis. The age, t, in years since the material was fired, is directly proportional to the number of fission tracks and can be readily calculated by

$$t = 1.2 \times 10^{16} \frac{N_t}{N_u} \qquad (17\text{--}7)$$

in which N_t/N_u is the number of naturally occurring tracks in a unit area relative to the number of ^{238}U atoms, and 1.2×10^{16} a constant related to the spontaneous fission rate of ^{238}U. The minimum detectable age is inversely proportion to the uranium content of the material. With a uranium concentration of 1 part per million, the minimum detectable age is 100,000 years; with a concentration of 1 per cent, the minimum detectable age is only 10 years. Thus this method can be used for dating recent materials such as uranium bearing glasses and pottery (Michels 1973).

Thermoluminescence

The thermoluminesence age dating technique was suggested in 1953 by Farrington Daniels and first used successfully in 1960 by George C. Kennedy and refined later during the 1960's by David Zimmerman and others at Oxford and the Universities of Pennsylvania and Wisconsin. It has become a highly useful and important method for age dating archeological materials and is applicable to geological materials as well. The thermoluminescence is based upon the radiation dose incurred by crystalline materials from the inclusion of trace amounts of uranium, thorium, and other naturally occurring radionuclides. This self-dosing produces a stored damage effect that results in the emission of light when the material is heated. The quantity of light emitted is proportional to the radiation history of the material, which in turn is a function of its age and the amount and kinds of radionuclides it contains.

By determining the amount of various radionuclides in the specimen, it is possible to calculate the energy deposited (i.e. the dose) in the material per unit time from the radioactive decay. Different radionuclides produce different energy depositions, so the contribution from each must be calculated. However, most of the dosage is from alpha particles from the nuclides in the various radiactive series. Typical internal dose rates from potassium and uranium and thorium plus daughters are in the neighborhood of 100 rad (1 Gy) per century, or about ten times ambient external levels. Buildup of dose begins with the time the material was fired in the case of pottery or other archeological artifacts. High temperatures anneal out any existing thermoluminescence. After the firing, the material will accumulate dose which is subsequently read in the form of light output upon heating. After reading, the material is annealed and exposed to a known calibration exposure, thus providing its own calibration reference. The age of the material in years is simply the product of the calculated annual dose and the calibration factor (i.e. thermoluminescence per unit dose) divided by the natural thermoluminescence. The thermoluminescence age dating method has been successfully used with archeological materials, largely pottery, dating from 5000 B.C. to 300 A.D.

The Oklo Phenomenon

Although the Oklo phenomenon is not an age dating technique, it nonetheless is an interesting and related natural effect. In 1972, H. Bouzigues noted that a sample of uranium ore from the Oklo mine in eastern Gabon contained a disproportionately small abundance of the 235 isotope— 0.7171 per cent as compared with the normal abundance of 0.7202 per cent. Subsequent analysis of other samples from the mine revealed much stronger depletion in the 235 isotope, down to as low as 0.44 per cent, or about three-fifths the normal abundance (Cowan 1977). The result was extraordinary, because prior to this discovery the isotopic composition of uranium had been thought to be constant in nature.

Further analyses of the ore revealed the presence of trace quantities of characteristic fission products leading to the conclusion that the Oklo mine had been the site of a natural self-sustaining chain reaction about two billion years ago. At that time, the abundance of the 235 isotope of uranium was about 3 per cent, based on radioactive decay calculation. (The 235 isotope has a half-life of 7×10^8 years relative to that of ^{238}U $(4.5 \times 10^9$ years). This is sufficient to produce a nuclear chain reaction under suitable conditions, i.e. sufficient percentage of uranium in the ore, appropriate geometry, and the presence of an appropriate neutron moderator. The first two conditions are easily met by the concentration of uranium in the rich Oklo mine; the latter could have been provided by the water of crystallization in a sedimentary ore of the type found in the Oklo mine, or by seepage of ground water. The estimated energy release of the reactor, determined from the fission products, was about 15,000 megawatt-years, equivalent to about four years of operation of a large light water reactor. Oklo was not a steady state constant operation, but probably occurred off and on over a period of 150,000 years. Since the inital discovery at the Oklo mine, similar discoveries of uranium slightly depleted in the 235 isotope have been made in sedimentary deposits of uranium in the Colorado plateau region.

In addition to its scientific significance, the Oklo observations may have enormous practical implictions. Most of the fission products and plutonium produced by the reactor have remain *in situ* or haver migrated little, thus providing the hope that long term disposal of high level radioactive wastes in stable geologic formations may be a feasible and environmentally safe method in the future.

References

1. Ahrens, L. H. 1956. "Radioactive Methods for Determining Geological Age", in *Physics and Chemistry of the Earth*, L. H. Ahrens, K. Rankama, and S. K. Runcorn, Eds., NcGraw-Hill, New York.
2. Aitken, M. J. 1974. *Physics and Archeology*, Second Ed., Oxford University Press, London.
3. Baxter, M. S. and A. Walton. 1971. "Fluctuations of Atmospheric Carbon-14 Concentrations during the Past Century", *Proc. Roy. Soc. London Ser. A* 321(1544):105.
4. Cowan, G. A. 1977. "A Natural Fission Reactor", *Scientific American*.
5. Duursma, E. K. 1972. "Geochemical Aspects and Applications of Radionuclides in the Sea", *Oceanography and Marine Biol. Ann. Rev.* , 10:137.
6. Faul, H., Ed. 1954. *Nuclear Geology A Symposium on Nuclear Phenomena in the Earth Sciences*, John Wiley and Sons, Ltd., New York.
7. International Atomic Energy Agency (IAEA). 1975. *The Oklo Phenomenon*, Proceedings of a Symposium, IAEA Publication IAEA-SM–204/4.
8. Kuroda, P. K. 1960. "Nuclear Fission in the Early History of the Earth", *Nature* 187:36. H. N. Michael and E. K. Ralph, Eds. 1971. *Dating Techniques for the Archaeologist*, MIT Press, Cambridge.
9. Michels, J. W. 1973. *Dating Methods in Archaeology*, Seminar Press, New York.

10. Rutherford, E. 1906. *Radioactive Transformations*, Archibald Constable and Son, London.
11. Rutherford, E. 1913. *Radioactive Substances and their Radiations*, Cambridge University Press.
12. Tite, M. S. 1972. *Methods of Physical Examination in Archaeology*, Seminar Press, New York.
13. York, D. and R. M. Farquhar. 1972. *The Earth's Age and Geochronology*, Pergamon Press, New York

Appendix: Radiological Units

TABLE A-1　Common Radiological Units.

Unit or Quantity	Symbol	Brief Description	Comment
Curie	Ci	3.7×10^{10} nuclear transformations per second	Unit of activity
Becquerel	Bq	1 nuclear transformation per second	SI Unit of activity
Roentgen	R	2.58×18^{-4} C/kg (photons in air)	Special Unit of exposure
Rad	rad	0.01 J/kg (100 erg/g)	Dose; applies to any radiation
Gray	Gy	1 J/kg ($= 100$ rad)	SI unit of dose
Dose Equivalent	H	dose $\times$ Q $\times$ any other modifying factors	Radiation protection
Quality Factor	Q	Biological effectiveness of radiation	Radiation protection
Rem	rem	rad $\times$ Q $\times$ any other modifying factors	Unit of dose equivalent
Seivert	Sv	Gy $\times$ Q $\times$ any other modifying factors	SI unit of dose equivalent

TABLE A–2   Relationship Between Old and SI Units.

Activity
1 Ci = 3.7 × 10^{10} Bq
Exposure
There is no SI unit for exposure. In the SI system, exposure in the special sense must be expressed in units of C/kg. 1 R = 2.58 × 10^{-4} C/kg.
Dose
100 rad = 1 Gy
Dose Equivalent
100 rem = 1 Sv

TABLE A–3   International Multiples and Submultiples.

Multiple	Prefix	Symbol
10^{18}	exa	E
10^{15}	peta	P
10^{12}	tera	T
10^{9}	giga	G
10^{6}	mega	M
10^{3}	kilo	K
10^{-1}	deci	d
10^{-2}	centi	c
10^{-3}	milli	m
10^{-6}	micro	μ
10^{-9}	nano	n
10^{-12}	pico	p
10^{-15}	femto	f
10^{-18}	atto	a

*Name Index

*Page numbers in italics refer to the complete Bibliographic information; page numbers in roman refer to citations in the text.

Subject Index

381